SO-AHS-488

CBMS
Conference Board of the Mathematical Sciences

Issues in Mathematics Education

Volume 8

Research in Collegiate Mathematics Education. IV

Ed Dubinsky
Alan H. Schoenfeld
Jim Kaput
Editors

Cathy Kessel, *Managing Editor*
Michael Keynes, *Assistant Managing Editor*

American Mathematical Society
Providence, Rhode Island
in cooperation with
Mathematical Association of America
Washington, D. C.

2000 *Mathematics Subject Classification.* Primary 00–XX, 97–XX.

ISBN 0-8218-2028-1
ISSN 1047-398X

Copying and reprinting. Material in this book may be reproduced by any means for educational and scientific purposes without fee or permission with the exception of reproduction by
services that collect fees for delivery of documents and provided that the customary acknowledgment of the source is given. This consent does not extend to other kinds of copying for general
distribution, for advertising or promotional purposes, or for resale. Requests for permission for
commercial use of material should be addressed to the Assistant to the Publisher, American
Mathematical Society, P. O. Box 6248, Providence, Rhode Island 02940-6248. Requests can also
be made by e-mail to reprint-permission@ams.org.
    Excluded from these provisions is material in articles for which the author holds copyright. In
such cases, requests for permission to use or reprint should be addressed directly to the author(s).
(Copyright ownership is indicated in the notice in the lower right-hand corner of the first page of
each article.)

© 2000 by the American Mathematical Society. All rights reserved.
The American Mathematical Society retains all rights
except those granted to the United States Government.
Printed in the United States of America.

∞ The paper used in this book is acid-free and falls within the guidelines
established to ensure permanence and durability.
Visit the AMS home page at URL: http://www.ams.org/

10 9 8 7 6 5 4 3 2 1   05 04 03 02 01 00

Contents

Preface

Welcome to the fourth volume of *Research in Collegiate Mathematics Education (RCME IV)*. Student learning and calculus are major themes in this volume. As in previous volumes, these are examined in a variety of ways. Seven of the eleven articles that comprise *RCME IV* concern different aspects of calculus. The first two give overviews of calculus reform in France and in the United States. The next two are small- and large-scale longitudinal comparisons of students who were enrolled in first-year reform and traditional courses. Four detailed studies of students' understandings of calculus and related topics follow. We then switch gears to focus more directly on relationships between instruction and students' understandings for courses other than calculus—abstract algebra and number theory—and finish with a cross-sectional study of a cross-cutting concept—quantifiers.

Calculus reform

Michèle Artigue gives an overview of calculus in France, including relevant research as well as a brief history of curricular change. Unlike the United States, France has a national program of study; teaching calculus at the high school level became widespread at the beginning of the 20th century. Alongside the national curriculum has been a coherent research enterprise which, for some decades, has explored mathematics teaching and learning at the upper high school (including calculus) and college levels. This research enterprise, much more homogeneous than in the U.S., is grounded in a particular sociocultural perspective, including the idea (called the "didactic contract") that students and teachers enter the classroom with a set of mutual, though often tacit, expectations, which play a strong role in shaping their classroom behavior. French cognitive studies have also focused on "epistemological obstacles," conceptual issues that have proven difficult both historically for the field and for individual students when learning the material. Artigue describes some of the theoretical frameworks used in research and some of the different kinds of student difficulties relevant to calculus which have been documented by research. This is followed by an account of the evolution of the teaching of calculus at the lycée level (grades 10–12).The syllabus changed in the 1960s and 1970s due to the influence of the Bourbaki. Another change occurred in 1982, this time influenced by the findings of mathematics education research, and the curriculum focused on approaches that were more intuitive than the formal approaches inspired by the Bourbaki. The situation in France may be of particular interest to readers from the United States because about 70% of French students take at least two years of calculus in high school, hence the French have been concerned far longer with the problem of how to make calculus accessible to the majority of students. (The following statistics give a sense of how many high school students in the United

States take calculus. In 1995 about 4% of grade 12 students took an Advanced Placement calculus exam. The National Science Foundation, extrapolating from a sample, indicates that about 11% of grade 12 students took a calculus course in 1993.) The French approach focusing on variation and approximation has had its successes, but some problems remain and new problems have arisen.

In the United States, calculus reform might be said to have begun with the Tulane conference of 1986. There were various reasons for concern. Calculus courses had high drop-out rates; the content of the courses wasn't adequate for further study of mathematics, science, and engineering; students didn't seem to understand that content well; and, unlike courses of the past, 1980s calculus courses did not appear to inspire many students to become mathematicians. After more than a decade, it is natural to ask whether reform has successfully addressed these concerns. However, any answer must take into account the different instantiations of calculus reform. They may include the use of technology, small groups during class, writing, or structured small-group sessions that supplement classes run in a traditional manner. Different textbooks and curriculum programs include these elements in different ways, along with treatments of topics that break with tradition. Betsy Darken, Robert Wynegar, and Stephen Kuhn give an overview of these different kinds of calculus reform and summarize the research on the effects of different elements of reform as well as the effects of different reform texts and programs. With the caveat that research on reform is limited, they conclude that calculus reform is doing no harm and may even be doing some good.

Darken and her colleagues contribute to the research on reform with a longitudinal comparison of students from first-year courses using a reform text (Ostebee and Zorn) and a traditional text. Students' course grades were compared for reform and traditional first-year calculus—and for "unreformed" second-year courses (Differential Equations and Multivariable Calculus). Students in the reformed first-year course withdrew less often although the grades for reform and traditional courses were not significantly different. Consistent with this, success rates (passing grades divided by all grades including withdrawals) were significantly different for the two courses. Student performance and retention were similar in the second-year courses.

Susan Ganter and Michael Jiroutek studied a different instantiation of calculus reform, a course that was traditional except for the addition of projects (written up every four weeks) and the replacement of one of four weekly class meetings by a computer lab. This reflects a conceptualization of reform as a change in teaching methods rather than a change in content or the way in which it is organized. Ganter and Jiroutek focus on two questions: Does the change in "delivery method" affect students' mastery of "basic" calculus skills—taking derivatives, finding equations of tangent lines, and the like? The answer in this case was yes, students in reform courses performed more poorly on tests of these skills than did their counterparts in traditional courses. Like Darken and her colleagues, Ganter and Jiroutek examined grades in later courses and found no significant differences for students from reform and traditional courses. Unlike the grades analyzed by Darken et al., these included grades in science as well as mathematics courses.

Calculus, Concepts, Computers, and Cooperative Learning (C^4L) is a reform calculus program that involves technology and has been extensively studied. Small-scale in-depth comparisons of the understandings of students taking C^4L and traditional courses have been made. But what about long-range issues like students' grades in later courses—or whether they take those later courses? Keith Schwingendorf, George McCabe, and Jonathan Kuhn address these questions with a large-scale study of the mathematics grades of 4636 Purdue University students who enrolled in first-year calculus in 1989, 1990, or 1991. Treatment A versus treatment B comparisons of such data often face methodological difficulties: in general—and in this case—students are not randomly assigned to courses. Schwingendorf et al. note that such difficulties may often be inevitable—hence methods for comparing calculus programs when students are not randomly assigned are needed. Their analysis controlled for factors that appeared to be important: predicted grade point average (a combination of entrance exam scores and high school grades), major, and gender. On average, C^4L students earned higher grades in calculus courses, took more courses beyond calculus and had slightly better grades in those courses.

Student understandings

What do students who succeed in calculus know? The next four articles give detailed accounts of students' understandings and abilities. Phenomena common to all four articles are: What each student knows or can do may differ considerably, even in a group of students receiving As or Bs. And such students may display surprising gaps in their knowledge.

The first two articles concern students' understandings of two important concepts in calculus: sequence and derivative. Michael McDonald, David Mathews, and Kevin Strobel use the Action-Process-Object-Schema (APOS) framework to analyze students' understanding of sequences. According to this framework, sequences can be understood as a process—a list of numbers, or as an object—in brief, a function whose domain is the integers. McDonald and his co-authors interviewed 21 students who would be considered successful according to standard criteria—with two exceptions they had earned As or Bs in second-semester calculus. One group of students had taken a traditional calculus course and the other a C^4L course. These students' interviews and written work suggest that the C^4L students are more likely to understand sequences as functions than traditionally taught students.

Michelle Zandieh describes a framework for analyzing student understanding of derivative and illustrates its use with a case study of nine high school students. Her framework is related to the PO of APOS: it consists of three layers of pairs where the first element of the pair is a process and the second is the object that is the result of that process. The first layer concerns ratio, the second, the limit of that ratio (derivative at a point), and the third, the collection of those limits (the derivative function). Each element of these pairs can be interpreted in different contexts: symbolically, graphically, in terms of rate (which leads to particular symbolic forms), and in terms of velocity (which leads to connections with physical experiences as well as with metaphorical uses of "velocity," "increase," and so on). Such a framework suggests just how complicated an adequate understanding of derivative might be. Zandieh uses her framework to analyze interviews of nine high school students enrolled in an Advanced Placement calculus course. Although these students attended the same class, their initial understandings of derivative fell in quite different categories of the framework. As the course progressed and students' understandings became more complete, they became more similar. The study suggests that: students' initial understandings of a particular topic may differ and students with similar initial understandings may not add to those partial understandings in the same ways.

Annie Selden, John Selden, Shandy Hauk, and Alice Mason provide data that remind us of some of the reasons for reform—specifically, that traditionally-taught students are not doing as well as we'd like. Selden et al.'s article is the third in a series of related studies. The previous studies analyzed traditionally-taught calculus students' responses to "moderately non-routine" calculus problems. The results were discouraging: Although responses on a test of related skills suggested that the students had the necessary skills, two-thirds of the students who earned As or Bs in calculus did not solve one problem and none of the students who earned Cs did. Do students develop the ability to solve such problems later in their mathematical lives? Perhaps. Selden et al. asked students who were finishing a traditional calculus/differential equations sequence to solve the same "moderately non-routine" calculus problems. Slightly more than half (16 of the 28 students) failed to solve any of the non-routine problems. Selden et al. analyze possible explanations. Was it absence of relevant knowledge? As in Selden et al.'s earlier studies the differential equations students were tested on the algebra and calculus necessary to solve the non-routine problems. Although results of the tests indicated that students were familiar with the calculus that could be used to solve the non-routine problems, solution methods tended to be algebraic. Selden et al. suggest that students may have the relevant knowledge but are not accessing it for "moderately non-routine" calculus problems—students do not have "problem situation images" that include "tentative solution starts," in the authors' terms. They describe a potential remedy: courses that include experiences fostering students' constructions of rich problem situation images that stimulate their recall of relevant knowledge.

William Martin also examined the effects of a previous course on students. His analysis suggests pedagogical implications. The question was: What is the effect of a college algebra course that incorporates graphing calculators? Martin interviewed 18 students toward the end of their first semester of calculus. Nine of the students had taken a college algebra course that included the use of graphing calculators. However, the course was not much changed from the traditionally taught college algebra course at the same university taken by the nine other students that Martin interviewed. Instructors had not received special instruction in the use of graphing calculators nor were their courses dramatically different from the traditionally taught courses. The students' performance was reminiscent of Selden et al.'s differential equations students: They performed well on routine pre-calculus tasks and had little success on conceptual items involving calculus. Although Martin found few statistically significant differences, he noticed an interesting trend. Students in both groups used inappropriate strategies apparently cued by superficial features of the tasks, but those in the graphing calculator group did so less often. Martin hypothesizes that, when faced with a problem traditional students may be more inclined to try standard procedures and that graphing calculator students may be more inclined to focus on understanding the problem. As Selden et al. point out, novices in a particular domain tend to focus on surface characteristics of a task. Martin's study suggests that the instruction that the graphing calculator students received may have helped them to focus on the structure of a task rather than its surface features.

Teaching

We now turn to studies concerning more detailed aspects of instructional design and their interaction with students' learning. John Hannah taught two instantiations of an abstract algebra course. His study focuses on his students' understanding of D_4 (the dihedral group of all symmetries of a square). In *RCME II*, Zazkis and Dubinsky discussed students' encounters with two representations of elements of D_4. These representations correspond to different ways of viewing the symmetries of a square: globally as transformations (reflections and rotations of the square) or locally as permutations (changes in position of its vertices). Computations with either representation should give the same result, but most of the students interviewed by Zazkis and Dubinsky got different results for compositions of the same elements, depending on whether they used transformations or permutations. These were connected with differences in notation (does composition go from left to right or right to left?) and visualization (does one focus on the vertices of the square moving against a fixed background or on the background positions receiving different labels?). That's a problem. What can instruction do to address it? For the first instantiation of his course, Hannah designed a system of labels and diagrams: for example, the four vertices of a square were labeled and the "background" against which the square had four positions labeled—both with numbers from 1 to 4. Despite this, in computing compositions of elements of D_4 using transformations and permutations, his students' responses were similar to those of Zazkis and Dubinsky. However, his students' suggestions that labels for the "background" be letters and those for vertices be numbers, led Hannah to a more successful refinement of the diagram–label system, which he used in the second instantiation of his course. But, although Hannah's students were more successful than those of the previous year in their computations with elements of D_4, he found that his students had difficulties in analogous situations: calculating compositions of transformations using three-dimensional models and calculating compositions of permutations using numerical arrays. Hannah's article suggests that instructors may find it helpful to attend to students' perceptions (which may be discovered from previous research and in more specific form from one's own students) and that consciousness of those perceptions may aid in redesigning instruction.

Rina Zazkis studied a very different group of students, prospective elementary teachers who were enrolled in her number theory course. She also used a different method, clinical interviews. She began her study with the intention of examining her students' understanding of three related terms (factor, multiple, and divisor), but found that connections with students' prior knowledge were impossible to ignore. Although the prospective teachers had studied factors and divisors for three weeks of Zazkis's course, in interviews they displayed incorrect understandings of these terms based on their pre-college schooling. This study suggests that teaching preservice teachers may be a very different enterprise than teaching children or mathematics majors, because prior knowledge may play a different role. As Zazkis puts it: "What often occurs in a content course for preservice elementary school teachers is not construction of new meanings or concepts, but reconstruction of previously constructed meanings." This poses a problem for research: How is learning different from re-learning? Can it be explained within existing theoretical frameworks? And it poses a problem for instruction: Can an undergraduate program help prospective elementary teachers re-construct meanings for terms learned during six years of elementary school?

Like Zazkis's preservice teachers, the science, mathematics, and mathematics education majors studied by Ed Dubinsky and Olga Yiparaki brought previous understandings to their interviews. Those interviews did not address a particular course topic, but rather a topic that occurs throughout mathematics—quantifiers, in particular, "there exists" and "for all," and alternations of the two. Undergraduate mathematics majors do often not study quantifiers explicitly as a separate topic, but they encounter them throughout their studies. Dubinsky and Yiparaki addressed the question of how students' interpretations of words used in everyday contexts, such as "all," "every," and "there is" might support their interpretations of similar terms used in mathematical statements. The findings were reminiscent of Zazkis's—in trying to interpret everyday and mathematical statements, the students were inclined to draw on their experiences and use context rather than syntax. This inclination seems to explain why interpreting mathematical rather than everyday statements was far more difficult for the students—they had far less mathematical than everyday experience and their everyday experiences had not provided them with a strong sense of syntax. At the end of each interview, the students were asked to play a game involving alternations of quantifiers for three of the statements. In general, this seemed to help the students. Both the findings, that students tend to rely on context rather than syntax and that instruction concerning syntax aids student understanding, suggest that instructors should not rely exclusively on analogies with everyday experience in helping students learn to interpret statements involving quantifiers. Instead, instruction might create experiences that allow students to learn about syntax.

All told, these papers show that a growing community of researchers is beginning to systematically gather and distill data regarding collegiate mathematics teaching and learning. We look forward to more reports in future volumes.

Ed Dubinsky
Alan Schoenfeld
Jim Kaput
Cathy Kessel

CBMS Issues in Mathematics Education
Volume **8**, 2000

Teaching and Learning Calculus:
What Can Be Learned from Education Research and Curricular Changes in France?[1]

Michèle Artigue

ABSTRACT. Research on mathematics teaching and learning has been stimulated by the persistent difficulties that students have when they are introduced to the conceptual field of analysis and from general feelings of dissatisfaction with calculus courses. After sketching the main research results, we analyze the methods adopted throughout the 20th century in preparing students for the baccalauréat in France[2] in order to overcome these difficulties. We also consider what the historical evolution of the curriculum can tell us about current teaching.

Introduction

It is not easy for students to get into the conceptual field[3] of elementary analysis (i.e., calculus). Education research developed in this area over the past 15 years shows this very clearly (Artigue & Ervynck, 1992; Farfán, 1993; Tall, 1991). Furthermore, this research has allowed us to better understand the nature of the difficulties and the obstacles students encounter, as well as the reasons for the failure, as much of traditional teaching strategies, which reduce analysis to algorithms for algebraic calculations, as of the theoretical approaches and formalisms developed in the context of modern mathematical reform.

Everywhere in the world new programs of study and new curricula are being set up in attempts to find ways to introduce this conceptual field that are, at the same time, rich in meaning and accessible. Intuitive approaches based on the use of information technology, calculators, and computers appear to be the most generally

---

[1]Translation by Ed Dubinsky of Enseñanza y aprendizaje del análisis elemental: ¿qué se puede aprender de las investigaciones didácticas y los cambios curriculares?, *Revista Latinoamericana de Investigación en Matemática Educativa*, 1998, 1, pp. 40-55.

[2]France has a national curriculum. In the French educational system, the secondary level begins with grade 6, when students enter what is called "collége," and lasts for four years. At grade 10, students enter one of several types of lycées for three years at the end of which time they can take an exam called the baccalauréat. If they pass, they can go on to the university. We will use this terminology throughout the article.

[3]By "conceptual field," educational researchers mean "the central ideas of a content area, understood conceptually, along with connections among them, so that the area is understood as a coherent whole."

©2000 American Mathematical Society

favored. What are their possibilities and their limitations? What can be learned from the experiences of countries where similar approaches were introduced a few years ago? In this study, we will try to reflect in some depth on such matters.

First, we will try to sketch the principal results that can be obtained from research in education. We will not try to be exhaustive; this sketch will only reflect a personal vision of the "state of the art," in which we will refer to the teaching processes in this field. Then, we will analyze methods of teaching and their evolution, in particular those in the French lycée, which will provide a good illustration of the general tendencies. In light of this case study, we will examine the possibilities and the limitations of the intuitive approaches that today are becoming generally used.

Student difficulties with the concepts of analysis

Research has shown the existence of strong and persistent difficulties. Those difficulties have various origins, but are mutually reinforcing and constitute an interrelated network. Nevertheless, in order to facilitate a synthesis we will group them in a few overlapping categories. These categories are the following:

- Difficulties connected with the mathematical complexity of the basic objects in this conceptual field: the real numbers, functions and sequences, which are objects continually in the mode of being mentally constructed by students when instruction in analysis begins.
- Difficulties connected with understanding the idea and the techniques of limit, which is the central concept in the field.
- Difficulties connected with the necessary break from modes of thought characteristic of algebraic manipulations.

Difficulties connected with the basic objects of the field. We cannot consider that the basic objects of analysis are new objects for students when they begin to study this material. In France, for example, irrational numbers and linear and affine functions are introduced in grades 8 and 9, and in grade 10, the concept of function becomes a central notion in the mathematical syllabus. Nevertheless, we cannot assume that these are already stable objects; on the contrary, analysis will play an essential role in their maturation and conceptualization.

Real numbers. Different studies show clearly that the concept of real numbers that students develop is not adequate for learning analysis (Robinet, 1986). Students' criteria for distinguishing among different sets of numbers remain weak and very dependent on the symbolic representations used (Munyazikwe, 1995). Beyond that, the increasing and barely controlled use of calculators in education tends to reinforce the idea that a real number is the same as a decimal, and one with less than 10 decimal places at that.

When teaching analysis begins, real numbers are algebraic objects. The students know very well that they are dense in themselves but, according to the context, students seem able to reconcile this property with the existence of a predecessor and successor for a given real number. For example, $0.999\ldots$ is often perceived as the number just before 1 and various questionnaires have shown that more than 40% of beginning university students in France believe that, if two numbers A and B satisfy the condition that $\forall n > 0, |A - B| < \frac{1}{n}$ then they are not necessarily equal but only very close, even infinitely close, that in a certain sense, one is the successor of the other. Students' understanding of the relationship between the real numbers and the real line also lacks coherence. Even when students claim *a priori* to accept the principle of a bijection between R and the real line, for example, they may or may not be convinced (depending on the context) that a specific number can be located on the real line (Castela, 1996).

Functions. For functions, the situation is even more complicated. It seems difficult to summarize in a few sentences the numerous and diverse results of research. Here we will limit ourselves to presenting only the major categories of difficulties identified in the research, categories that, once again, cannot be considered to be independent.

Difficulties related to the identification of what is actually a function and the realization that sequences of numbers are also functions. It is well established that the criteria used by students to check the functional character of a mathematical object do not necessarily correspond to the formal definition of function, even when the students can correctly quote this definition (Vinner & Dreyfus, 1989). These criteria depend more on the examples the students are most familiar with, examples that acquire the status of prototypes; also on associations such as the function/formula association or the function/curve association. Thus, the same mathematical object can be considered as a function or not according to the form of its symbolic representation. For example, the function f defined by $f(x) = 2$ is not recognized as a function because the given algebraic expression does not depend on x, but if it is introduced by means of its graphical representation, it will be recognized as such because it will be represented by a straight line. Such phenomena have led some researchers to differentiate between on the one hand what they call the "concept definition" and, on the other hand, what they call the "concept image" (Tall & Vinner, 1981).

Difficulties in going beyond a purely process conception of function to be able to relate flexibly the process and object conceptions in order to develop a sophisticated understanding (Tall & Thomas, 1991). In fact, research shows the qualitative jump that exists between the conceptual levels of function: process and object (Dubinsky, 1991; Dubinsky & Harel, 1992; Sfard, 1992). We can relate this jump to difficulties encountered by beginners when they have to consider as equal functions defined by equivalent but different processes, or when they have to work, not with specific functions, but with functions defined by some arbitrary property. Working in analysis becomes very difficult if the students can only rely on a process conception. This work, in fact, needs to consider functions as objects that can be included in higher level processes such as, for example: integration and differentiation and it also needs to consider not only particular objects, but classes of functions defined by specific properties, e.g., continuous, C^1, or Riemann integrable.

Difficulties in relating different symbol systems (Duval, 1995) used in representing and working with functions. These difficulties have been well investigated (Romberg, Carpenter, & Fennema, 1994) as much for the processes of translation from one symbol system to another, especially in the difficulties in translating from a graphical to an algebraic system (Dagher, 1996; Schoenfeld, Smith, & Arcavi, 1993) as for the difficulties connected with the simultaneous use of information that concerns different notions within the same system such as, for example, the graph of a function, its derivative, or its primitives. Moreover, research has explained very well how, in this domain, standard teaching practices tend to reinforce the difficulties by the way graphical representations are handled and by the status accorded to graphical reasoning.

Difficulties in going beyond numerical and algebraic modes of thought. This category of difficulties appears less in the education research literature, perhaps because our students are rarely allowed to choose the modes of thought they use. Nevertheless, it is an essential difficulty. For Euler, at least from the time of his famous book *Introductio in analysis infinitorum*, analysis is a mathematical field organized around the function concept, processes of variation, and functional thinking. Current research in France (Pihoué, 1996) tends to show that, when they enter 11th grade, students, who have been hearing about functions for three years at least, do not really see what is the interest in and value of functional thinking. For the majority, the function model remains essentially something related to the didactic contract, that is, the often implicit understanding between teacher and students that certain things are part of the education process, whether or not their purpose is clear.

Difficulties connected with the concept of limit. The difficulties that students encounter when they come in contact with the field of analysis are not restricted to those just mentioned. Difficulties associated with the notion of limit have been investigated extensively (cf. Cornu, 1991, for a first collection). For this particular domain, it is necessary to mention the role played by the notion of *epistemological obstacle* introduced by the philosopher Bachelard (see Cornu, 1983; Schneider, 1991; Sierpińska, 1985, 1988). For Bachelard, scientific knowledge is not developed in a continuous process, but rather results from the rejection of previous forms of knowledge that become epistemological obstacles. Researchers following Bachelard hypothesize that, in mathematics as well, some learning difficulties, especially the most persistent, result from forms of knowledge that have been, at some time, coherent and useful for students in their social life or at school. They also hypothesize that these epistemological obstacles are found both in the historical development of the concept and in current learning, in spite of obvious cognitive and cultural differences, as if they were a part of the genesis of the concept. Therefore, they make full use of historical analysis.

Regarding limits, there seems to be agreement at least on the following as epistemological obstacles:

- The common meaning of the word limit, which induces persistent conceptions of the limit as an unreachable barrier, or as the last term of a process, or which tends to restrict convergence to monotone convergence.
- The over-generalization of finite processes to infinite processes; in other words, the application of Leibniz's principle of permanence.
- The strength of a geometry of forms that impedes a clear identification of the objects involved in the limit process and the underlying topology. This, moreover, makes it difficult to understand the subtlety of the interplay between numerical and geometric contexts which is fundamental to understanding limits.

The endurance of such epistemological obstacles is confirmed by the difficulties that even advanced students typically have when asked the following, non-standard question: Why is it that the same method that consists of cutting a sphere into short cross-sections, approximating each section by a cylinder, approximating the sphere by the union of small cylinders, and then passing to the limit, gives the correct answer when applied to calculate the volume of the sphere, but gives an incorrect answer when used to calculate its area? Since the set of cylinders approximates the sphere geometrically, most students cannot understand how the limiting values for all numerical magnitudes associated with the cylinders do not necessarily equal the corresponding magnitudes for the sphere.

As research has shown, all of these obstacles are also found in the historical development of the concept, in spite of the cognitive and cultural differences mentioned earlier.

In the education literature concerning limits, the search for epistemological obstacles has played an important role, but we can't think that the difficulties students have can be reduced to them. The concept of limit, like that of function, has two dimensions: one a process and the other an object. The possibility of managing these two dimensions effectively requires cognitive processes, called encapsulation (according to the theory elaborated by Dubinsky) or condensation and reification (according to the theory elaborated by Sfard). The complexity of those processes is well understood today. This fact contributes to explaining why, in all countries, students find it so hard to identify $0.999\ldots$ and 1: the first symbolic representation being clearly of process type and the second of object type. In order to consider the two as equal, it is necessary to avoid falling into the trap of these symbolic differences and be able to see beyond the process described by $0.999\ldots$ and be aware of the number created by this process and distinct from it.

Another important category of difficulties comes from the characteristics of the formal definition of the limit concept, at least in the standard analysis that is taught today: the definition is complex and requires inverting the direction of the function process that gives to the variable x the value of the function $f(x)$. But more than these formal characteristics, there is an essential point: between an intuitive conception of limits and a formal conception, there is a fundamental qualitative jump, also attested to by the history of the concept. The formal concept of limit is a concept that, partially, breaks with previous understandings of this notion. When it appears on the mathematical scene, its role as a unifying concept, as an intrument for establishing a foundation for the field of analysis, as a "proof generated concept" in the sense of Lakatos (1976), is perhaps more important than its role as a productive tool for solving problems. We find here an epistemological dimension of the concept whose didactic transposition (that is, the transformation of an idea from its formulation in the field itself to a form convenient for teaching and learning it) in teaching is not easy. In fact, investigators such as Robert (Robert & Robinet, 1966) are convinced now that such epistemological characteristics cannot be transposed using a-didactic processes (Brousseau, 1986), that is, by confronting students with appropriate problems to solve. They hypothesize that an effective transposition needs specific mediations of a meta-mathematical nature. These investigators do not refer to Vygotsky but we can find certain similarities with the distinctions introduced by Vygotsky between different types of concepts, regarding their different modes of formation.

Difficulties connected with the necessary break with algebraic thinking. Mathematical activity in analysis is supported by facility with algebraic computations but, at the same time, in the introduction to what we may call, for lack of a better phrase, analytic thinking, it is necessary to take a certain distance from algebraic thinking. In France, Legrand (1993) has contributed to calling attention to this problem. As he has shown clearly, the break between algebraic and analytic thinking is organized according to various dimensions. The most important are the following:

First, to enter the world of analysis and become an effective analyst, it is necessary to enrich one's vision of the notion of equality and develop new methods for proving equalities. It is interesting to note that a similar reconstruction of the notion of equality in the transition from numerical thinking to algebraic thinking was revealed by education research.

Briefly, in algebra, when one wants to show that two expressions, $a(x)$ and $b(x)$ are equal, the standard strategy is the following: transform one or both expressions by successive equivalences until you obtain two expressions that are clearly equal, or transform the difference (respectively, the quotient) in the same way until you obtain 0 (respectively, 1). In analysis, if we don't just restrict ourselves to the algebraic part, this strategy will often not be achievable, or at least, it will not be very economical because we don't see the objects of analysis as algebraic objects and because we often work with local properties. That's why we have to develop a vision of equality associated with the idea of "locally infinite nearness," that is, if, for a given pair if numbers A, B, $\forall \epsilon > 0$, $d(A, B) < \epsilon$, then $A = B$. This new relation with equalities gives, in fact, a predominant role to inequalities and to the mode of local reasoning by sufficient conditions.

For example, if you want to show that there exists a neighborhood of x_0 in which $a(x) < b(x)$, you rarely try to solve this inequality as you would in algebra. The transformation introduces successive expressions $a_1(x), a_2(x), \ldots, a_n(x)$, reducing the neighborhood as necessary to guarantee that locally $a(x) < a_1(x) < a_2(x) < \ldots < a_n(x)$ and continuing until you obtain a certain neighborhood of x_0 on which $a_n(x) < b(x)$. Each step of the process requires difficult decisions: you have to accept a certain loss of information about $a(x)$, but not too much because you want to remain less than $b(x)$ so you have to combine these decisions with the subtle play of the neighborhoods.

An awareness of all these changes, and of the corresponding increase in technical difficulty of the mathematical work, helps us to understand better the distance that separates the ability to formulate the formal definition of limit, to illustrate it with examples and counterexamples, to represent it graphically from technical mastery of this definition, that is, the ability to use it as an operative tool for solving problems.

To clarify this first part, we will mention another dimension of the algebra–analysis break. Entering the world of analysis also obliges students to reconstruct already familiar mathematical objects, but in other worlds. The notion of tangent provides a prototypical case of such a construction. From the teaching in the lycée, the students first encounter this notion in the context of the circle. The tangent is a geometric object that possesses certain specific properties:

- it does not cross the circle,
- it touches it at a single point,
- at the point of contact, it is perpendicular to the radius.

All of these ideas are global and have nothing to do with the idea of common direction. Furthermore, to help students be aware of the abstract nature of geometrical objects, teachers emphasize that even if, visually, the circle and the tangent appear to coincide locally, in reality, there is only one common point. This geometric conception can be generalized and applied to other curves, for example, parabolas. It leads to the algebraic conception of the notion of tangent as a line which has a double intersection with the curve, which is an operative result for algebraic curves. Clearly, there is no direct relationship between this tangent, and the definition in analysis, which is characterized by a local property: the line with which the curve tends to coincide locally (affine approximation of order 1) and whose slope is given by the value of the derivative of the associated function.

A recent study by Castela (1995) shows that, at least in France, the educational system is not sensitive to this problem and leaves to individual student work the necessary reorganization of the conceptions. This has the consequence that, when they have passed their baccalauréat, the majority of students find themselves in an intermediate phase, without having a coherent global conception of the tangent, even if they are successful with the usual exercises. The same study shows that, when teaching deals with it, the reorganization occurs and is stable.

Evolution of the French programs

The reform of 1902: A pragmatic and algebraic approach to analysis. As in other countries, it was at the beginning of the 20th century, with the reform of 1902, that teaching analysis at the lycée level become widespread in France (Artigue, 1996). The introduction was helped by the most famous mathematicians of the era such as Poincaré, Borel, and Hadamard, among others, and it was clearly successful as shown by various indicators, for example, the ICME study organized 10 years later (Beke, 1914).

According to the mathematicians who participated in this reform, teaching analysis at the lycée level should be rigorous, free of all metaphysics (e.g., all kinds of infinitesimals), but at the same time, it should be accessible to the students and useful for mathematics as well as the physical sciences. The following declaration that Poincaré made in his famous lecture on mathematical definitions (Poincaré, 1904) and the report of the ICME study illustrate these positions clearly:

> It is no doubt difficult for a teacher to teach what does not completely satisfy her or him, but the satisfaction of the teacher is not the only object of teaching; one should be concerned primarily with the mind of the student and what one wants to do with it.

> Our principal task is to introduce the ideas of differential and integral calculus in an intuitive manner, starting with geometrical and mechanical considerations, and progressively rise to the necessary abstraction. All of our statements have to be true, but we don't have to state the whole truth.

Mathematicians like Poincaré were convinced that it was possible to develop a coherent curriculum in analysis based on these principles without encountering any important difficulties. Concerning the limit concept, we can read the report of the ICME study already cited: "The concept of limit comes up so frequently in the lycée and even at the lower levels (infinite decimals, area of a circle, geometric series, etc.) that its general definition should not cause any difficulty."

In fact, then, what was taught was a standard calculus, but it is necessary to realize that the contributions of this calculus as much to solving the classical problems as well as the level of rigor that could be achieved, made its advantages clear to all.

The reform of the 1970s and the reform of modern mathematics: Towards a formal approach to calculus. The curriculum was not changed very much until the beginning of the 1960s. At this time, structuralism came to be predominant and the positivism that had inspired the reform of 1902 was discredited. Mathematicians had discovered the power of algebraic structures and the issue of foundations became more and more important. In France, it was the golden age of the Bourbaki, created in 1937 with the goal of revising the differential and integral calculus course in undergraduate mathematics. The teaching of analysis at the lycée level was perceived as obsolete, not very rigorous, and not concerned with the central ideas of the field.

The revision of the secondary curriculum began in 1960, when a less empirical and pragmatic concept of calculus was introduced with a more complete focus on the fundamental concepts: functions, limits, continuity and on its structural dimension. In the same period, quantifiers, elements of set theory and algebraic structures entered explicitly into the syllabuses.

This was definitely a revision, but not a revolution. A careful study of the textbooks shows that, at this point, it was in a transition phase, the novelties introduced did not transform the old organization, the old world.

The reform of modern mathematics definitely turned a page at the beginning of the seventies. Analysis cannot be considered a mainstay of the reform, but the spirit of modern mathematics came to have a profound influence on its teaching which became essentially formal and theoretical, concentrating more on questions of definition and foundations than on the solution of problems. This type of teaching would be rejected shortly by the general reform of modern mathematics.

The rejection of modern mathematics and the introduction of intuitive approaches. The last important reform resulted from the rejection we have just mentioned which occurred in 1982. It was supported by reflection and experimental work conducted within the purview of the Mathematics Education Research Institutes (IREM from its initials in French) which had been created in the seventies. The reform was based on a conception of mathematics as a constructed science, depending on historical and cultural contexts for its development, an idea that was becoming prevalent. Moreover, it tried to find a more adequate balance between the logic of mathematical knowledge and the logic of the cognitive development of the student. It also tried to find a more satisfactory equilibrium between the "tool" and "object" dimensions of analysis according to the distinction introduced by R. Douady in 1986, that is, between internal development and the mathematical structuring of the fundamental concepts in the field on one hand, and on the other, its utilization as instruments for solving internal or external problems in the field of mathematics. The proposals of the commissions formed by representatives of the IREMs about analysis published in 1981 reflect these ambitions. They are as follows:

- Modify the relations between theory and applications and organize the syllabuses around the solution of problems that are wide-ranging and representative of the epistemology of the field, such as problems of approximation and optimization. Analysis is thus seen, according to Dieudonné, as a field in which the central processes are "approximate, underestimate, and overestimate."

- Find a better balance between qualitative and quantitative analysis, giving more importance to problems of a quantitative type (for example, problems concerning rate of convergence, numerical approximations of integrals, or solutions of equations, etc.), making use of pocket calculators.
- Give particular importance to simple and typical examples that then can serve as references, and avoid too much early interest in pathological situations.
- Theorize only as necessary with reduced levels of formalization that are accessible to students.
- Pay attention to the relations between teaching and learning in the constructivist theoretical framework.

The proposals that were made reflected directly on the new curriculum published in 1982. The general strategy that was used to introduce various key notions illustrates this clearly:

- Take advantage of typical simple behavior, both numerical and graphical, with the aid of calculators for the numerical part.
- Exploit these explorations to produce quantitative definitions adapted to the simplest cases and work with them.
- Let the students be aware of the limitations of the first approach by introducing more complicated cases.
- Introduce more general and qualitative definitions.

For example, the concept of derivative is introduced by means of the idea of first order expansion, exploring numerically and graphically the local behavior (at 0) of simple functions permitting estimates of the form $|f(x+h) - f(x) - f'(x)h| \leq Mh^2$. Afterwards, functions that do not admit such simple estimates are introduced and lead to the general definition of first order expansion.

Formalization is downplayed. A formal definition is given only for the limit at 0, and the comments that accompany the syllabus suggest that the teacher should be very careful about introducing it and should use it sparingly both in class and in the exercises. As one might expect, this isolated definition, whose role is diluted, was to disappear in the next version of the curriculum three years later, and the paragraph of the syllabus dedicated to limits is titled, very modestly, "the language of limits."

Mathematical activity is organized around solving problems: optimization problems, approximation of numbers and of functions, discrete and continuous variational models, etc. The notion of derivative, above all that of derived function, which is an essential tool for solving such problems becomes the central notion.

The logical order, limits-continuity-derivative, is broken. A minimal intuitive language of limits is introduced as a basis for the introduction of derivation, and then the notion of derived function becomes the keystone of the edifice. The notion of continuity almost disappears since, with the definition of limit that was chosen, any function that has a limit at a point in its domain of definition is necessarily continuous at this point.

The influence of analysis is already evident in the syllabus at Grade 10, a year before it is taught officially, as we see from the following extract:

- Upper and lower estimates of a function on an interval.

- Search for maxima and minima associated with elementary optimization problems.
- Rate of change: finding upper and lower estimates of the rate; inequalities of the form, $|f(y) - f(x)| \le M|y - x|$ for arbitrary x, y. Geometric interpretation.
- Use of the variations of a function to study equations of the form $f(x) = b$ and inequalities.

This curriculum is ambitious. As we pointed out earlier, it tries to emphasize, from the beginning of instruction, the epistemological value of analysis as a field where the processes of approximation play an essential role; it tries to organize a progressive introduction of students to thisfield simultaneously at technical and conceptual levels. In such an introduction, the relations among algebraic symbolic representations, graphs and techniques play an important role. This is indicated explicitly in comments in the syllabuses. The analysis that is taught is not a formal analysis, it is an intuitive and experimental approach to the field. Furthermore, the goal of not being limited to algebraic manipulations is noted clearly.

The (national) curriculum was modified in 1985, 1990, and 1993 so as to adapt it to the growing democratization of teaching at the lycée level, but its spirit has been maintained, at least in terms of official statements. That is why we can now measure the long term effects of nearly 15 years of such intuitive and experimental approaches.

Potentialities and limitations of intuitive approaches

First, we must point out some obvious successes. We will limit the discussion to three that apparently are particularly important.

The intuitive approaches have made this field accessible, up to a certain point, to all students. This success has not been considered to be of small importance, above all, when we take into consideration the fact that today the overwhelming majority (about 70%) of the students enter either a traditional or vocational lycée and therefore have two years of analysis.

The students come into contact very quickly with some central problems of this field such as variation, optimization, and approximation. The analysis that is taught is not restricted to its algebraic part and, following the syllabuses, the textbooks try to give a real importance to the numerical and graphical aspects of the concepts and techniques and to the quantitative dimension of analysis.

Calculators and even graphing calculators are standard tools for students, doubtless because, for the last 15 years, any kind of calculator can be used in national and regional examinations. The use of calculators has clearly led to making viable the numerical and graphical approaches envisioned in the syllabuses.

Nevertheless, in spite of these successes, one cannot think that we have found the royal road to teaching analysis. Some of the important problems have not been solved and new problems arise. Once again, this discussion will concentrate on those that are particularly outstanding.

The use of calculators has obvious limitations and furthermore, their real integration into the life of the classes continues to be unsatisfactory. As has already been mentioned, calculators, and even graphing calculators have been widespread in education at the baccalauréat level. In 1981, the Minister of

Education took the decision to authorize all kinds of calculators in the baccalauréat examinations and this situation remains in effect today. Therefore, French students can get their baccalauréat diploma with a graphing calculator and even a TI-92. The syllabuses themselves specify that baccalauréat students should have calculators and programming calculators at the lycée and they should learn to use them, in particular to study functions and sequences. These decisions were taken to promote the institutional use of these instruments and to contribute to overcoming certain resistance anticipated from teachers. But now, 15 years later, we cannot consider that the desired integration into teaching practices has been achieved. The majority of teachers continue to consider calculators to be "private" to the students and have not taken charge of the necessary instrumental learning. Fifteen years after the fact, they continue to be more concerned with the difficulties that arise from their introduction than the help they provide for the work of the teacher. Recent research published in the IREMs (Trouche, 1996) shows the negative effects of their uncontrolled use in teaching on conceptions developed by the students, for example, on those that concern the concept of limit, numerical approximation, or the interpretation of graphical representations produced by the machines. Moreover, we are realizing more and more that effective teaching of analysis with calculators requires specific learning and specific knowledge (Artigue, 1996). The education system does not easily recognize this fact and has little taste for dedicating the necessary time and energy to this learning.

The difficulties of viability of the approximation dimension in analysis. In recent years we have seen the difficulties that the education system has had with the approximation component of analysis. As was mentioned earlier, in order to develop this dimension, it is necessary to move a certain distance from the usual modes of thinking and introduce complex techniques whose mastery can be thought about only in long range terms. Teachers have serious difficulties in organizing and preserving an "ecological niche" for such mathematical practices, because they cannot avoid the competition between the techniques of approximation and algebraic techniques which are much easier to handle and learn (Artigue, 1993). Therefore, in spite of the guidance given in the text of the syllabuses, we notice a growing disequilibrium between approximation and algebraization.

The difficulties of viability of teaching centered on solution of wide ranging problems of epistemological significance. The curriculum of 1982 was intended to organize teaching analysis around wide ranging problems of epistemological significance. We note, on the one hand, once again a growing breach between these goals continually expressed in the syllabuses and, on the other hand, the organization and contents of the textbooks. The most widespread recent textbooks drift towards a weakly structured accumulation of limited problems whose solution is decomposed into so many subproblems that the students, merely executing tiny tasks, lose all awareness of the global problem situation. Perhaps it is possible to see here a more or less conscious adaptation of the educational system to changes introduced by the growing democratization of teaching for the baccalauréat. But, more than this, such an evolution shows clearly the strength of the processes of didactic transposition that shape the real curriculum (Chevallard, 1985) and always tend to bring the functioning of the system back to a previous equilibrium.

The difficulties that result from the growing lack of structuring. Once again, these difficulties are clearly noted on reading the most prominent recent textbooks. The status of objects, of concepts, of assertions are left hanging. Formal definitions have been rejected and replaced by more or less precise statements in natural language. Actually, these expressions only have the appearance of natural language; they don't have anything to do with the everyday language of our students. And they don't lead to an effective control over their practice. Moreover, because quantifiers are sometimes placed at the beginning of the sentence, sometimes at the end, these formulations do not necessarily help the students understand the subtle play of quantifications in definitions. Theorems are accepted on the basis of some explorations and not always mentioned as such. On reading these textbooks, one gets the uncomfortable feeling that the coherence induced by the logical structures of mathematical knowledge has disappeared without being replaced by any other solid form of coherence.

Conclusion

All evidence from education research shows that it is not easy for students to enter the field of analysis when it is not reduced to its purely algebraic version, but rather seeks the development of modes of thought and techniques that are fundamental to it.

The general introduction to analysis with the reform of 1902 gave to teaching in the lycée efficient tools for solving classical problems in mathematics as well as in the physical sciences. This introduction was clearly a success but its intentions remained limited and what was taught in this period was essentially a differential and integral calculus, that is, the most accessible part of analysis. With the reforms of the sixties, the curriculum in analysis was given new goals: analysis was made independent of algebra and its dimension as object emerged from its dimension as instrument. Shortly thereafter, the reform of modern mathematics imposed a formal vision on the field in which procedures based on fundamentals tended to become dominant. With the counterreform of 1982, this formal vision was rejected. A new organization of this conceptual field around problems of variation and approximation constituted in its history emerged and intuitive and experimental approaches were envisioned.

These intuitive and experimental approaches were imposed progressively. Today they appear as the only reasonable entrée, especially because courses in analysis are not now reserved for a mathematical or social elite. But, it must be admitted that learning introductory analysis has not been made miraculously easy, nor has the teaching been completely satisfactory. Intuitive and experimental approaches permitted the solution of some teaching and learning problems, but in the long run, they tended to generate others, as the French example shows. These intuitive approaches must be better controlled if we want to make sure that the ease of the first contacts does not generate serious obstacles in future learning. Current education research in France is increasingly devoted to these problems, which, 15 years after the counterreform, are becoming critical and concern the lycée as much as the transition from the lycée to the university.

The study of the processes of didactic transposition also shows the difficulties encountered when we try to take advantage of results or experiments that locally have had success in order to organize substantial and more global curricular changes.

The cognitive and epistemological focuses dominant today are clearly insufficient for taking that advantage. It is necessary to integrate approaches to the field of education that allow us better to take into account the role played by laws and institutional and cultural forces in the problems of learning and teaching.

References

Commission interIREM Université. (1990). *Enseigner autrement les mathématiques in DEUG SSM première année.* IREM de Lyon.

Artigue, M. (1991). Analysis. In D. Tall (Ed.), *Advanced mathematical thinking* (pp. 167-198). Dordrecht: Kluwer Academic Press.

Artigue, M. & Ervynck, G. (Eds.). (1994). *Proceedings of Working Group 3 on students'difficulties in calculus.* ICME 7. Université de Sherbrooke.

Artigue, M. (1993). Enseignement de l'analyse et fonctions de référence. *Repères IREM, 11*, 115-139.

Artigue, M. (1995). La enseñanza de los principios del cálculo: problemas epistemológicos, cognitivos y didácticos. In P. Gomez (Ed.), *Ingeniería didáctica en educación matemática* (pp. 97-140). México: Grupo Editorial Iberoamericano.

Artigue, M. (1996). Réformes et contre-réformes dans l'enseignement de l'analyse au lycée 1902-1994. In B. Belhoste, H. Gispert & N. Hulin (Eds.), *Les sciences au lycée—Un siècle de réformes des mathématiques et de la physique en France et à l'étranger* (pp. 197-217). Paris: Ed. Vuibert.

Beke, E. (1914). Rapport général sur les résultats obtenus dans l'introduction du calcul différentiel et intégral dans les classes supérieures des établissements secondaires. *L'Enseignement Mathématique, 16*, 246-284.

Bachelard, G. (1937). *La formation de l'ésprit scientifique.* Paris: J. Vrin.

Brousseau, G. (1986). Fondements et méthodes de la didactique des mathématiques. *Recherches en Didactique des Mathématiques, 7*(2), 33-116.

Castela, C. (1995). Apprendre avec et contre ses connaissances antérieures—un exemple concret: celui de la tangente. *Recherches en Didactique des Mathématiques, 15*(1), 7-47.

Castela, C. (1996). *La droite des réels en seconde: Point d'appui disponible ou enjeu clandestin?* IREM de Rouen.

Chevallard, Y. (1985). *La transposition didactique.* Grenoble: La Pensée Sauvage.

Commission interIREM Analyse (de.). (1981). *L'enseignement de l'analyse.* IREM de Lyon.

Cornu, B. (1983). *Apprentissage de la notion de limit: Conceptions et obstacles.* Doctoral dissertation, Université de Grenoble.

Cornu, B. (1991). Limits. In D. Tall (Ed.), *Advanced mathematical thinking* (pp. 153-166). Dordrecht: Kluwer Academic Press.

Dagher, A. (1996). Apprentissage dans un environnement informatique: Possibilité, nature, transfert des acquis. *Educational Studies in Mathematics, 30*(4), 367-398.

Douady, R. (1986). Jeux de cadre et dialectique outil-objet. *Recherches en Didactique des Mathématiques, 7*(2), 5-32.

Dubinsky, E. (1991). Reflective abstraction in advanced mathematical thinking. In D. Tall (Ed.), *Advanced mathematical thinking* (pp. 95-123). Dordrecht: Kluwer Academic Press.

Dubinsky, E. & Harel, G. (Eds.). (1992). *The concept of function: Some aspects of epistemology and pedagogy* (MAA Notes no. 25). Washington, DC: Mathematical Association of America.

Duval, R. (1995). *Semiosis et pensée humaine-registres sémiotiques et apprentissages intellectuels.* Berne: Ed. Peter Lang.

Farfán, R. M. (1993). *IV Seminario Nacional de Investigación en Didáctica del Cálculo.* Monterrey: Cinestav-IPN, Mexico.

Lakatos, I. (1976). *Proofs and refutations: The logic of mathematical discovery.* Cambridge: Cambridge University Press.

Legrand, M. (1993). Débat scientifique en cours de mathématiques et spécificité de l'analyse. *Repères IREM 10*, 123-159.

Munyazikwiye, A. (1995). Problèmes didactiques liés aux écritures de nombres. *Recherches en Didactique des Mathématiques, 15*(2), 31-62.

Pihoué, D. (1996). *L'entrée dans la pensée fonctionnelle en classe de seconde.* DEA, Université Paris 7.

Poincaré, H. (1904). Les definitions en mathématiques. *L'Enseignement Mathématique, 6*, 255-283.

Robert, A., & Robinet, J. (1996). Prise en compte du meta en didactique des mathématiques. *Recherches en Didactique des Mathématiques, 16*(2), 145-176.

Robinet, J. (1986). Les réels: quels modèles en ont les élèves. *Cahier de Didactique des Mathématiques, 21*, IREM Paris 7.

Romberg, T., Carpenter, T., & Fennema, E. (Eds.). (1994). *Integrating research on the graphical representation of functions.* Hillsdale, NJ: Lawrence Erlbaum Associates.

Schoenfeld, A., Smith, J., & Arcavi, A. (1993). Learning, the microgenetic analysis of one student's evolving understanding of a complex subject matter domain. In R. Glaser (Ed.), *Advances in Instructional Psychology* (Vol. 4, pp. 55-175). Hillsdale, NJ: Lawrence Erlbaum Associates.

Schneider, M. (1991). Un obstacle épistémologique soulevé par des découpages infinis de surfaces et de solides. *Recherches en Didactique des Mathématiques, 11*(2,3), 241-294.

Sfard, A. (1992). On the dual nature of mathematical conceptions: Reflections on processes and objects as different sides of the same coin. *Educational Studies in Mathematics, 22*, 1-36.

Sierpińska, A. (1985). Obstacles épistémologiques relatifs à la notion de limite. *Recherches en Didactique des Mathématiques, 6*(1), 5-67.

Sierpinska, A. (1988). Sur un programme de recherche lié à la notion d'obstacle épistémologique. *Actes du Colloque: Construction des savoirs: obstacles et conflits.* Montréal: CIRADE.

Tall, D. (Ed.). (1991). *Advanced mathematical thinking.* Dordrecht: Kluwer Academic Press.

Tall, D., & Thomas, M. (1991). Encouraging versatile thinking in algebra using the computer. *Educational Studies in Mathematics, 22*, 125-147.

Tall, D., & Vinner, S. (1981). Concept image and concept definition in mathematics with particular reference to limits and continuity. *Educational Studies in Mathematics, 12*(2), 151-169.

Trouche, L. (1994). Calculatrices graphiques: la grande illusion? *Repères IREM*, 14, 39-55.

Trouche, L. (1996). *A propos de la apprentissage des limites de fonctions dans un "environnement calculatrice": étude des rapports entre processus de conceptualisation et processus d'instrumentation.* Doctoral dissertation, Université de Montpellier 2.

Vinner, S. & Dreyfus, T. (1989). Images and definitions in the concept of function. *Journal for Research in Mathematics Education, 20*(4), 356-366.

UNIVERSITÉ DE PARIS VII, DENIS DIDEROT, PARIS, FRANCE
E-mail address: Michele.Artigue@gauss.math.jussieu.fr

CBMS Issues in Mathematics Education
Volume **8**, 2000

Evaluating Calculus Reform:
A Review and a Longitudinal Study

Betsy Darken, Robert Wynegar, and Stephen Kuhn

For more than fifteen years, members of the mathematics community have debated ways of improving the calculus curriculum; labored to implement new methods, technologies, and textbooks; and occasionally evaluated what has been accomplished with these changes. Unfortunately, the latter aspect of curricular change—which is the most difficult—has received the least attention. In 1995, the MAA report *Assessing Calculus Reform Efforts* lamented "the dearth of literature on evaluation of learning in undergraduate mathematics courses" (Tucker & Leitzel, p. 32), particularly in regard to calculus reform efforts. Because considerably more research has been published since this report, we will summarize the literature on the effectiveness of efforts to analyze and improve undergraduate mathematics education in general, and the calculus curriculum in particular. However, as this review will show, longitudinal studies of student performance in subsequent courses are still rare. To contribute to this critically important area of evaluation, we next describe a comparative longitudinal study of student performance through two years of calculus, including the performance of students switching from reform to traditional calculus, or vice versa, during the first year. As the number of such studies increases, the mathematics community will become more knowledgeable about the effects of curricular changes and be able to make more informed decisions.

1. Research Literature Review

Although anecdotal and reflective articles on undergraduate mathematics education have become quite common, research in this area is still limited. Research studies prior to the mid 1980's, which were primarily observational, revealed that even good students understood considerably less about key mathematical concepts than professors may have hoped. Orton (1983), for instance, found that almost all students did not understand integration as a limit of a sum, nor did they understand the symbols of differentiation. More recent research ranges from investigations of thought processes to practical studies of the effectiveness of various teaching methods, reflecting the interests both of researchers seeking epistemological and psychological foundations and of practitioners keen on identifying curricular and instructional improvements. For example, Tall (1991) analyzed the way in which human beings make generalizations, and hypothesized the existence of a "generic extension principle," in which people implicitly assume that irrelevant similarities among examples are essential components of the principle at hand. As a case in point, he found that students presented with examples of limits in which the terms

Calculus project supported by NSF grant DUE-1953285.

©2000 American Mathematical Society

never equal the limit tend to assume that this characteristic is generalizable to all limits. As this example illustrates, learners seem to develop a more complete understanding of complex concepts only through an evolutionary process. Moreover, this process may vary from learner to learner. For instance, in a study of the manner in which students developed deeper levels of understanding of the derivative, Zandieh (1997, and this volume) found that none of the nine students in her case study learned aspects of this concept in the same order.

In regard to identifying characteristics of effective instruction, Good, Mulyran, and McCaslin (1992) warned that a common pitfall of reform movements has been to assume that a single type of instruction will solve a general educational problem. Instead, a key conclusion to be drawn from research, mainly conducted at the K–12 level, is that the method emphasized is not as important as the quality of planning and instruction. The Research Advisory Committee of the NCTM (1996) has also advised reformers to examine innovations critically and be prepared to make adjustments based on experience. In addition, Hiebert (1999) has warned that debates about implications of research are often unproductive unless there is agreement about the goals and priorities of instruction. These points were illustrated recently in a discussion of teaching purposes and practices in calculus (Krantz, 1999; Tucker, 1999) by two well-known mathematicians and successful teachers, one heavily involved in calculus reform and the other identified with traditionalist teaching. Both clearly use well planned and carefully implemented teaching strategies, and because of their open-mindedness as well as cautiousness about new approaches, their teaching practices have evolved with experience (albeit in different ways).

Similarly, particular goals and teaching strategies may be interpreted or implemented very differently by different teachers. Because it is important to take such differences into account when considering research related to the calculus reform movement, main issues addressed by this research are listed below, along with their various interpretations and complications.

Conceptual understanding. "Conceptual understanding" is usually assumed to be self-explanatory, and is often contrasted to the results of drill work, the learning of procedures, and performance of computations. However, the boundary between conceptual and procedural understanding is hardly well-defined, the measurement of conceptual understanding has not been systematized, and, as the MAA's Task Force on the NCTM *Standards* remarked, albeit with respect to the K–12 curriculum, "[t]he challenge, as always, is balance" (Ross, p. 253).

Problem solving. Although this term has multiple meanings, it is usually associated with nonstandard problems. However, measurement of the ability to solve such problems is complicated by the fact that a problem which is nonstandard to one student may be standard to another.

The use of technology. "Technology" has been used to refer to graphing calculators, computer algebra systems like *Mathematica*, a variety of software packages, and calculator-based laboratories. "Use of technology" may range from making calculators or computers available to requiring their regular use. In addition, there are many variations on what technology is used *for*, from making computations to using graphing capabilities and simulations to enhance conceptual understanding.

Small groups. The use of small groups, also known as cooperative or collaborative learning, varies considerably. The authors of *A Practical Guide to Cooperative*

Learning in Collegiate Mathematics (Reynolds et al., 1995) remarked that "implementations of cooperative learning run the gamut from very loosely organized to highly structured classroom settings" (p. 3). Variations include: students in large lecture halls being given time to compare answers to a problem; individuals working with each other during a required class; students meeting in voluntary study sessions; groups organized by ability, heterogeneously or homogeneously; temporary or semester-long groups; unstructured or structured activities; group activities either supplemental to a lecture or as the main form of instruction; and groups engaged in major projects or brief classroom interactions.

Use of writing. The use of writing in reform mathematics has often been closely tied to developing conceptual understanding, with assignments including the writing of explanations, justifications, and proofs. However, writing assignments have also included lecture summaries, "math autobiographies," and essays addressing math anxiety (see Sterrett, 1990). In addition to this range of interpretations, other sources of variation include the extent to which writing is required, instructors' expectations, grading methods, and so on.

1.1. Reform in Undergraduate Education.

Research studies directly related to the effectiveness of college level reform efforts are relatively limited. The strongest conclusions that can be drawn from this literature are not surprising: first, students have a very difficult time solving non-routine problems and understanding concepts such as functions, limits, and rates of change (e.g., Ferrini-Mundy & Graham, 1991; Selden et al., this volume; Thompson & Thompson, 1996; White & Mitchelmore, 1996); second, there are no panaceas; and third, improving a curriculum is challenging, with "the devil in the details." It is also reasonable to conclude, based on the aggregated results of these studies, that certain curricular changes have been efficacious at some institutions. Unfortunately, the overall picture may quite possibly be skewed by selective reporting, as successful reform practitioners may have been more inclined to report results than unsuccessful ones. In addition, the studies are of varying quality, ranging from a significant number of excellent studies to reports with incomplete information. Such facts must be taken into consideration in drawing conclusions and determining what needs to be done to assess curricular changes more thoroughly.

This literature review focuses primarily on studies of courses using reform calculus texts, but first it is useful to note studies which have investigated issues connected to reform in college level mathematics. A comprehensive review of the research literature from 1975 to 1991 on teaching and learning college mathematics, conducted by Becker and Pence (1994), addressed many such issues, including research on problem solving, concept formation and misconceptions, as well as the use of computers in calculus. For instance, considerable evidence was found to support the value of explicit instruction on problem solving techniques, along with overwhelming evidence that, without instruction, students are poor at solving problems. A comprehensive report by Ganter (1999) on the effects of calculus reform is forthcoming. Preliminary results from this study, along with a number of other studies, are reviewed briefly below.

Technology. Kaput and Thompson (1994) provide a review of research on the use of technology through the early 1990's, ending with a lament about the paucity of technology-related articles in the *Journal for Research in Mathematics Education*, the premier journal for research in mathematics education. The number of studies on this issue, especially dissertations, has greatly increased in the last ten years. In these studies, the effectiveness of graphing calculators (e.g., Alkhateeb, 1995; Cassity, 1997; Hylton-Lindsay, 1997; Kinney, 1997; Lauten et al., 1994, Penglase & Arnold, 1996; Quesada & Maxwell, 1994; Stick, 1997; Wilson & Krapfl, 1994) and computers or computer software packages (Alarcon & Stoudt, 1997; Ayers, Davis, Dubinsky, & Lewin, 1998; Connors, 1995; French, 1997; Hare, 1996; Heid, 1988; Mayes, 1995; Padgett, 1994; Palmiter, 1991) has received the most attention. In her comprehensive review, Ganter (1999) found that 50% of the institutions conducting studies on the impact of technology reported improvements in conceptual understanding and other areas without loss of computational skills, while another 40% reported that students in classes with technology had done at least as well as those in traditional courses. An inspection of various reports support these findings. In particular, simply providing access to technology is usually unproductive (Alkhateeb, 1995); on the other hand, if the use of technology is integrated into a course, it has often led to improved conceptual understanding and maintained computational skills (e.g., Connors, 1995; Cooley, 1997). This has been the case with a number of sustained calculus reform programs that placed a primary emphasis on the use of technology, including the University of Connecticut project (Hurley, Koehn, & Ganter, 1999). In the latter, which also incorporated group problem-solving sessions, the overall mean score on a common final examination of students in a computer integrated calculus course was higher than that of traditionally taught students for all seven semesters included in the study. This examination included both conceptual and procedural questions. In addition, students from the computer-integrated sections took significantly more key post-calculus major courses than the traditional students (4.95 vs. 3.50), with no significant difference in the mean grades of the two groups. An earlier, less extensive study at Dartmouth (Baumgartner & Shemanske, 1990), where the computer package True BASIC used at Connecticut and other schools was developed, found no differences in performance between reform and traditional calculus students on a traditional final examination.

There are also well-documented reports of improved results based on the use of graphing calculators. For instance, Hylton-Lindsay (1997) found that business precalculus students using this technology outperformed traditional students and used better self-regulation. Hollar and Norwood (1999) found that an intermediate algebra class using graphing calculators and a strong graphing approach had a significantly better understanding of functions than a traditional class, while performing equally well on a common final examination of traditional algebra skills. However, an uneasiness exists among calculus instructors about the use of technology. According to a survey of 89 instructors who taught calculus using technology (Rochowicz, 1996), agreement about significant benefits of the use of technology was usually coupled with concern that students might not do as well without technology.

Use of small groups. There is an increasing amount of research on the effects of students working together at the college level (Bonsangue, 1994; Conrad, 1994; Davidson, 1990; Dees, 1991; Dubinsky, Mathews, & Reynolds, 1997; Ganter, 1994; Keeler & Voxman, 1994; Norwood, 1995; Urin & Davidson, 1992). As mentioned earlier, research in this area includes a broad category of activities, and the results of these studies are mixed. However, there is significant evidence that carefully structured small group learning has been quite effective in some settings. Norwood

(1995), who tracked 178 students from a developmental algebra course emphasizing problem solving and cooperative learning, found that 70% of the students in the treatment completed the subsequent college algebra course, compared to 46% of the 212 control students. These impressive results reflect those reported by Treisman (1985) and Bonsangue (1994) on the effective use of carefully structured workshops for minority calculus students. Treisman's methods have also been used effectively by others (Murphy, Stafford, & McCreary, 1998). Like Treisman, Bonsangue studied students who participated voluntarily in a special weekly workshop restricted to minority students, in which students worked collaboratively on challenging calculus problems. The mean grade for participating students was almost a letter grade higher than for minority non-workshop students, with almost twice as many still enrolled in mathematics or mathematics related fields after three years of enrollment. Interviews indicated that the workshop was a critical part of students' academic and social development. Citing psychological and other research, Bonsangue argued that the key to his students' success was identification with a group that was involved in meaningful academic activities and structured around cooperation rather than competition.

Supplemental Instruction. Consistent improvements have also been found with the use of Supplemental Instruction (SI), a relatively well-defined system which utilizes mainly voluntary, small group, cooperative learning sessions supervised by students or professionals trained in conducting such sessions. SI is usually attached to courses with traditionally low success rates. Burmeister, Kenney, and Nice (1996) presented summary data for 45 different institutions that showed significantly higher course grades for SI students compared to non-SI students in college algebra (2.21 vs. 1.98), calculus (2.28 vs. 1.83) and statistics (2.49 vs. 2.32). Kennedy and Kallison (1994) reported similar results in two studies of calculus classes, noting that SI appeared to help lower-achieving students most. Although SI and the programs cited above relied mainly on voluntary participation, the size of the effects strongly suggests that cooperative learning in carefully structured environments is effective.

Writing. Little research has been done on the use of writing in mathematics. A number of dissertations and unpublished reports (Beidleman, Jones, & Wells, 1995; Bolte, 1997: Hirsch & King, 1983; Isom, 1997; Porter, 1996; Wells, 1995) have documented mixed results. Significant results in some studies, such as Isom's finding of improvement in analytical abilities, suggest that there may be some potential for the use of writing to improve student achievement and understanding.

1.2. Research on Reform Calculus Texts. Most quantitative investigations of reform calculus materials involve the most popular reform text and the second best selling calculus book on the market, *Calculus* by Hughes-Hallett et al. (1998) (hereafter referred to as CCH), originally produced by the Calculus Consortium based at Harvard University. Several other studies have been conducted using other texts, including *Calculus from Graphical, Numerical, and Symbolic Points of View* (OZ) by Ostebee and Zorn (1997a), the third most popular reform text and the subject of the study described later in this article; *Calculus: Modeling and Application* (CALC) by Smith and Moore (1996) of Project CALC at Duke University; *Calculus, Concepts, Computers, and Cooperative Learning* (C^4L), an unpublished set of materials by Dubinsky and Schwingendorf (1991) and later Dubinsky, Schwingendorf, and Mathews (1995); and *Calculus and Mathematica* (C&M) (Davis, 1994), a text originally developed at the University of Illinois. Although a number of other reform texts have appeared on the market, these are the only ones for which we have found several studies of student performance.

These texts all claim to emphasize conceptual understanding, incorporate the use of technology, include varied and more challenging problems, and use a variety of representations (the "Rule of Four"). However, differences occur in both content and emphasis. CCH offers "a focused treatment of a limited number of topics" (Hughes-Hallett et al., 1998, p. viii). OZ limits the number of proofs and concentrates on "helping students understand concretely what theorems say, why they're reasonable, and why they matter" (Ostebee & Zorn, 1997a, p. xii), and expects that most instructors will incorporate group work and writing assignments. CALC uses only limited lecturing, emphasizing group projects involving real world problems and written reports. C&M is probably the most technology-intensive, with students exploring concepts and formulae using *Mathematica*. According to Park and Travers (1996), students are typically in the computer laboratory 6 to 12 hours per week. C^4L is probably the most unusual. Its central philosophy is that students should construct their own understanding of mathematical concepts, and it uses a mathematical programming language to facilitate this goal. Since these texts are different from each other, we have taken care to identify which text or texts were used for each study. However, this review is organized according to key categories such as conceptual understanding rather than texts because we found this made the results more understandable. We also found that within most of these categories the results were relatively similar across texts.

Our primary purpose in focusing on studies involving the above reform texts is to summarize the evidence available regarding the most well-defined, the most common, and the most comprehensive means of implementing calculus reform. The effectiveness of these texts is of keen interest to practitioners. Of course, as research related to the implementation of the NCTM *Standards* (1989) clearly shows (Ma, 1999; Stigler & Hiebert, 1999), and as anecdotal reports of calculus reform reconfirm (e.g., Brown, 1996), teachers—not textbooks—hold the key to changing educational practices. However, textbooks can be a significant guide to those intent on change. Schneider (1995), who studied the question of CCH implementation at the University of Texas at Austin, found that the CCH curriculum was for the most part implemented as intended. A much broader 1996 survey by Lauten (1996) of 117 institutions using CCH revealed that reform accompanied the use of the text, although in various forms.

Most studies of reform materials were aimed at answering the key question for practitioners, "Are reform materials 'better' than traditional materials?" Definitions of "better" included better student grades; higher performance on common exams; higher enrollment rates and better grades in successive courses; greater student interest, confidence, and involvement in mathematics; better conceptual understanding and problem solving skills; and better performance and retention among special groups of students, including minorities, women, and those with lower aptitude or weaker backgrounds. Despite a keen awareness that changing goals for calculus calls for a concomitant change in measures of achievement (e.g., Roberts, 1996), traditional performance measures were the most common means of assessment. These included semester grades, performance on common final exams, and grades in subsequent courses. The latter two measures are especially critical for comparative purposes because higher grades in a course do not in and of themselves indicate a text's superiority. Other measures of assessment included attitude tests, in-depth interviews, and pre- and post-tests. Most of the studies reviewed here included statistics or statistical test results. Studies which failed to include adequate information regarding statistical significance are clearly indicated.

End of Term Results: Final grades. Because a high failure rate in calculus courses was one of the original catalysts for reexamining the calculus curriculum, student grades in reform calculus have been of considerable interest. First, we will consider a number of studies using final grades to make comparisons, then studies using common final examinations. The former are subject to the charge that better grades may simply be due to grade inflation, although users have often claimed that reform courses are at least as demanding as traditional courses (e.g., Bookman & Blake, 1996). Another potential problem, at least for cross-institutional comparisons, is that reports of rates of success (grades of C or above) almost invariably fail to state whether or not withdrawals are included in the base. Because many mathematics courses are notorious for their high numbers of withdrawals, the exclusion of the latter can make quite a difference in reported success rates. In any case, comparisons of final grades within institutions, in combination with the results of other studies, provide useful information about the impact of calculus reform. Studies which compared final grades are listed in Table 1. Roman numerals listed with each text refer to the semester of calculus, while Hindu numerals refer to quarters.

Table 1. End of Term Grades in First Year Calculus Courses

Study	N Trad.	N Reform	Final Course Grades Traditional	Reform	Significance
C^{4}L: I					
Monteferrante, 1993	109	60	37% ABC	72% ABC	$p < 0.001$
Schwingendorf, 1999	≈ 4600	≈ 200	$C^4L >$ Traditional		not reported[a]
CCH: I					
Baxter et al., 1997	1122	1147	51.6% ABC	63.6% ABC	$p < .0001$
Ratay, 1993	≥ 250	≥ 250	2.08	2.67	not reported[b]
Mittag & Collins, 1999	549	62	2.96[c]	2.98[c]	not reported
Johnson, 1995	≈ 110	≈ 110	62% ABC	67% ABC	not reported
Brunett, 1995	41	31	similar	similar	–
Luke & Muller, 1994	≥ 10 sections		similar	similar	–
C^{4}L: II					
Monteferrante, 1993	57	57	54% ABC	68% ABC	$p = 0.06$
CCH: 2					
Ratay, 1993	≥ 250	≥ 250	2.09 GPA	2.45 GPA	not reported[b]
CCH: 3					
Ratay, 1993	≥ 250	≥ 250	1.98 GPA	2.48 GPA	not reported[b]
CCH: II					
Baxter et al., 1997	large	large	$GPA_{trad} + .26 = GPA_{CCH}$[d]		$p < .05$
Johnson, 1995	> 110	> 110	71% ABC	80% ABC	not reported
Luke & Muller, 1994	≈ 10 sections		similar	similar	

[a]Information about this C^{4}L study is based on an abbreviated report that implied that these results were significant. See Schwingendorf et al. (this volume) for more details.

[b]Ratay's differences are probably significant due to large sample sizes.

[c]Mittag and Collins' study was restricted to Calculus I students enrolled in Calculus II, so that only passing Calculus I grades (A through D) were reported.

[d]Baxter et al. adjusted Calculus II grades according to Calculus I grades, which, according to the authors, may have resulted in a smaller adjusted difference between Calculus II grades.

For the first term of calculus, none of eight studies showed reform students with significantly lower average grades, and four probably found significantly higher grades for reform students. Two of these studies (Baxter, Majumdar, & Smith, 1998; and Ratay, 1993) compared two successive years of large first year classes that had similar scores on standardized tests and other predictors of performance. Although no statistical test results were reported in Ratay's study, these are almost undoubtedly significant given the large differences and sample sizes. Another study favoring reform, by Monteferrante (1993), had the advantage of comparing reform and traditional Calculus I and II classes all taught by the same instructor over four years.

Of the four studies which reported similar grades in Calculus I, the study by Luke and Muller (1994), which involved at least 10 instructors and a campus-wide implementation, pointed out that the CCH final exam was much more challenging than the traditional final exam. In contrast, Brunett (1995) reported that traditional students performed better on five common problem solving questions than CCH students, although final grades were similar.

Of the six studies of grades in the latter part of first year calculus, none reported significantly higher average grades for traditional students and four found statistically higher grades for reform students, including several very impressive differences. A fifth study by Johnson (1995), which did not report significance and relied on relatively small samples, found higher reform grades. The only study which reported similar grades (Luke & Muller, 1994) included the comment that the CCH final examination was much more challenging than the traditional final examination.

The evidence presented above, with most of the reports showing higher reform calculus grades and none reporting significantly higher traditional grades, supports the claim that reform calculus grades are at least as high as traditional grades and higher in some situations. It would be useful to conduct a multi-institutional survey of grades in reform and traditional calculus to establish much more definitive conclusions regarding the issue of comparative success rates.

End of Term Results: Common examinations. We found eight studies, four of which were contained in newsletters published by commercial publishers, which compared reform and traditional students' performance on common examinations or parts of examinations (see Table 2).

Six studies that compared performance on traditional problems found no differences between reform and traditional students, with three conducting statistical tests to confirm this result. Of the seven studies comparing graphical understanding, conceptual understanding, or problem solving ability, four found that reform students performed significantly better than traditional students, two found results favorable to reform students without reporting significance, and one found that traditional students performed better. One of the results favorable to reform, the C&M study by Park and Travers (1996) is particularly noteworthy as the only study which utilized pre- and post-test achievement test scores adjusted for prior mathematical knowledge. On the other hand, Holdener's results at the Air Force Academy (1997), which reported the average reform grade on twelve common conceptual questions as 89.9% and the average traditional grade as 71.8%, must be interpreted in light of the fact that the second year C&M calculus students were selectively chosen from the upper third of the prior semester's Calculus II classes.

In an attempt to adjust for this fact, Holdener used GPAs, standardized test scores, and historical data to predict that the C&M group would out-perform the control group by about 5% on the common final exam. In fact the average overall score of the C&M group was 11% higher than the control group, possibly indicating an effect due to the use of C&M.

Table 2. Common Final Exam Grades in Calculus Courses

Study	Text	N Trad.	N Reform	Traditional/ Procedural	Graph, concept, problem solving
				Type of Exam Question	
Penn, 1997	CCH: I	620	122	NSD[a]	Reform > Trad[b]
Park & Travers, 1996	C&M: II	42	26	NSD	Reform > Trad[b]
Holdener, 1997	C&M: mvc	133	32	NSD	Reform > Trad[c]
Tidmore, 1994	CCH: I	288	126	similar	Reform > Trad
Hershberger & Plantholt, 1994	CCH: I	49	44	similar	Reform > Trad
Lefton & Steinbart, 1995	C&M: I	67	37	similar	–
Judson, 1994	OZ: I	unknown	2 sections	–	Reform > Trad[b]
Brunett, 1995	CCH: I	41	31	–	Trad > Reform[b]

[a]NSD stands for "no significant difference."
[b]Reported to be significant.
[c]Significant, but reform students were from the upper third of calculus II students.

In the third study with a significant difference favoring reform, conducted at the Naval Academy, Penn (1994) reported total scores to be significantly better for CCH students in both the standard 4-hour calculus course and the 5-hour version taken by weaker students, with no significant difference in the 3-hour course for students with prior calculus experience. In the fourth study, Judson (1994) found that reform students' grades were 17% higher than those for traditional students.

Two other studies reporting positive results for reform, by Hershberger and Plantholt (1994) and Tidmore (1994), unfortunately did not report statistical tests for significance. The first of these studies had the commendable feature of controlling for the effect of instructors by using two instructors who each taught a section of CCH and a section of traditional calculus for two semesters. Tidmore's study compared averages within three groups, with students grouped according to standardized or placement test scores, and found CCH scores to be higher than non-CCH in all three groups.

Only one statistical study reported results favorable to the traditional method. This previously mentioned dissertation by Brunett (1995), which focused on a comparison of problem solving abilities of 72 students, found that a traditional class performed significantly better than a CCH class on five common final exam problems developed to test problem-solving ability.

Another study, an unpublished report (Silverberg, n.d.) not included in Table 2, described a phenomenon occasionally mentioned in the descriptive literature: the mean final exam grades of three C^4L sections taught in three successive semesters went from much worse, to the same as, to better than traditional students' mean grades. According to the author, this improvement was due to working out a variety of kinks in the new program, most related to its use of the specialized C^4L programming language.

An unusual follow-up study was conducted by Bookman and Friedman (1994), who tested CALC and traditional students one to two years after they had taken a year of calculus. Traditional students performed better, although not statistically better, on ten standard calculus skills problems, a result that inspired a revision of CALC to include more skills practice. In contrast, CALC students performed better, but not significantly better, on an essay question about the concept of the derivative, although both groups performed equally poorly on ten short answer conceptual questions. Regarding the latter results, Bookman and Friedman commented that fatigue may have been a factor because these questions were at the end of a test that lasted several hours.

The preponderance of the evidence reported above suggests that students using reform calculus texts usually perform as well as traditional students on traditional test questions, and perhaps better than traditional students on conceptual or graphical questions. This conclusion is strengthened by the fact that the two studies which contained the strongest evidence of superior CCH performance involved large numbers of students, while the only study with negative results for a reform class involved only a few sections. However, given that these studies are limited in number and observational in nature, a broader cross-institutional study would be very useful here.

Subsequent Course Performance. Obviously student performance in subsequent courses is a matter of critical concern in the hierarchical mathematics curriculum. Unfortunately such longitudinal studies are not only difficult to conduct but often have inconclusive results due to small sample sizes. Six longitudinal studies were found in the literature, with results for mathematics courses summarized in Table 3.

Table 3. Reform and Traditional Students in Subsequent Math Courses

Study	Text	Trad. Calc. II	Diff. Eqns.	Multivar/ Vector Calc.	Lin. Alg.
Mittag & Collins, 2000	CCH	R > T[a,b]	–	–	–
Armstrong et al., 1994	CCH	T > R[b,c]	–	NSD[d]	NSD
Johnson, 1995	CCH	T > R[b]	(T > R)[e]	–	(T > R)[e]
Baxter et al., 1998	CCH	–	NSD	T > R[b]	–
Schwingendorf, 1999	C[4]L	–	NSD[f]	NSD[f]	NSD[f]
Bookman & Friedman, 1999	CALC	–	FD[g]	FD[g]	FD[g]

[a]Reform > Traditional.
[b]Statistically significant.
[c]Traditional > Reform.
[d]NSD stands for "no significant difference."
[e]Sample sizes very small; no statistical test conducted.
[f]No significant difference for unspecified post-calculus mathematics courses.
[g]"Few differences" in subsequent mathematics related courses.

Three studies statistically compared CCH and traditional Calculus I student performance in traditional Calculus II. Johnson (1995) found that only 55% of 47 CCH Calculus I students made C or above in Calculus II, compared to 81.1% of 185 traditional students. Armstrong et al. (1994) also found results in favor of traditional students ($p < .05$). In contrast, Mittag and Collins (2000) found a statistical difference in grade distribution favoring 62 CCH students over 549 traditional students ($p = .047$). However, the distributions of A, B, and C grades were very similar, with identical overall success rates of 56%. The difference occurred among

D grades and "other" grades, presumably mostly withdrawals. Among traditional
students, 11% made D's and 22% made "other"; among CCH students, 24% made
D's and 12% made "other." This behavior had the interesting effect of yielding
a lower—although not significantly lower—Calculus II GPA for CCH Calculus I
students (2.13) compared to traditional Calculus I students (2.20), which indicates
that simply comparing GPA's without considering withdrawal rates may be misleading. Nonetheless, the overall evidence bodes ill for CCH Calculus I students
taking a traditional Calculus II course.

Johnson also compared reform and traditional Calculus I student performance
in CCH Calculus II. Success rates were not significantly different, with 92% of 25
traditional Calculus I students making C or above, compared to 80.6% of the 67
CCH Calculus I students.

With regard to student performance in second year mathematics courses, results are meager. The largest and most extensive longitudinal study was conducted
by Baxter et al. (1998), whose results in other areas have been reported earlier. An
entire class of over one thousand first-year CCH students were compared with the
previous year's first-year students, who had about the same average ACT score.
Although no significant differences were found in Differential Equations, there were
lower grades for CCH students in Vector Calculus. The authors conducted further
analyses that indicated that CCH students performed about the same as traditional students in Vector Calculus sections in which the styles of homework were in
some ways similar to CCH, but significantly worse in sections that were completely
traditional. These researchers also found a few other statistically significant differences after examining a multitude of science courses. These included first-semester
physics, in which the CCH students averaged half a grade higher with a 20% higher
success rate (grades of at least C), and Physics II, in which CCH students did less
well than traditional students.

In a second study, Bookman and Friedman (1994) found few differences in
grades for most subsequent mathematical courses. However, in a comparison of
subsequent science courses, a significant difference of 0.2 between average GPAs
was found which favored traditional students, apparently due to differences in less
mathematically dependent courses such as Introductory Biology. The authors spec-
ulated that this may have been due to the propensity of premedical students to
choose the traditional calculus series. The third longitudinal study by Schwingendorf (1999) briefly reported finding no significant differences, without mentioning
specific courses. The fourth, by Armstrong, Garner, and Wynn (1994), briefly summarized three sets of comparisons of subsequent grades for three types of Calculus
I students, including CCH students, students using *Calculus Using Mathematica*
(based at the University of Iowa), and traditional students. The only statistically
significant difference, which occurred in only one of the three studies, favored *Calculus Using Mathematica* students in their subsequent mathematics courses ($p < .05$).
Wisely, the authors warned that this result should be treated cautiously due to such
problems as small sample sizes. They reported no significant differences in a number
of other courses, including linear algebra and multidimensional calculus.

Three more longitudinal studies did not report any significant differences. Johnson (1995) reported that the success rates of less than 20 CCH students were lower
than those of 81 traditional students in Differential Equations and Linear Algebra,

but sample sizes were too small to make any inferences. An unpublished dissertation by Alexander (1997) reported that subsequent grades of a large number of CCH students at the University of Arizona may have been somewhat better than those of traditional students. A similar study, conducted at three other CCH institutions by Alexander and Madden (1998), reported that, at two of these schools, there was a very mild trend favoring CCH students. The study also reported that women and minorities using CCH were more likely to enroll in and receive higher grades in subsequent courses, but that the overall numbers of such students were too small to make inferences. The firmest conclusion reached by the authors was that CCH students were not at a disadvantage relative to traditional students in subsequent courses.

Several tentative conclusions are suggested by these studies. First, due to the mixed evidence regarding performance of reform calculus students in subsequent traditional mathematics courses, close attention should be paid to such transitions. This is especially true in the case of CCH students enrolling in traditional Calculus II. Second, there is a pressing need for more longitudinal studies, especially with regard to the performance of students moving from reform to traditional courses.

Conceptual Understanding. A fascinating development in calculus reform has been the shift from an initial effort to concentrate on fewer topics, in an *implicitly teacher-oriented* curriculum, to a focus on developing deeper understanding in an *explicitly student-oriented* curriculum. As Schoenfeld put it in 1995, " 'understanding the concepts' has taken on a richer meaning than might have been suspected by many of the participants in the Tulane conference" (1995, p. 4). Conceptual understanding has become the focal point of most reform efforts, relative to which reform activities are valued. This apparently major difference in emphasis between traditional and reform calculus indicates that comparative studies limited to grade distributions and traditional exam results are not in and of themselves adequate for evaluating reform efforts.

Evidence has already been presented regarding comparative performance on conceptually-oriented final examination questions. Ironically, further in-depth evidence pertaining to this key feature of reform is hard to come by. While improved understanding is a common claim in anecdotal reports (e.g., Gordon, 1997; Judson, 1994) and research on student understanding has resulted in an expanding number of tightly focused theoretical studies of specific concepts (Williams, 1991; Zandieh, 1997), only six comparative studies have been found which studied conceptual understanding in detail. The largest and most thorough was conducted by Park and Travers (1996), who found that 26 C&M students demonstrated better understanding of conceptual interrelationships than 42 traditional students. They also found, via interviews of 12 C&M and 22 traditional students, that C&M students demonstrated clearer understandings of the nature of the derivative and integral. Similarly, Allen (1995) concluded that six CCH students had a better understanding of continuity, differentiation, and integration, although six traditional students to whom they were compared performed better on limits. In a less formal and even smaller study, Roddick (1995) conducted a case study of seven students in an engineering mechanics course, and found that the four C&M students were more likely to solve problems from a conceptual viewpoint than the three traditional students.

Two additional studies found no significant differences. In a comparison between 14 students using an unspecified reform text and 14 using a traditional text,

Williams (1994) found no significant differences in understanding of the function concept. In fact, both groups showed poor understanding. Likewise, Soto-Johnson (1996) found no significant differences among an unspecified number of CALC, C&M, and traditional students on conceptual understanding of infinite series, although there may have been some evidence that CALC students had a better understanding of the relationships among differentiation, integration, and series.

In contrast to the five studies above, the sixth study of conceptual understanding favored traditionally taught students. Meel (1998) conducted an in-depth study of the understandings of third semester honors calculus students, 16 of whom had completed a year of C&M and 10 of whom had completed a year of traditional calculus. He found that traditional students performed significantly better on conceptually oriented test questions and tasks presented without figures. There were no differences in regard to differentiation and integration items.

Unfortunately, as is typical of clinical research, all of the studies above involved small or even very small sample sizes. So the results are quite tentative, with a mild trend in favor of reform. As better tools are developed and used for assessing conceptual understanding in the classroom, more definitive evaluations of conceptual understanding in reform and traditional calculus programs can be conducted. The recent MAA publication *Student Assessment in Calculus* (Schoenfeld, 1997) and other publications on assessment may expedite such work.

One final study related to conceptual understanding was conducted by Porzio (1994), who studied 100 students from three types of calculus programs. This study found that C&M students were better than the other two groups (a graphics calculator group and a traditional group) on their ability to use and make connections between numerical, graphical, and symbolic representations in calculus, providing some evidence that direct teaching of multiple representations of various concepts can be effective. The assumption that becoming more familiar with multiple representations is important for gaining better understanding has received some research support (e.g., Shoaf-Grubbs, 1994).

Problem Solving. Problem solving, an ability considered to be closely associated with conceptual understanding and regarded by many as the *sine qua non* of mathematics education, has also received considerable attention in the reform movement. Research in this area includes such works as *Mathematical Thinking and Problem Solving* (Schoenfeld, 1994). A summary of the literature relevant to mathematical thinking and problem solving (Schoenfeld, 1992) was published in 1992. Only three studies have been found which investigated problem solving in the context of reform calculus texts. Bookman and Friedman (1999) found that 88 CALC students performed significantly better than 247 traditional students on a problem-solving test administered over two successive years at Duke University. Meel (1998) found that C&M students were more successful and flexible in solving realistic problems. This study is also the one referred to earlier which reported that traditional students outperformed C&M students on a test of conceptual understanding, indicating that perhaps problem solving and conceptual understanding may not be as closely linked as some may think. On the other hand, the third study, the previously cited study by Brunett (1995), found that a class of 31 traditional students outperformed a class of 41 CCH students on a common problem-solving portion of their final exams. Brunett also found that the traditional students used significantly more problem solving techniques related to solving maximum/minimum problems. Obviously a great deal more research is needed in this area.

Attitude and Interest. Student attitudes toward and interest in mathematics have also received considerable attention in calculus reform. Many anecdotal accounts (e.g., Alkhateeb, 1995; Lock, 1994) report very positive improvements in these areas, but there have been other reports indicating major dissatisfaction, especially in the first year of implementation (Alarcon, 1997; Silverberg, n.d.). Unfortunately, research on attitudes of students using reform texts seems to be very limited, with our search yielding only five studies. Three studies, by Allen (1995),f Bookman and Friedman (1999), and Park and Travers (1996), found better attitudes toward mathematics among reform students. Allen based this conclusion on a case study involving only 12 CCH students, while the results of Park and Travers were based on a larger sample size of 68 C&M students. Bookman and Friedman's study was particularly interesting because he measured CALC and traditional students' attitudes one to two years after they had taken a year of calculus. Two other studies, by Soto-Johnson (1996) and Alexander (1997), both apparently studying only two or three classes, found no significant differences in attitude between traditional and CCH, CALC, and C&M students.

Although various measurements of students' affective characteristics are certainly of interest, they may well present too simplistic a reading of a complex phenomenon. After all, the feelings and opinions of teachers and professors about reform and traditional calculus are often quite mixed, especially in this period of transition, and it is likely that the same can be said for calculus students. Meel (1998) found such mixed reactions in interview responses of C&M honors and traditional honors students regarding the strengths and weaknesses of their respective programs. It would also be useful if future studies of affective characteristics separated students' likes or dislikes from their willingness to work hard and continue their studies in mathematics. That these are two separate issues is shown by international studies comparing Japanese and American secondary students, in which American students indicated a more positive attitude toward mathematics than their Japanese counterparts, while the latter demonstrated far greater mathematical achievement (McKnight et al., 1987).

Retention. We found seven reports on student retention, most very abbreviated but all describing higher retention among reform students. Unfortunately, none reported any statistical results, although some of the differences are so large that significance is virtually assured. Such is the case with Monteferrante's study (1993), which reported startlingly different attrition rates from Calculus I to Calculus II: 5% for C^4L and 41%–59% for traditional students. Unfortunately, these results are difficult to interpret since "attrition rate" is not well defined in the study. Lefton and Steinbart (1995) also found large differences, with 58% of C&M Calculus I students going immediately into second semester calculus, compared to 39% of traditional students. Hershberger and Plantholt (1994) reported that 10% to 15% more CCH students continued to Calculus II. Johnson (1995) found that 63% of the CCH students took Calculus II compared to 56% of the traditional students. However, the latter study also reported minor differences in enrollment rates in Differential Equations and Linear Algebra, both favoring traditional students. Bookman and Friedman (1999) reported some improvement in continuation rates for CALC students from Calculus I to Calculus II and from there into more advanced mathematics courses. Schwingendorf (1999) found that C^4L students took more calculus and post-calculus courses. Phillips (1995) also found an initial increase in post Calculus I mathematics enrollments of 8.1% for CCH students, in particular increases of 6% to 10% from Calculus I to Calculus II during the first three years of CCH implementation. However, a decline occurred after the mathematics department switched to OZ. Phillips remarked that other changes at the school, notably an increase in science majors, may have had more of an effect on enrollments than changes in the calculus. Overall, these results provide evidence that, on average, the use of reform texts and methods increases retention.

With regard to the performance of weaker students in reform calculus, Ratay (1993) found a gap of almost a full letter grade in favor of "weaker" CCH students with Math SAT scores between 500 and 540 compared to similar traditional students. There were also moderate to small gaps in favor of CCH among students with higher SAT scores. On the other hand, Armstrong (1997) who studied the characteristics of successful CCH and traditional students at a large number of community colleges, found no support for the claim that students with weak precalculus backgrounds, as measured by students' algebraic ability, can succeed in CCH calculus with the aid of technology. Both of these studies were reasonably large but in very different settings (respectively, the U.S. Merchant Marine Academy and a number of community colleges), a fact that may explain these results. In particular, "weaker" students at community colleges would probably not even have been admitted to the Merchant Marine Academy. Armstrong also commented that there was some evidence that non-Asian minorities performed less well in reform calculus.

Conclusion. Obviously more research needs to be conducted on the effects of calculus reform. Yet the limited research reviewed here provides some evidence that the use of reform texts and methodologies can have a positive impact. Although results are not consistent, the fact that there are very few studies reporting negative effects may indicate that the calculus reform movement is adhering to the advice of at least one critic to "First, do no harm" (Askey, 1997). There are enough studies reporting positive results to infer that certain changes may even be doing some good, which is certainly good news to the mathematics community. Of course, since almost all studies reviewed here are observational rather than experimental, causality cannot be proven. However, as in the case of the relationship between the use of tobacco and the occurrence of cancer, an accumulation of carefully collected and well-reported observational evidence can become quite convincing. It is important, therefore, to continue collecting evidence. The longitudinal study which follows serves this purpose.

2. Longitudinal Study

The primary purpose of this study was to compare the performance of students using the Ostebee and Zorn reform first year calculus text (OZ) to those using traditional texts. Comparisons were made by using final grades in Calculus I, Calculus II, and second year calculus. There were no common final exams. The drawback of using grades to measure achievement was mitigated to some extent by the use of large numbers of students and instructors in first year calculus, as well as by comparisons of subsequent course performance. A secondary purpose was to determine if the use of two textbooks was deleterious to students who "crossed over" from one type to another as they advanced from Calculus I to Calculus II.

2.1. Program Description. Since 1991, several sections in each semester of first year calculus at the University of Tennessee at Chattanooga (UTC) have been taught using *Calculus from Graphical, Numerical and Symbolic Points of View* by Ostebee and Zorn (1997a). UTC is a public metropolitan university of about 8500 students, mostly commuters, with an average ACT of 21.9 for first year students, a mandatory placement system, and a relatively small number of mathematics and engineering majors. The OZ text was chosen because of its similarity in content with traditional texts. Features which distinguish the OZ text from traditional texts include the use of group projects, much greater use of graphs to illustrate concepts, more emphasis on numerically approximating convergent improper integrals and infinite series, significant use of technology, and reduced attention to complex symbolic integration problems. In general there is a much greater emphasis on conceptual understanding and less emphasis on mechanical operations than in a traditional textbook. Another distinguishing feature, based on anecdotal evidence, may be that OZ students work harder than traditional students.

2.2. Samples. The experimental sample used in Calculus I consisted of calculus students who took OZ Calculus I during the first eight semesters it was offered, from 1992 to 1997. A total of eight instructors volunteered to use the reform textbook in 20 of approximately 50 sections of Calculus I offered during these semesters. Students who enrolled in traditional sections of Calculus I during this same period were chosen as the comparison group. Students were not randomly assigned to nor solicited for the different types of Calculus I, nor was any differentiation made between these types in the schedule of classes. A comparison of high school grade point averages and ACT Mathematics scores found no significant differences between the two groups of students.

In Calculus II, the second semester of first year calculus, four instructors used the reform textbook in 14 sections of the course for six semesters between 1993 and 1998. For comparisons within Calculus II, a comparison group was chosen to be the 227 students enrolled in 10 traditional sections of Calculus II taught during these same semesters. A comparison of high school grade point averages and ACT Mathematics scores found no significant differences between these two groups.

In sequential investigations, students' backgrounds were designated as either reform or traditional according to their last successful attempt in the previous course. The experimental sample consisted of all students who had completed OZ calculus and the control sample consisted of students in the database who had taken traditional calculus during the same years as OZ students.

2.3. Results. Calculus I Comparisons. A comparison of the first-time grades of reform and traditional Calculus I students showed that the success rate (number of ABC divided by number of ABCDFW) of 55.5% for 321 reform students was significantly higher than the comparable rate of 46.2% for 450 traditional students ($p = 0.01$). If W's are disregarded, success rates of 68.2% vs. 60.5% for OZ and traditional students respectively were still significantly different ($p = 0.05$). The difference in withdrawal rates of 18.7% and 23.6% respectively was not quite significant ($p = 0.11$), nor was the difference in GPAs of 2.06 and 1.89 ($p = 0.13$).

Calculus II Comparisons. In Calculus II, a number of large and significant differences were obtained: the success rate of 64.5% for 197 OZ students was much higher than the rate of 43.8% for 146 traditional students. In addition, the OZ withdrawal rate of 14.2% was less than half the traditional W rate of 30.1%, and the OZ GPA of 2.32 was almost half a letter grade higher than the 1.93 GPA for traditional students. All of these differences were significant (all p-values < 0.01).

Crossover Results. An analysis of the performance of Calculus II students according to their Calculus I backgrounds showed surprisingly close results within each type of Calculus II, as shown in the next two tables. Table 4 summarizes results for students in traditional Calculus II and Table 5 summarizes results for students in OZ Calculus II.

Table 4. First-Time Traditional Calc II Grades by Calc I background

	Total Enroll	A	B	C	D	F	W	Avg. Grade	Success Rate	% Grade D or above	W Rate
Reform Calc. I	23	2	5	4	1	4	7	2.00	48%	52%	30%
Trad. Calc. I	76	7	16	13	11	10	19	1.98	47%	62%	25%
Total	99	9	21	17	12	14	26	1.99	47%	60%	26%

Table 5. First-Time OZ Calc II Grades by Calc I background

	Total Enroll	A	B	C	D	F	W	Avg. Grade	Success Rate	% Grade D or above	W Rate
Reform Calc. I	77	17	19	22	8	8	3	2.39	75%	86%	4%
Trad. Calc. I	46	7	15	10	3	4	7	2.46	70%	76%	15%
Total	123	24	34	32	11	12	10	2.42	73%	82%	8%

Tables 4 and 5 show that the performances of students within traditional Calculus II and within OZ Calculus II were very similar regardless of Calculus I background. None of the differences in Table 4 are significant. In Table 5, the only significant difference is between withdrawal rates of 4% (OZ) and 15% (traditional) ($p = 0.026$). In addition, the sharp differences described earlier between the two types of Calculus II are even wider for the more restricted samples, as evidenced by the striking contrast of results in Table 5 with those in Table 4.

The two tables above include students who repeated Calculus I at least once, so that some students were exposed to both types of Calculus I. This fact may have had a confounding effect on the results. For this reason a second set of comparisons, restricted to students who passed Calculus I on their first attempts, is summarized in Table 6. As before, all differences in performance *between* OZ and traditional Calculus II are highly significant, while the differences *within* each type of Calculus II are nonsignificant. However, it is interesting to note that both traditional and reform Calculus II GPAs and success rates are higher for students from OZ Calculus I, although none of the differences are significant. The lack of significance may be due to smaller samples and the consequent lack of statistical power.

Table 6. Calc II Grades of Students Completing Calc I on First Attempt

	Reform Calculus II				Traditional Calculus II				Total
Calculus I	Total	GPA	C or above	%W	Total	GPA	C or above	%W	
OZ	61	2.52	82%	2%	19	2.29	58%	26%	80
Traditional	33	2.45	73%	9%	57	2.02	53%	18%	90

Performance in Second Year Calculus. OZ and traditional Calculus II students were also compared by analyzing their subsequent performance in Differential Equations and Multivariable Calculus. In Differential Equations, the GPA of 2.40 for 48 students from OZ Calculus II was identical to the GPA of 2.40 for 80 traditional students, and the grade distributions were very similar. In Multivariable Calculus, the results were also very similar: 26 students from OZ Calculus II had a GPA of 2.5, compared to 2.54 for 89 students from traditional Calculus II.

Retention. About the same percentages of reform and traditional students who completed Calculus I enrolled in Calculus II. There was a small but not statistically significant gap in completion rates for Calculus II: 27.0% of the 396 students who ever enrolled in OZ Calculus I made at least a C in Calculus II, compared to 22.0% of the 696 control students. No significant differences were found between enrollment rates of OZ and traditional Calculus II students in second year calculus.

Summary of Results. The results presented in this longitudinal study support the following conclusions.

(1) There is strong evidence that OZ Calculus I students have a higher success rate and a lower withdrawal rate than traditional Calculus I students, although not a higher GPA.

(2) There is good evidence that OZ Calculus II students have a much higher success rate, a much lower withdrawal rate, and higher grades than traditional Calculus II students, regardless of Calculus I background.

(3) There is strong evidence that students who switch from one type of calculus sequence to another in first year calculus perform at about the same level as those who do not switch.

(4) There is no significant evidence that reform and traditional students perform differently or have different retention rates in second year calculus courses.

Discussion. One of the current concerns of the mathematics community has been to make calculus "a pump, not a filter," for mathematics and related majors. The mildly improved success rates in OZ Calculus I and the considerably improved success rates in OZ Calculus II suggest that this sequence of reform courses may be achieving this purpose. This conclusion is especially likely in light of the low success rates in traditional Calculus II, regardless of Calculus I backgrounds. These results add to the evidence already cited in the literature review that the use of reform texts has a positive effect on success rates in first year calculus. In particular, these results are similar to studies by Baxter et al. (1998) and others cited in the literature review which found that reform Calculus I students performed significantly better in reform Calculus II than traditional Calculus I students performed in traditional Calculus II. However, it bears repeating that neither the study reported above nor those reviewed were experimental in design, so that results must be treated cautiously, especially in light of the limited number of Calculus II instructors involved in the study above.

The performance of students in subsequent courses in the calculus sequence are of critical importance in evaluating the effectiveness of the OZ text. In this regard, the evidence presented above strongly indicates that OZ Calculus I students are at least as well prepared for Calculus II, either traditional or reform, as traditional Calculus I students. These findings are at variance with evidence reported in the review that CCH Calculus I students may have been at a disadvantage in traditional

Calculus II. Midyear crossover performance may be an instance in which the use of different reform texts may produce different results, and may confirm the opinion of those at UTC who selected OZ as the reform text which would cause the least problems for cross-over students.

With regard to second year calculus, the evidence indicates that OZ students perform as well as traditional students in both Multivariable Calculus and Differential Equations. These longitudinal findings are consistent with most of the reports discussed earlier. Bookman and Friedman (1999), Armstrong et al. (1994), and Schwingendorf (1999) also found no significant differences in subsequent courses, and Baxter et al. (1998) found no significant differences between large samples of CCH and traditional students in Differential Equations. However, Baxter et al. (1998) did find a significant difference in favor of traditional students in traditional Vector Calculus, and Johnson (1995) reported that CCH students had a lower, but probably not statistically lower, success rate than traditional students in both Differential Equations and Linear Algebra.

These mixed results may be due to the use of limited numbers of second year students and possible major variations in second year courses, as well as unknown effects of the lack of experimental design. Nonetheless, studies such as the one conducted at UTC and those reviewed above provide data points regarding the long term effectiveness of calculus reform, from which a clear pattern may eventually be discerned. It may also be argued that the traditional second year mathematics curriculum is not appropriate, either for meeting more general curricular goals or for judging the effectiveness of a reformed first year curriculum. These possibilities have not escaped the attention of reformers, and several reform multivariable calculus texts are already on the market (McCallum et al., 1997; Ostebee & Zorn, 1997b; Smith & Moore, 1997). Obviously much more longitudinal research will be needed to evaluate these developments.

Concluding Remarks. A critical step in monitoring curricular changes in sequential courses is to observe how well students perform in both revised and subsequent courses. Although we have pointed out that improved performance lacks meaning without clear curricular goals and standards, it is obvious that acceptable student performance in current and subsequent courses is certainly a necessary condition for the acceptance of revised sequential courses. The literature review and the longitudinal study of OZ students presented above provide some evidence that reform calculus courses are meeting this basic criterion. However, more longitudinal studies are crucial in order to gain additional evidence about the effects of curricular changes.

3. References

Alarcon, F. E. & Stoudt, R. A. (1997). The rise and fall of a *Mathematica*-based calculus curriculum reform movement. *Primus, 7,* 73-88.

Alexander, E. H. (1997). An investigation of the results of a change in calculus instruction at the University of Arizona. *Dissertation Abstracts International.*

Alexander, E. H. & Madden, D. J. (1998). Reform through grade analysis. *Focus on Calculus, 15,* 6.

Alkhateeb, H. M. (1995). The effect of using graphics calculators on students' attitudes towards mathematics and students' achievement in introductory calculus. *Dissertation Abstracts International.*

Allen, G. H. (1995). A comparison of the effectiveness of the Harvard calculus series with the traditionally taught calculus series. *Dissertation Abstracts International.*

Armstrong, G., Garner, L., & Wynn, J. (1994). Our experience with two reformed calculus programs. *Primus, 4*, 301-311.

Armstrong, S. (1997). A multivariate analysis of the dynamics of factors of social context, curriculum, and classroom process to achievement in the calculus at the community college. *Dissertation Abstracts International.*

Askey, R. (1997). What do we do about calculus? First, do no harm. *American Mathematical Monthly, 104*, 738-743.

Ayers, T., Davis, G., Dubinsky, E., & Lewin, P. (1998). Computer experiences in learning composition of functions. *Journal for Research in Mathematics Education, 19*, 246-259.

Baumgartner, J. & Shemanske, T. (1990). Teaching calculus with True BASIC. In *Priming the calculus pump: Innovations and resources* (MAA Notes no. 17, pp. 33-50). Washington, DC: Mathematical Association of America.

Baxter, J., Majumdar, D., & Smith, S. (1998). Subsequent-grades assessment of traditional and reform calculus. *Primus, 8*, 317-330.

Baxter, J. (1999). Assessing reform calculus. *Focus on Calculus, 16*, 4.

Becker, J. & Pence, B. (1994). The teaching and learning of college mathematics: current status and future directions. In J. Kaput & E. Dubinsky (Eds.), *Research issues in undergraduate mathematics learning: Preliminary analyses and results* (MAA Notes no. 33, pp. 5-16). Washington, DC: Mathematical Association of America.

Beidleman, J., Jones, D., & Wells, P. (1995). Increasing students' conceptual understanding of first semester calculus through writing. *Primus, 5*, 297-316.

Bolte, L. A. (1997). *Assessing mathematical knowledge with concept maps and interpretive essays.* Unpublished manuscript.

Bonsangue, M. (1994). An efficacy study of the calculus workshop model. In E. Dubinsky, A. Schoenfeld, & J. Kaput (Eds.), *Research in collegiate mathematics education I* (pp. 117-137). Providence, RI: American Mathematical Society.

Bookman, J. & Friedman, C. (1994). A comparison of the problem solving performance of students in lab based and traditional calculus. In E. Dubinsky, A. Schoenfeld, & J. Kaput (Eds.), *Research in collegiate mathematics education I* (pp. 101-116). Providence, RI: American Mathematical Society.

Bookman, J. & Blake, L. (1996). Seven years of Project CALC at Duke University: Approaching steady state? *Primus, 6*, 221-234.

Bookman, J. & Friedman, C. P. (1999). The Evaluation of Project CALC at Duke University, 1989-1994. In B. Gold, S. Keith, & W. Marion (Eds.), *Assessment practices in undergraduate mathematics* (MAA Notes no. 49, pp. 253-256). Washington, DC: Mathematical Association of America.

Brown, M. (1996). Planning and change: The Michigan Calculus Project. In A. W. Roberts (Ed.), *The dynamics of change* (MAA Notes no. 39, pp. 52-58). Washington, DC: Mathematical Association of America.

Brunett, M. R. (1995). A comparison of problem-solving abilities between reform calculus students and traditional calculus students. *Dissertation Abstracts International.*

Burmeister, S., Kenney, P., & Nice, D. (1996). Analysis of effectiveness of Supplemental Instruction (SI) sessions for college algebra, calculus, and statistics. In J. Kaput, A. H. Schoenfeld, & E. Dubinsky (Eds.), *Research in collegiate mathematics education II* (pp. 145-154). Providence, RI: American Mathematical Society.

Cassity, C. (1997). Learning with technology: research on graphing calculators. In *Proceedings of Selected research and development presentations at the 1997 National Convention of the Association for Educational Communications and Technology.*

Connors, M. A. C. (1995). An analysis of student achievement and attitudes by gender in computer-integrated and non-computer-integrated first year college mainstream calculus courses. *Dissertation Abstracts International.*

Conrad, J. (1994). An experiment using group worksheets in college algebra. *Primus, 4,* 264-272.

Cooley, L. A. (1997). Evaluating student understanding in a calculus course enhanced by a computer algebra system. *Primus, 7,* 308-316.

Davidson, N. (1990). *Cooperative learning in mathematics: A handbook for teachers.* Menlo Park, CA: Addison-Wesley.

Davis, W. (1994). *Calculus and Mathematica.* Reading, PA: Addison Wesley-Longman, Inc.

Dees, R. L. (1991). The role of cooperative learning in increasing problem-solving ability in a college remedial course. *Journal for Research in Mathematics Education, 22,* 409-421.

Dubinsky, E., Mathews, D. & Reynolds, B. (Eds.). (1997). *Readings in cooperative learning for undergraduate mathematics* (MAA Notes no. 38). Washington, DC: Mathematical Association of America.

Dubinsky, E. & Schwingendorf, K. (1991). *Calculus, concepts and computers: Some laboratory projects for differential calculus.* Unpublished manuscript.

Dubinsky, E., Schwingendorf, K. E., & Mathews, D. M. (1995). *Calculus, concepts and computers* (2nd. ed.). New York: McGraw-Hill.

Ferrini-Mundy, J. & Graham, K. (1991). An overview of the calculus curriculum reform effort: Issues for learning, teaching, and curriculum development. *American Mathematical Monthly, 98,* 627-635.

French, D. M. (1997). *A comparison study of computer assisted instruction using interactive software versus traditional instruction in a college precalculus course.* Ed. D. dissertation, Temple University.

Ganter, S. (1994). The importance of empirical evaluations of mathematics programs: a case from the calculus reform movement. *Focus on Learning Problems in Mathematics, 16,* 1-19.

Ganter, S. (1997). The impact of ten years of calculus reform on student learning and attitudes. *Association for Women in Science Magazine, 26*(6), 10-15.

Ganter, S. (1999). An evaluation of calculus reform: A preliminary report of a national study. In B. Gold, S. Keith, & W. Marion, (Eds.), *Assessment practices in undergraduate mathematics* (MAA Notes no. 49, pp. 233-236). Washington, DC: Mathematical Association of America.

Good, T. L., Mulyran, C. & McCaslin, M. (1992). Grouping for instruction in mathematics: A call for programmatic research on small-group processes. In D. A. Grouws (Ed.), *Handbook of research on mathematics teaching and learning* (pp. 165-196). New York: MacMillan.

Gordon, S. (1997). Out of the mouths of babes: Students' questions and comments in reform courses. *Primus, 7*, 25-34.

Hare, A. C. (1996). *An investigation of the behavior of calculus students working collaboratively in an interactive software environment.* Unpublished doctoral dissertation, American University.

Heid, L. R. (1988). Resequencing skills and concepts in applied calculus using the computer as a tool. *Journal for Research in Mathematics Education, 19*, 3-25.

Hershberger, L. D. & Plantholt, M. (1994). Assessing the Harvard Consortium Calculus at Illinois State University. *Focus on Calculus, 7*, 6-7.

Hiebert, J. (1999). Relationships between research and the NCTM *Standards. Journal for Research in Mathematics Education, 30*, 3-19.

Hirsch, L. R. & King, B. (1983). *The relative effectiveness of writing assignments in an elementary algebra course for college students.* Paper presented at the Annual Meeting of the American Educational Research Association.

Holdener, J. (1997). *Calculus and Mathematica* at the U.S. Air Force Academy: Results of an anchored final. *Primus, 7*, 62-72.

Hollar, J. & Norwood, K. (1999). The effects of a graphing approach intermediate algebra to students' understanding of function. *Journal for Research in Mathematics Education, 30*, 220-226.

Hughes-Hallett, D., Gleason, A. M. et al. (1998). *Calculus* (2nd ed.). New York: Wiley.

Hurley, J., Koehn, U., & Ganter, S. L. (1999). Effects of calculus reform: Local and national. *American Mathematical Monthly, 106*, 800-811.

Hylton-Lindsay, A. A. (1997). *The effect of the graphing calculator on metacognitive aspects of student performance in pre-calculus for business.* Ed. D. dissertation, Columbia University Teachers College.

Isom, M. A. (1997). *The effect of a writing-influenced curriculum on student beliefs about mathematics and mathematics achievement.* Unpublished doctoral dissertation, University of Northern Colorado.

Johnson, K. (1995). Harvard Calculus at Oklahoma State University. *American Mathematical Monthly, 102*, 794-797.

Judson, P. (1994). Calculus reform blooms in Texas. *In General Terms* 2, 3.

Kaput, J. J. & Thompson, P. W. (1994). Technology in mathematics education research: the first 25 years in the JRME. *Journal for Research in Mathematics Education, 25*, 676-684.

Keeler, C. M. & Voxman, M. (1994). The effect of cooperative learning in remedial freshman level mathematics. *AMATYC Review, 16*, 37-44.

Kennedy, P. A. & Kallison, J. M. (1994). Research studies on the effectiveness of Supplemental Instruction in mathematics. *New Directions in Teaching and Learning, 60*, 75-82.

Kinney, D. P. (1997). *The effect of graphing calculator use and the Lesh translation model on student understanding of the graphical relationship between function and derivative in a nonrigorous calculus course.* Unpublished doctoral dissertation, University of Minnesota.

Krantz, S. G. (1999). You don't need a weatherman to know which way the wind blows. *American Mathematical Monthly, 106*, 915-918.

Lauten, A. D. (1996). *Profiles of reform in the teaching of calculus: a study of the implementation of materials developed by the Calculus Consortium based at*

Harvard (CCH) curriculum project. Unpublished doctoral dissertation, University of New Hampshire.

Lauten, A. D. et al. (1994). Student understanding of basic calculus concepts: interaction with the graphics calculator. *Journal of Mathematical Behavior, 13,* 225-237.

Lefton, L. E. & Steinbart, E. M. (1995). *Calculus & Mathematica*: An end-user's point of view. *Primus, 5,* 80-96.

Lester, F. K. (1994). Musings about problem-solving research. *Journal for Research in Mathematics Education, 25,* 660-675.

Lock, P. F. (1994). Reflections on the Harvard calculus approach. *Primus, 4,* 229-234.

Luke, R. & Muller, A. (1994). The impact of CCH at a comprehensive community college. *Focus on Calculus, 6,* 7.

Ma, L. (1999). *Knowing and teaching elementary mathematics: Teachers' understanding of fundamental mathematics in China and the United States.* Mahwah, NJ: Lawrence Erlbaum Associates.

Mayes, R. L. (1995). The application of a computer algebra system as a tool in college algebra. *School Science and Mathematics, 95,* 61-68.

McCallum, W. G., Hughes-Hallett, D., Gleason, A. M. et al. (1997). *Multivariable calculus.* New York: Wiley.

McKnight, C. S., Crosswhite, E. J., Dossey, J. A., Kifer, E., Swafford, J. O., Travers, K. J., & Cooney, T. J. (1987). *The underachieving curriculum: Assessing U. S. school mathematics from an international perspective.* Champaign, IL: Stipes Publishing Company.

Meel, D. (1998). Honors students' calculus understanding: Comparing *Calculus & Mathematica* and traditional calculus students. In A. Schoenfeld, J. Kaput, & E. Dubinsky (Eds.), *Research in collegiate mathematics education III* (pp. 163-197). Providence, RI: American Mathematical Society.

Mittag, K. C. & Collins, L. B. (2000). Relating Calculus I reform experience to performance in traditional Calculus II. *Primus, 10,* 82-94.

Monteferrante, S. (1993). Implementation of Calculus, Concepts and Computers at Dowling College. *Collegiate Microcomputer, 11*(2), 47-50.

Murphy, T. J., Stafford, K. L. & McCreary, P. (1998). Subsequent course and degree paths of students in a Treisman-style workshop calculus program. *Journal of Women and Minorities in Science and Engineering, 4,* 381-396.

National Council of Teachers of Mathematics. (1989). *Curriculum and evaluations standards for school mathematics.* Reston, VA: Author.

Norwood, K. S. (1995). The effects of the use of problem solving and cooperative learning on the mathematics achievement of under-prepared college freshmen. *Primus, 5,* 229-252.

Orton, A. (1983). Students' understanding of integration. *Educational Studies in Mathematics, 14,* 1-18.

Ostebee, A. & Zorn, P. (1997a). *Calculus from graphical, numerical and symbolic points of view* (Vols. 1 & 2). Orlando, FL: Saunders College Publishing.

Ostebee, A. & Zorn, P. (1997b). *Multivariable calculus from graphical, numerical and symbolic points of view* (Preliminary Edition). Orlando, FL: Saunders College Publishing.

Ostebee, A. & Zorn, P. (1997c). Pro choice. *American Mathematical Monthly, 104*, 728-730.

Padgett, E. (1994). *Calculus I with a laboratory component*. Ed.D. dissertation, Baylor University.

Palmiter, J. R. (1991). Effects of computer algebra systems on concept and skill acquisition in calculus. *Journal for Research in Mathematics Education, 22*, 151-156.

Park, K. & Travers, K. (1996). A comparative study of a computer-based and a standard college first-year calculus course. In J. Kaput, A. H. Schoenfeld, & E. Dubinsky (Eds.), *Research in collegiate mathematics education II* (pp. 155-176). Providence, RI: American Mathematical Society.

Penglase, M. & Arnold, S. (1996). The graphics calculator in mathematics education: a critical review of recent research. *Mathematics Education Research Journal, 8*, 58-90.

Penn, H. L. (1994). Comparison of test scores in Calculus I at the Naval Academy. *Focus on Calculus, 6*, 6-7.

Phillips, A. (1995). *Persistence: MAT 131 → MAT 132*. SUNY Stony Brook Department of Mathematics Report.
www.math.Sunysb.edu/~tony/calc/finrep/finrep.html.

Porter, M. K. (1996). *The effects of writing to learn mathematics on conceptual understanding and procedural ability in introductory college calculus*. Unpublished doctoral dissertation, Syracuse University.

Porzio, D. T. (1994). *The effects of differing technological approaches to calculus on students' use and understanding of multiple representations when solving problems*. Unpublished doctoral dissertation, The Ohio State University.

Quesada, A. R. & Maxwell, M. E. (1994). The effects of using graphing calculators to enhance college students' performance in precalculus. *Educational Studies in Mathematics, 27*, 205-215.

Ratay, G. M. (1993). Student performance with calculus reform at the United States Merchant Marine Academy. *Primus, 3*, 107-111.

Research Advisory Committee of the NCTM. (1996). Justification and reform, *Journal for Research in Mathematics Education, 27*, 517-520.

Reynolds, B. E. (1995). *A practical guide to cooperative learning in collegiate mathematics* (MAA Notes no. 37). Washington, DC: Mathematical Association of America.

Roberts, A. W. (Ed.). (1996). *Calculus: The dynamics of change* (MAA Notes no. 39). Washington, DC: Mathematical Association of America.

Rochowicz, J. A. (1996). The impact of using computers and calculators on calculus instruction: Various perceptions, *Journal of Computers in Mathematics and Science Teaching, 15*, 423-435.

Roddick, C. S. (1995). *How students use their knowledge of calculus in an engineering mechanics course*. Paper presented at the Annual Meeting of the North American Chapter of the International Group for the Psychology of Mathematics Education.

Ross, K. A. (1998). Doing and proving: The place of algorithms and proofs in school mathematics. *American Mathematical Monthly, 105*, 252-255.

Schneider, E. M. (1995). *Testing the rule of three: A formative evaluation of the Harvard based calculus consortium curriculum.* Unpublished doctoral dissertation, University of Texas, Austin.

Schoenfeld, A. H. (1992). Learning to think mathematically: Problem solving, metacognition, and sense making in mathematics. In D. Grouws (Ed.), *Handbook of research on mathematics teaching and learning* (pp. 334-370). New York: MacMillan.

Schoenfeld, A. H. (Ed.). (1994). *Mathematical thinking and problem solving.* Hillsdale, NJ: Lawrence Erlbaum Associates.

Schoenfeld, A. H. (1995). A brief biography of calculus reform. *UME Trends, 6*(6), 3-5.

Schoenfeld, A. H. (Ed.). (1997). *Student assessment in calculus: A report of the NSF Working Group on Assessment in Calculus* (MAA Notes no. 43). Washington, DC: Mathematical Association of America.

Schwingendorf, K. E. (1999). Assessing the effectiveness of innovative educational reform efforts. In B. Gold, S. Keith, & W. Marion (Eds.), *Assessment practices in undergraduate mathematics* (MAA Notes no. 49, pp. 249-252). Washington, DC: Mathematical Association of America.

Shoaf-Grubbs, M. M. (1994). The effect of the graphing calculator on female students' spatial visualization skills and level-of-understanding in elementary graphing and algebra concepts. In E. Dubinsky, A. H. Schoenfeld, & J. Kaput (Eds.), *Research in collegiate mathematics education I* (pp. 169-194). Providence, RI: American Mathematical Society.

Silverberg, J. A. (n.d.) *A longitudinal study of an experimental vs. traditional approach to teaching first semester calculus.* Unpublished manuscript.

Smith, D. A. & Moore, L. C. (1996). *Calculus: Modeling and applications.* Lexington, MA: D. C. Heath & Co.

Smith, D. A. & Moore, L. C. (1997). *Calculus: Modeling and applications, multivariable chapters* (final preliminary edition). Boston, MA: Houghton Mifflin.

Soto-Johnson, H. (1996). *Technological vs. traditional approach in conceptual understanding of series.* Unpublished doctoral dissertation, University of Northern Colorado.

Sterrett, A. (Ed.). (1990). *Using writing to teach mathematics* (MAA Notes no. 16). Washington, DC: Mathematical Association of America.

Stick, M. E. (1997). Calculus reform and graphing calculators: A university view. *Mathematics Teacher, 90*, 356-360, 363.

Stigler, J. & Hiebert, J. (1999). *The teaching gap.* New York: Free Press.

Tall, D. (Ed.). (1991). *Advanced mathematical thinking.* Dordrecht: Kluwer Academic Publishers.

Thompson, A. G. & Thompson, P. W. (1996). Talking about rates conceptually, part II: Mathematical knowledge for teaching. *Journal for Research in Mathematics Education, 27*, 2-24.

Tidmore, E. (1994). A comparison of calculus materials used at Baylor University. *Focus on Calculus, 7*, 5-6.

Treisman, P. U. (1985). *A study of the mathematical performance of black students at the University of California.* Unpublished doctoral dissertation, University of California, Berkeley.

Tucker, A. C. & Leitzel, J. R. C. (1995). *Assessing calculus reform efforts: A report to the community.* Washington, DC: Mathematical Association of America.

Tucker, T. W. (1999). Reform, tradition, and synthesis. *American Mathematical Monthly, 106*, 910-914.

Urion, D. K. & Davidson, N. A. (1992). Student achievement in small-group instruction versus teacher-centered instruction in mathematics. *Primus, 2*, 257-264.

Wells, P. J. (1995). *Conceptual understanding of major topics in first semester calculus: a study of three types of calculus courses at the University of Kentucky.* Unpublished doctoral dissertation, University of Kentucky.

White, P. & Mitchelmore, M. (1996). Conceptual knowledge in introductory calculus. *Journal for Research in Mathematics Education, 27*, 79-95.

Williams, C. G. (1994). *Using concept maps to determine differences in the concept image of function held by students in reform and traditional calculus classes.* Unpublished doctoral dissertation, University of California, Santa Barbara.

Williams, S. R. (1991). Models of limit held by college calculus students. *Journal for Research in Mathematics Education, 22*, 219-236.

Wilson, M. R. & Krapfl, C. M. (1994). The impact of graphics calculators on students' understanding of function. *Journal of Computers in Mathematics and Science Teaching, 13*, 252-264.

Zandieh, M. J. (1997). *The evolution of student understanding of the concept of derivative.* Unpublished doctoral dissertation, Oregon State University.

DEPARTMENT OF MATHEMATICS, UNIVERSITY OF TENNESSEE AT CHATTANOOGA, CHATTANOOGA, TN 37403

E-mail address: Betsy-Darken@utc.edu

DEPARTMENT OF MATHEMATICS, UNIVERSITY OF TENNESSEE AT CHATTANOOGA, CHATTANOOGA, TN 37403

E-mail address: Robert-Wynegar@utc.edu

DEPARTMENT OF MATHEMATICS, UNIVERSITY OF TENNESSEE AT CHATTANOOGA, CHATTANOOGA, TN 37403

E-mail address: Stephen-Kuhn@utc.edu

CBMS Issues in Mathematics Education
Volume **8**, 2000

The Need for Evaluation in the Calculus Reform Movement
A Comparison of Two Calculus Teaching Methods

Susan L. Ganter and Michael R. Jiroutek

ABSTRACT. It has become increasingly clear that a much greater effort to evaluate projects in calculus reform is necessary. This paper presents one such study that tracks the progress of students in reform calculus courses as compared to traditionally-taught courses, as well as the progress of these two groups in subsequent courses. Results indicate that, although the students in the reform course who completed the first year of calculus performed less well on traditional measures than the students in the traditional course, they had equivalent performance levels in subsequent courses from several disciplines.

Introduction

A report released five years ago by the Mathematical Association of America entitled *Assessing Calculus Reform Efforts* (Tucker & Leitzel, 1995) indicated that more than 500 mathematics departments at post-secondary institutions nationwide were implementing some level of calculus reform. These "reformed" courses were, at that time, affecting an estimated 150,000 calculus students each year, or approximately 33% of the students enrolled in calculus. Since then, efforts to make changes in calculus have continued to grow as well as the likelihood that calculus reform will continue to affect an even greater percentage of the next generation (Foley & Ruch, 1995; George, 2000; Lightbourne, 2000; Roberts, 1996; Schoenfeld, 1996; Smith, 2000).

Why has so much effort been put into changing calculus? Perhaps because studies show that students passing "traditional" calculus courses tend to lack "basic skills," not to mention their even poorer understanding of how to apply the ideas of calculus to real problems (Frid, 1994; Ganter, 1997, 1999; McCallum, 2000; NSF, 1987, 1991; Smith, 2000). To compound this problem, the amount of time students devote to calculus homework and students' perceived value of mathematics has been steadily decreasing over the past two decades (AMATYC, 1995; Andrews, 1995; Bookman & Friedman, 1996, 1998; Ganter, 1997; Mathews, 1995). The 1986 Tulane Conference helped to launch the calculus reform movement by challenging the mathematics community to change the structure of calculus courses. Participants identified five major problems with calculus at that time:

©2000 American Mathematical Society

- too few students successfully complete calculus
- many calculus courses involve implementing algorithms with no conceptual understanding
- many faculty are frustrated with the constant struggle to help students learn only a small part of calculus
- calculus has become an unmotivated "filter" for many disciplines
- most calculus courses involve little, if any, use of technology. (Lightbourne, 2000; Tucker & Leitzel, 1995)

Specifically, the importance of conceptual understanding and the need to change current modes of instruction to include technology and active student learning were emphasized. Instructors were asked to focus on numerical, graphical, and modeling problems through the use of computers, open-ended projects, writing, applications, and cooperative learning. Cooperation throughout the mathematical community, as well as broad dissemination of the resulting ideas and materials, were viewed as critical to the success of calculus reform (Tucker & Leitzel, 1995).

The result has been a significant increase in the use of alternative teaching strategies and technology in many reform courses (AMATYC, 1995; Crockett & Kiele, 1993; Ellis, 2000; Farr & VanGeel, 1994; Foley & Ruch, 1995; Ganter & Kinder, 2000; Hurley, Koehn, & Ganter, 1999; McCallum, 2000; NSF, 1996; Roberts, 1996; Smith, 1996). Many other conditions in the late 1980s also promoted the widespread growth of calculus reform, including the increasing availability of technology, the publication of the *Curriculum and Evaluation Standards for School Mathematics* (NCTM, 1989), and the reinstitution of the Division of Undergraduate Education (DUE) at the National Science Foundation (Tucker & Leitzel, 1995).

The Need for Evaluation. Although calculus reform has received a great deal of financial support from the National Science Foundation and other granting agencies, very few of the original projects included a plan to evaluate outcomes as part of their work. Therefore, the research on student learning in reform calculus is very limited (Asiala, Brown, DeVries, Dubinsky, Mathews, & Thomas, 1996; Bookman & Friedman, 1994; Ganter, 1999; Schoenfeld, 1996; Tucker & Leitzel, 1995). In fact, very few mathematicians involved in calculus reform have training in the design of studies to measure learning in mathematics. Although some reform projects have designed and implemented some measures of evaluation (e.g., U.S. Military Academy [see West, 1994], Project CALC [see Bookman, 2000; Bookman & Friedman, 1994, 1996, 1998], C^{4}L [see Mathews, 1995], University of Connecticut [see Hurley et al., 1999], and others [see Roberts, 1996]), most studies have been very limited in scope, in part because of limited resources for evaluation. Consequently, there is still relatively little information on the impact of the reform efforts on student learning (AMATYC, 1995; Asiala et al., 1996; Bookman, 2000; Frid, 1994; NSF, 1996; Tucker & Leitzel, 1995). It has become increasingly clear that a much greater effort to evaluate these projects is indeed necessary, as the mathematics community now has little evidence to justify the expenditure of enormous amounts of time, energy, and financial resources.

In addition, the efforts in calculus have encouraged change in other mathematics courses, including precalculus, differential equations, and linear algebra. These courses have increasingly adopted the philosophies begun in calculus, including the use of computer algebra systems, applied projects, and a move to more group discussion instead of lecture in class (AMATYC, 1995; Dossey, 1998; Gordon, 2000;

Tucker & Leitzel, 1995; Smith, 1996). It is hard to support such a move without data that justify the increased budget and workload on faculty that such programs involve.

Understandably, many in the mathematics community are skeptical of curricular reform, since their questions about its success often cannot be addressed with anything more than anecdotes from those teaching the courses. Many not involved in the reform are unconvinced by claims that simply help those who have invested time and resources to justify their efforts. There is a growing concern that these reformed courses may not be recognized by some as "real" calculus and that this is simply another educational experiment (Ganter, 1999; Keith, 1995; McCallum, 2000). Therefore, it is imperative that thorough evaluation plans be designed, implemented, and disseminated in order to significantly increase the amount of evidence about the areas of success (and failure) of the calculus reform movement.

What is Evaluation? Project evaluation uses student performance data, as well as other evidence, to make judgments about the progress and, ultimately, success of an educational innovation. Such information is often used to make decisions about the future of such projects, including continuation of funding or modifications in procedure (Asiala et al., 1996; George, 2000; NCTM, 1995; NSF, 1996; Stevens, Lawrenz, & Sharp, 1993). Many such questions can be answered by assessing students' progress through the affected courses. However, an evaluation plan should be designed to incorporate multiple sources of quantitative and qualitative data, including data from assessment instruments developed specifically for the project as well as those already in place. These data need to be carefully collected, analyzed, and interpreted in order to make valid inferences that will result in important programmatic decisions (Ganter & Kinder, 2000; NCTM, 1995; Schoenfeld, 1996; Stevens et al., 1993).

Evaluation differs from assessment because it is a process of collecting information, not simply the result of a single measure (Eisner, 1996; Slavin, 1999; Stevens et al., 1993). Therefore, it can affect not only an individual student or classroom, but also the general instructional practices in a department, school, or region. Complete data from all students involved in the project are not necessary to make programmatic decisions provided one uses a carefully selected sample that is representative of all student participants. Assessment tasks that involve traditional measures such as short-answer and multiple choice questions can be used to effectively gain information about basic skill levels of students. However, long-range evaluation plans should incorporate whenever possible a variety of measures, such as interviews, classroom observations, and holistic assessment techniques, as each assessment task may favor certain types of teaching methods (Asiala et al., 1996; Bookman, 2000; Eisner, 1996; NCTM, 1995; Schoenfeld, 1996).

Evaluation and Calculus Reform. The advent of the calculus reform movement in 1986 was the result of a long history of discontent with calculus in the mathematics community. As much as a century before, mathematicians were expressing concern about the appropriate presentation of topics in calculus (e.g., Durell, 1894). Discussions about undergraduate mathematics continued into the 20th century, with calculus returning to the forefront in the 1950s and 1960s in response to Sputnik. Methods of presentation and organization of topics in the course were challenged much as they have been in recent years, with experiments that included self-taught and discovery method courses (Baum, 1958; Cummins, 1960; McKeachie, 1954), use of computers and other technologies (Buck, 1962; CUPM, 1969; Davis, 1966; Dyer-Bennett, Fuller, Seibert, & Shanks, 1958; May, 1964), variations in class size and organization using small groups (Levi, 1963; May, 1962; McKeen, 1970; Stockton, 1960; Turner, Alders, Hatfield, Croy, & Sigrist, 1966), the implementation of learning theories from other educational settings (Holder, 1967; Larsen, 1961; Moise, 1965), and changes in the course content (Monroe, 1966). However, evaluation of the impact on student learning was sparse and mostly in the form of doctoral dissertations (e.g., Davidson, 1970; Levine, 1968; Monroe, 1966; Shelton, 1965; Stannard, 1966).

Although the need to reform calculus received less attention in the two decades prior to 1986, much relevant research conducted during this time would pave the way for the current reform efforts. Mathematicians commonly found that students could easily respond to standard questions posed in a predictable way, but when pushed to reveal any deeper understanding of the concepts, they often had difficulties (Tall, 1990). This troubling realization prompted an increasing interest in understanding the cognitive development of students in mathematics at all levels, including calculus. For example, it has been known for some time that many of the alternative methods that have been embraced by the recent reform efforts in calculus do enhance student learning in mathematics, including small group activities (Davidson, 1985, 1990; Kroll, 1989; Slavin, 1983, 1985; Urion & Davidson, 1992) and the use of technology (Held, 1988; Johnson, Johnson, & Stanne 1986; Tall, 1990; Tall et al., 1984; Webb, Ender, & Lewis, 1986; Yackel, Cobb, & Wood, 1991). However, research on the teaching and learning of calculus as influenced by the reform movement is very new, with only a few studies that investigate student achievement, the general state of calculus instruction, or the influences of non-traditional curricula and teaching strategies in calculus on student learning (Asiala et al., 1996; Bookman, 2000; Foley & Ruch, 1995; Frid, 1994; Ganter, 1999; Smith, 2000). Many studies that are available have been conducted only very recently and as graduate research, with the results published as theses and dissertations and often not widely disseminated (e.g., Ahmadi, 1995; Alexander, 1997; Allen, 1995; Aspinwall, 1994; Atkins, 1994; Barton, 1995; Newell, 1994; Padgett, 1994; Rochowicz, 1993; Schneider, 1995; Simonsen, 1995; Wells, 1995; Williams, 1994). This is similar to the trend that was seen thirty years earlier, where studies to evaluate the impact of reform in action are not supported to the extent necessary to adequately inform and affect progress. Consequently, there is a great need for information on various instructional formats and their subsequent effects on learning.

In addition, the mathematics community needs to develop an understanding of the perspectives, methodologies, and findings of research on calculus learning, as it is important to discover the "right mix" of content and pedagogy (Ferrini-Mundy & Lauten, 1994; Ganter, 1997; Tucker & Leitzel, 1995). Concrete information about what and how students learn, as well as decisions about program improvements, cannot be made reasonably with only anecdotes and success stories. The value of changes made to a course with such impact as calculus is very difficult to quantify; such decisions cannot be made without an enormous amount of data collected over a long period of time that include information on students before, during, and after the course. Ultimately, the true goal of any calculus course is to ensure that students completing the course will be able to adequately analyze and solve problems involving calculus, whether those problems be in physics, engineering, or other mathematics courses.

Reformed calculus courses have also received very skeptical reviews from colleagues in other disciplines who are not convinced that the decreased attention to algorithmic skills is a good thing (Buccino, 2000; Cipra, 1996; McCallum, 2000; Mumford, 1997; Smith, 1996; Tucker & Leitzel, 1995). Similar arguments were made for rejecting the New Math movement of the 1960s; i.e., it did not support anything that looked even remotely like "mere rote learning" (Andrews, 1995). In addition, many students and faculty complain that the reformed courses involve significantly greater amounts of time and effort than their more traditional counterpart (Bookman & Friedman, 1996, 1998; Ganter, 1997; Tucker & Leitzel, 1995). Comments such as "shouldn't calculus be getting easier with technology, rather than harder?" are often seen on student evaluations (Keith, 1995, p. 6). It is critical to analyze whether these increased efforts truly help students understand calculus and its applications. Although instructors feel that reform students are doing better in terms of passing rates, general performance, retention rates, out-of-class studying, and enrollment in upper-level mathematics courses, there is very little solid evidence to support these instructors' convictions or their belief that reform students are learning and retaining more with improved conceptual understanding (Bookman, 2000; Bookman & Friedman, 1996; Buccino, 2000; Ganter, 1999; Tucker & Leitzel, 1995).

The following study was conducted to address some of these issues by investigating the progress of students in reformed calculus courses as compared to traditionally-taught courses at a small private university in New England. In addition, a subset of students from both types of courses were tracked through several calculus-based, technical courses in the subsequent year.

Although the information to be presented does not give a complete evaluation of the project's impact, the student performance data for traditional measures that will be reported, analyzed, and discussed yield enlightening results that can be used to guide further study. Results from individual attitudinal interviews conducted with reform and traditional students after the calculus course will also be presented.

Background

The Calculus Reform Project. The project for this study was first implemented in the fall of 1990 with support from the National Science Foundation. The goal of the project was to develop a modular approach to calculus through computer labs and extensive projects that utilize a computer algebra system (*Maple*). The project heavily emphasized applications, mathematical problem solving, and an early introduction of functions of several variables. Sections of the course involved in the project used the same textbook as the traditional sections; however, the reform sections also utilized computer labs and applied projects that were written as part of the project. All lab and project work was completed in groups of 2–4 students, with one of the four weekly class meetings being in the computer lab. The three class meetings outside the lab setting were not structured as part of the project and were conducted as deemed appropriate by the individual instructors. Students were given one week to complete each computer lab, while the time allotted to projects was approximately four weeks. In a typical seven-week term, students were required to complete four computer lab assignments and one project that were common to all pilot sections. Both pilot and non-pilot sections utilized the same final exam; however, other tests and homework assignments varied for each section as determined by the individual instructor. Distribution of grades for all assignments, including the labs and projects, also varied for each instructor, with the percentage of final grades allotted to labs and projects ranging from 50% to 70% in the pilot classes.

The pilot course placed a heavy emphasis on the importance of technical writing and communication through the labs and projects. Students were not only asked to solve applied problems that required modeling and analysis, but also to discuss their solutions in the context of a technical report. Project reports were completed only after several drafts were reviewed by the instructor and discussed with the student project group.

By academic year 1993–94, the year in which the data collection began, the projects and labs were being piloted in approximately 40% of the calculus courses on the "normal" track (i.e., students in these courses enrolled in Calculus I at the start of the academic year). Results from the data collection were to be used by the department to determine the level at which the labs and projects would be implemented in future calculus courses. In addition, the initial report of the 1993–94 study was distributed and reviewed by other departments campuswide to solicit input for the improvement of the project.

The Study. The university at which this study was conducted utilizes a calendar consisting of four seven-week terms (A, B, C, D) in an academic year. Therefore, data were collected for the two groups in 1993–94 for Calculus I (derivatives), II (integrals), III (sequences and series), and IV (multivariate) over the course of these four terms, A93, B93, C94, and D94, respectively. Comparisons were made at the beginning and end of each of these four terms to produce a picture of the two groups' progress in the entire calculus sequence while ensuring that the populations remained comparable throughout the year. Data were also collected for a subset of the same two groups of students in several courses taught during academic year 1994–95 that require calculus as a prerequisite; comparisons were made between the two groups in the subsequent courses (grouped by discipline).

In order to compare the reformed sections of calculus to the traditional ones in a meaningful way, a scientific method for measuring the knowledge gained by students in each type of course needed to be developed. Specifically, students selected experimental ("pilot") and control ("traditional") sections with no knowledge of the experiment. The participants were then analyzed by course type on several variables to establish "equivalence" between the groups. Only after establishing these two "identical" groups was it possible to correlate any differences in course performance with the treatment. Students were encouraged to maintain their pilot or non-pilot status in subsequent terms and, although some students transferred from pilot to non-pilot after the first term, all students who completed the pilot Calculus IV course in D94 had been pilot students for all four terms of the experiment.

Several tests were used to establish the homogeneity of the two groups upon entering Calculus I in A93. Tests included comparisons for age, gender, and mathematical competence as measured by a pre-test administered prior to A93. Homogeneity of the two groups was examined prior to each of the four terms by re-analyzing these results, taking into consideration any changes in the enrollment

status of students in the experiment. This was done to ensure that the two populations for each term were still equivalent as measured prior to A93, implying that
changes in the groups' enrollments at the beginning of each term did not affect the
reliability of comparisons between the two groups.

Analysis

Data Instruments. The pre-test used to establish the homogeneity of the two
groups was the Calculus Readiness Exam, available from the Mathematical Association of America. This standardized exam contains 25 multiple-choice questions
on mathematical concepts up to, but not including, calculus. Topics on the test
include, but are not limited to, logarithmic and exponential functions, trigonometric functions, graphs of functions, and solutions for algebraic equations. The test
was administered once immediately prior to term A93; test scores were recorded for
each student as the number correct out of a possible 25. Each problem was worth
one point, with no partial credit given.

For each of the four courses, a special final exam was developed and administered to all students in order to quantify and compare the mathematical progress
of both groups (see Appendix). Common content in the pilot and non-pilot courses
ensured that all students had experience with the same mathematical concepts,
even though the method of delivery varied between the two groups. Therefore, it
was possible to give the two groups the same exam in each of the four terms. Exams were graded using a rubric developed by both pilot and non-pilot instructors.
Each question was graded by one instructor for all pilot and non-pilot sections to
maintain consistency in scoring across sections.

Each of the four common exams consisted of six to nine questions on various
mathematical topics from the course. This made it possible to compare the pilot
and non-pilot students not only on total exam scores, but also on each major topic
covered. The questions on these exams were written in a traditional format; i.e.,
students were not required to utilize any of the computer or writing skills emphasized in the pilot courses. This was intentional, as the project seeks to establish
that students in reform courses are still able to perform basic calculus skills, a concern throughout the mathematics community about reform courses since they place
less emphasis on traditional skills and more emphasis on writing, communication,
and the use of technology. If the pilot students can perform at levels equal to or
higher than their traditional counterparts on traditional assessment measures, then
it can be argued that their educational experience is greatly enhanced by the reform
effort, given the other practical skills gained in such a course.

A standardized, multiple-choice calculus exam developed by the Educational
Testing Service (ETS) was given at the end of Calculus IV in addition to the
common final exam for the course. The ETS exam covers much of the material
taught in the four calculus courses, making it similar to a cumulative final for the
entire academic year. Therefore, the purpose of the ETS exam was to compare
the two groups on the cumulative knowledge retained from all four terms of the
calculus course.

Data were collected during academic year 1994–95 for subsequent courses in
which many students from the 1993–94 study were enrolled. These courses included
Advanced Calculus, Differential Equations, Linear Algebra, General Chemistry,
Organic Chemistry I and II, Chemical Thermodynamics, and two entry-level courses

in Electrical and Computer Engineering. Because of the difficulty in implementing a standardized exam in several courses from multiple departments, course grades were used to evaluate performance. Ordinarily, biases in awarding course grades would make it impossible to use them in comparisons of academic progress. However, the common exams from Calculus IV in D94 implied that the entering abilities of students in these courses could be monitored. In addition, the subsequent courses that were examined were each taught by only one faculty member, making the comparison of pilot and non-pilot students on course grades within each course a valid measure of success. To obtain the course grades, signed release forms had to be obtained from the students; therefore, the sample size for this portion of the study was limited to 40 ($n = 15$ for the pilot, $n = 25$ for the non-pilot).

Three pilot and two non-pilot students also participated in individual interviews approximately nine months after completing Calculus IV. These interviews were designed to obtain student reactions to the pilot program. All interviews were conducted in a consistent format that included a brief statement of purpose, instructions for the interview, and a series of ten questions about the pilot calculus program.

Method of Analysis. Due to the short, discrete range for the scores on exam questions from the common finals, normality was frequently a problem. Thus the NPAR1WAY procedure in SAS was utilized in the analysis of exam data to enable the use of non-parametric statistical tests; i.e., tests that do not require that the data be approximately normally distributed. Since the purpose of this study was to determine any significant differences in the knowledge retained by students in the pilot and non-pilot calculus courses, the analysis needed to test

$$H_0 : \mu_{\mathrm{p}} - \mu_{\mathrm{np}} = 0 \quad \text{vs.} \quad H_a : \mu_{\mathrm{p}} - \mu_{\mathrm{np}} \neq 0$$

where μ_{p} is the mean of the scores for each exam question or one of the exam totals for the students enrolled in a pilot course and μ_{np} is the mean for students in a non-pilot course on the same measures. Testing this hypothesis revealed any statistically significant differences in the test scores of the two student populations. An analysis of variance on ranks was performed and several statistics (including normal theory z-scores) were computed based on the empirical distribution function and certain rank scores of the response variables (e.g., final exam question scores) across a one-way classification (see SAS, 1990). The Wilcoxon Rank-Sum Test was used for this analysis. Means and p-values on each question for each population will be reported and discussed.

Course grades for the subsequent courses were assigned numerical values for the purpose of the analysis, with A = 4, B = 3, C = 2, and below C = 1.[1] In order to obtain samples large enough to determine significant differences between the pilot and non-pilot students, course grades were grouped by subject (mathematics, chemistry, and electrical engineering) across academic terms. Although this requires the combination of data from different instructors, the results will still be reliable since the ratio of pilot to non-pilot students enrolled in each of the courses was approximately equal.

---

[1]The grading system at this university does not distinguish between grades of D and F.

Results

Homogeneity of the Populations. The initial analysis focuses on the homogeneity of the two populations under consideration; i.e., were the students enrolled in either the non-pilot or the pilot course at an advantage when the courses began? The NPARlWAY procedure established no difference in the gender distribution of the two groups (p-value $= 0.3480$) and although a significant difference was seen in the mean age of the two groups (p-value $= 0.0251$), the mean ages for the two groups were 18.28 and 18.68 years for the pilot and non-pilot students, respectively. This difference will be considered insignificant for the purpose of this study, since the statistical significance of this difference was simply due to the small range of values within the populations, not a large actual difference in ages. Results of the NPARlWAY procedure for the pre-test for each term can be seen in Table 1.

Table 1. Homogeneity as Measured by Calculus Readiness Test
(administered August 1993).

| | sample size[a] | | mean | | |
term	pilot	non-pilot	pilot	non-pilot	p-value[b]
A93	125	174	16.01	15.52	0.2778
B93	72	190	16.15	15.57	0.4516
C94	58	199	16.48	16.28	0.9811
D94	56	158	16.50	16.48	0.9207

[a]Samples do not include students who were unable to attend the exam that subsequently enrolled in the course.
[b]p-value ≤ 0.05 indicates a significant difference.

The table indicates no significant difference between the two groups on this exam, suggesting an approximately equal understanding of precalculus concepts upon entering Calculus I in A93. Although the data used for the above comparisons are limited, it is felt that they provide enough information to assume that the two populations began each of the four calculus courses with equivalent mathematical backgrounds and characteristics as measured prior to enrolling in Calculus I in A93.

Homogeneity between the two samples studied in the 1994–95 subsequent courses was determined by comparing scores on the ETS Standardized Calculus Exam and the Calculus IV common final exam, both administered at the end of D94. It is necessary to determine any significant differences between the two groups so that progress in the subsequent courses can be measured and compared. Since the sample size is small ($n = 40$), the assumption is made that the variances of the two independent samples are equal and that the data come from a normal distribution. By making these assumptions, a two-sample t-test with equal variances can be used. Results from the t-tests for both exams can be seen in Table 2.

Table 2. Homogeneity for Subsequent Courses (measured May 1994).

| | sample size | | mean | | |
Exam	pilot	non-pilot	pilot	non-pilot	p-value[a]
ETS	15	25	15.80	19.52	0.2180
cumulative common final	15	25	66.07	80.17	0.0529

[a]p-value ≤ 0.05 indicates a significant difference.

The results indicate that performance between the two groups was statistically equivalent on the ETS exam (p-value $= 0.2180$); however, the non-pilot students performed marginally significantly higher on the common final exam (p-value = 0.0529). Although conclusive results cannot be implied from such a small sample, we will see that a similar analysis on the entire Calculus IV population ($n = 253$) yields similar results (see Table 6). Therefore, the results from this sample appear to be representative of those from the entire Calculus IV population and it will be assumed that the pilot students entered the subsequent courses in 1994–95 with marginally significantly lower achievement on traditional Calculus IV material than the non-pilot students, given the previous caution of inference from small sample sizes.

A term 1993. The common final exam for term A93 consisted of eight questions covering the following topics: equations of tangent lines, calculation of derivatives, limits of rational functions, absolute maximum and minimum, continuity of piecewise functions, rate of change, graph of velocity function, and general concepts. Table 3 displays the means and p-values for each of the eight questions and the total score on the final exam for each of the two student groups.

Table 3. Common Final for A Term 1993.

question topic	mean pilot	mean non-pilot	p-value[a]
tangents	7.16	7.31	0.6728
derivatives	18.47	18.66	0.4337
max/min	4.10	4.53	0.2766
continuity	6.95	6.54	0.2010
rate of change	6.32	5.23	0.0295
velocity	7.95	7.94	0.7637
general	4.56	4.53	0.8943
total	65.69	64.96	0.8128
sample size	146	204	

[a]p-value ≤ 0.05 indicates a significant difference.

Note that the non-pilot sample size was 204 for all variables, while the corresponding pilot sample size was 146. Only Question 6 (rate of change) resulted in a significant difference between the two groups (p-value = 0.0295), with the pilot students outperforming the nonpilot students. Therefore, with the exception of Question 6, the null hypothesis, H_0, is not rejected in favor of the alternative for all exam questions and total exam scores, implying that the pilot and non-pilot students performed equally well on the mathematical topics included on the common final exam at the end of term A93.

B term 1993. The common final exam for term B93 consisted of seven questions covering the following topics: derivatives of functions involving natural logarithms and exponentials, integrals, solids of revolution, graphing, areas, calculating maximums and minimums on a graph, and area approximation. Table 4 displays the means and p-values for each of the seven questions and the total score on the final exam for each of the two student groups.

Table 4. Common Final for B Term 1993.

question topic	mean pilot	mean non-pilot	p-value[a]
deriv: ln/exp	12.06	12.40	0.9749
integrals	19.49	20.06	0.5902
solid of rev.	9.39	9.55	0.5623
graphing	5.39	6.16	0.1503
area	7.45	7.55	0.7482
max/min	6.47	7.78	0.0286
area approx.	.58	6.89	0.6181
total	65.29	69.46	0.1516
sample size	85	209	

[a]p-value ≤ 0.05 indicates a significant difference.

Note that the non-pilot sample size was 209 for all variables, while the corresponding pilot sample size was 85. Although there is a large difference in the sample sizes of the pilot and non-pilot groups for this term (as well as subsequent terms), three factors proved these differences to be unimportant: the robustness of the Wilcoxon test; the fact that the pilot group, although small in comparison to the non-pilot group, is still quite large; and the small variances of the two groups in each term, implying this was not a concern. Only Question 6, dealing with the calculation of maximums and minimums on a graph, resulted in a significant difference between the two groups, with the non-pilot students outperforming the pilot students. Therefore, with the exception of Question 6, the null hypothesis, H_0, is not rejected in favor of the alternative for all exam questions and total exam scores, implying that the pilot and non-pilot students performed equally well on the mathematical topics included on the common final exam at the end of term B93.

C term 1994. The common final exam for term C94 consisted of nine questions covering the following topics: ordinary differential equations (ODEs), areas, integrals, derivatives and integrals of inverse sines, Taylor polynomials, Taylor errors, limits, convergence, and Taylor series. Table 5 displays the means and p-values for each of the nine questions and the total score on the final exam for each of the two student groups.

Table 5. Common Final for C Term 1994.

question topic	mean pilot	mean non-pilot	p-value[a]
ODEs	7.93	6.44	0.0208
area	8.71	8.62	0.6784
integral	8.34	8.45	0.1615
inverse sine	2.83	3.94	0.0172
Taylor poly.	9.14	10.11	0.0427
Taylor error	4.74	5.12	0.5701
limits	7.76	8.67	0.0081
convergence	6.28	7.38	0.0009
Taylor series	5.62	6.08	0.7257
total	61.34	64.15	0.0870
sample size[b]	68	236	

[a]p-value ≤ 0.05 indicates a significant difference.
[b]Includes mid-year transfer students from other institutions.

Note that the non-pilot sample size was 236 for all variables, while the corresponding pilot sample size was 68. Here, Questions 1, 4, 5, 7 and 8 all resulted in a significant difference between the two groups, with the pilot students outperforming the non-pilot students on Question 1 and the non-pilot students doing significantly better on Questions 4, 5, 7 and 8. Therefore, the null hypothesis, H_0, is not rejected in favor of the alternative only for exam questions 2, 3, 6, 9, implying that the non-pilot students performed better on several of the mathematical topics included on the common final exam at the end of term C94. However, the total exam scores are still statistically equivalent for the two groups.

D term 1994. The common final exam for term D94 consisted of six questions covering the following topics: second partials, first partials, gradients, cross products, dot products, and maximums and minimums of multivariate functions. Table 6 displays the means and p-values for each of the six questions and the total score on the final exam for each of the two student groups.

Table 6. Common Final and ETS Exam for
D Term 1994.

question topic	mean		p-value[a]
	pilot	non-pilot	
2nd partials	7.38	9.76	0.0019
1st partials	10.36	12.48	0.0001
gradient	12.82	14.47	0.0010
cross product	8.05	13.45	0.0001
dot product	10.97	11.23	0.9613
max/min	9.94	11.05	0.0282
total	59.32	72.35	0.0001
ETS Exam	39.39	44.06	0.0001
sample size	66	187	

[a]p-value ≤ 0.05 indicates a significant difference.

Note that the non-pilot sample size was 187 for all variables, while the corresponding pilot sample size was 66. Only Question 5, dealing with dot products, resulted in a non-significant difference between the two groups. The non-pilot students outperformed the pilot students on all other questions, as well as the total score. Therefore, the null hypothesis, H_0, is rejected in favor of the alternative for exam questions 1, 2, 3, 4, 6 and the total exam scores, implying that the non-pilot students performed better on nearly all the mathematical topics included on the common final exam at the end of term D94.

The final comparison between these groups is the scores on the ETS exam. This standardized test was taken only at the end of term D94 in order to compare the cumulative performance of the pilot and non-pilot students after four terms of calculus. As seen in Table 6, the non-pilot students significantly outperformed the pilot students on this standardized exam.

1994–95 Subsequent Courses. Analysis of students' performance in subsequent courses was observed using course grades in nine courses grouped into three disciplines: chemistry, electrical engineering, and mathematics. Results of t-tests for each of the three disciplines yielded no significant differences between the pilot and non-pilot students in any of the three disciplines (see Table 7). As was previously discussed, the pilot students entered these subsequent courses at a significant disadvantage as measured by the common final exams from Calculus IV at the end of the previous academic year.

Samples include combined enrollment in each discipline. Further, sample sizes reflect the fact that many students in the follow-up study were enrolled in more than one of the courses and therefore were counted once for each of the courses they attended.

Table 7. Subsequent Course Grades.[a]

discipline	sample size		mean		p-value[b]
	pilot	non-pilot	pilot	non-pilot	
Chemistry	21	21	2.43	2.86	0.1031
Elec. Eng.	11	8	2.09	2.88	0.0782
Mathematics	18	31	2.44	2.84	0.1060

[a]A = 4.0, B = 3.0, C = 2.0, below C = 1.0.
[b]p-value ≤ 0.05 indicates a significant difference.

Individual interviews with pilot and non-pilot students in February 1995 yielded a wide variety of responses. However, several points were agreed upon by all students interviewed from both groups. First, all believed that the topics from calculus were important for success in their subsequent coursework. All students also felt that open-ended projects and opportunities to work in groups aid in learning and are important for continued academic achievement. Computers were believed to be an important tool for learning by all students. However when specifically asked about the benefits of the pilot calculus program, the opinions of the pilot and non-pilot students differed dramatically. Non-pilot students expressed serious concerns about the ability of the pilot program to prepare students for subsequent coursework and spoke very negatively about the program even though they had never been enrolled in the pilot course; conversely, the pilot students felt that they had benefited greatly from the program and that the pilot calculus course prepares students better for their subsequent coursework than more traditional methods.

Discussion and Issues for Further Research

The purpose of this study was to evaluate the impact of a reform calculus course on student achievement as traditionally measured in college calculus. The subsequent impact of reform calculus on courses requiring calculus as a prerequisite was also examined. A review of the data reveals an increasingly negative trend in the ability of the pilot students to perform traditional tasks while enrolled in the calculus sequence as measured by common final exams and standardized tests. However, this is certainly to be expected given the significantly lower emphasis placed on such activities in the pilot courses. For example in Calculus IV, the only course that demonstrated significantly different overall performance levels for the two groups, the percentage of the course grade devoted to both the common final exam and the ETS exam was only 15% *combined* in the pilot sections, as compared to approximately 30% in the traditional sections. As a result, the pilot students devoted very little study time to these exams, as revealed in course evaluations. Efforts were instead put into computer lab reports and the open-ended project, which comprised as much as 70% of the course grade; similar trends were observed in Calculus I–III.

Is the decrease in study time for traditional skills and increased time for open-ended problems and technical writing a negative outcome of such a curriculum? Maybe not, since this significant difference in performance on traditional skills seemed to have very little bearing on success in subsequent courses from several disciplines. Perhaps the issue to be investigated further is not how to maintain performance levels on these traditional skills, but rather the value of these skills for subsequent work in comparison with those skills obtained through the reform curriculum. In addition, the individual student interviews revealed a positive attitude toward the pilot program and mathematics in general from the pilot students. How much will this attitude contribute to their level of commitment in the total undergraduate experience, especially in the areas of science, mathematics, engineering, and technology? Issues such as these, as well as methods for eliminating the misconceptions about the value of alternative delivery methods in calculus both in students and faculty, are critical areas for further study.

Specifically, an important part of the implementation process for new teaching methods in undergraduate mathematics is open lines of communication about the philosophies and research that support such change. This communication must occur not only within mathematics departments, but also between faculty from many disciplines and, perhaps most importantly, between faculty and students. Because many of our students have grown up in an educational system that does not utilize the techniques that undergraduate mathematics reform is embracing, they often resist and even become hostile to such change. This hostility, as reflected in the large number of students who transferred from pilot to non-pilot sections after the first term, is quite understandable given their success with the traditional system, the extra work that most reform curricula require, and the uncertainty of faculty who are teaching under circumstances that are often foreign to them. Both students and faculty need to understand the importance of experimentation and change in our educational system.

Although this study had many limitations inherent to data collected at one institution, several important observations surfaced that warrant careful further study. Perhaps the most important of these is the need to investigate the purpose of the calculus course in the preparation of undergraduate students for other courses and for a technological society. The skills that we as mathematicians deem critical to the mathematical development of our students may in fact be less important, given the changes implied by increasingly sophisticated technologies. Certainly, the students in the reform calculus courses did not suffer from the experience; in fact, one would be remiss to overlook the many skills that were developed in these students through the pilot course. It is well-documented that several experiences deemed essential by employers of college graduates are technical writing, computer literacy, group work, and the ability to solve open-ended problems (Davis, 1993, 1994, 2000; Dudley, 1997; Ganter & Kinder, 2000; NSF, 1996; SIAM News, 1995). Clearly, the reform course offers multiple opportunities for these experiences that often are not offered in a more traditional setting.

It will take considerable effort, experimentation, evaluation, and re-examination of our goals to find the appropriate combination of projects, group work, computer labs, and traditional skills for our calculus students. Indeed, the program studied has already undergone several revisions since these data were collected in an effort to find that ideal combination. In fact, the process of experimentation, evaluation, and re-grouping must be the norm in undergraduate mathematics education. Only through such continuous change can our educational system produce citizens prepared to face an increasingly technological, ever-changing society.

References

Ahmadi, D. C. (1995). *A comparison study between a traditional and reform Calculus II program.* Unpublished doctoral dissertation, The University of Oklahoma.

Alexander, E. H. (1997). An investigation of the results of a change in calculus instruction at the University of Arizona. *Dissertation Abstracts International.*

Allen, G. H. (1995). A comparison of the effectiveness of the Harvard Calculus series with the traditionally taught calculus series. *Dissertation Abstracts International.*

American Mathematical Association of Two-Year Colleges (1995). *Crossroads in mathematics: Standards for introductory college mathematics before calculus.* Memphis, TN: Author.

Andrews, G. E. (1995). The irrelevance of calculus reform: Ruminations of the sage-on-the-stage. *UME Trends, 6*(6), 17, 23.

Asiala, M., Brown, A., DeVries, D., Dubinsky, E., Mathews, D., & Thomas, K. (1996). A framework for research and curriculum development in undergraduate mathematics education. In E. Dubinsky, A. H. Schoenfeld, & J. Kaput (Eds.), *Research in collegiate mathematics education II* (pp. 1–32). Providence, RI: American Mathematical Society.

Aspinwall, L. N. (1994). *The role of graphic representation and students' images in understanding the derivative in calculus: Critical case studies.* Unpublished doctoral dissertation, The Florida State University.

Atkins, C. D., Jr. (1994). *A study to produce guidelines for evaluating calculus reform projects.* Unpublished doctoral dissertation, North Carolina State University.

Barton, S. D. (1995). *Graphing calculators in college calculus: An examination of teachers' conceptions and instructional practice.* Unpublished doctoral dissertation, Oregon State University.

Baum, J. D. (1958). Mathematics, self-taught. *American Mathematical Monthly, 65*(9), 701–705.

Bookman, J. (2000). Program evaluation and undergraduate mathematics renewal: The impact of calculus reform on student performance in subsequent courses. In S. L. Ganter (Ed.), *Calculus renewal: Issues for undergraduate mathematics education in the next decade* (pp. 91–102). New York: Kluwer Academic/Plenum Publishers.

Bookman, J. & Friedman, C. P. (1994). A comparison of the problem solving performance of students in lab based and traditional calculus. In E. Dubinsky, A. H. Schoenfeld, & J. Kaput (Eds.), *Research in collegiate mathematics education I* (pp. 101–116). Providence, RI: American Mathematical Society.

Bookman, J. & Friedman, C. P. (1996). *Student attitudes and calculus reform* (included in Project CALC information package). Unpublished.

Bookman, J. & Friedman, C. P. (1998). Student attitudes and calculus reform. *School Science and Mathematics* (March), 117–122.

Buccino, A. (2000). Politics and professional beliefs in evaluation: The case of calculus renewal. In S. L. Ganter (Ed.), *Calculus renewal: Issues for undergraduate mathematics education in the next decade* (pp. 121–146). New York: Kluwer Academic/Plenum Publishers.

Buck, R. C. (1962). Teaching machines and mathematics programs: Statements by R. C. Buck. *American Mathematical Monthly, 69*(6), 561–564.

Cipra, B. (1996). Calculus reform sparks a backlash. *Science, 271*, 901–902.

Crockett, C. & Kiele, W. (1993). A comprehensive game plan for calculus reform: Thoughts grown through experience. *Primus, 3*(4), 355–370.

Committee on the Undergraduate Program in Mathematics. (1969). Calculus with computers. *Newsletter of the CUPM*, 4.

Cummins, I. (1960). A student experience-discovery approach to the teaching of calculus. *Mathematics Teacher, 53*(3), 162–170.

Davidson, N. A. (1970). *The small group-discovery method of mathematics instruction as applied in calculus*. Unpublished doctoral dissertation, The University of Wisconsin, Madison.

Davidson, N. A. (1985). Small group learning and teaching in mathematics: A selective review of the research. In R. Slavin (Ed.), *Learning to cooperate, cooperating to learn* (pp. 211–230). New York: Plenum Press.

Davidson, N. A. (Ed.). (1990). *Cooperative learning in mathematics: A handbook for teachers*. Menlo Park, CA: Addison-Wesley Publishing Company.

Davis, P. W. (1993). Some glimpses of mathematics in industry. *Notices of the American Mathematical Society, 40*(7), 800–802.

Davis, P. W. (1994). *Mathematics in industry: The job market of the future* (1994 SIAM Forum final report). Philadelphia: Society for Industrial and Applied Mathematics.

Davis, P. W. (2000). Calculus renewal and the world of work. In S. L. Ganter (Ed.), *Calculus renewal: Issues for undergraduate mathematics education in the next decade* (pp. 41–52). New York: Kluwer Academic/Plenum Publishers.

Davis, T. A. (1966). An experiment in teaching mathematics at the college level by programmed instruction. *American Mathematical Monthly, 73*(6), 656–659.

Dossey, J. (Ed.). (1998). *Confronting the core curriculum: Considering change in the undergraduate mathematics major* (MAA Notes no. 45). Washington, DC: Mathematical Association of America.

Dudley, U. (1997). Is mathematics necessary? *College Mathematics Journal, 25*(5), 364.

Durell, F. (1894). Application of the new education to the differential and integral calculus. *American Mathematical Monthly, 1*(1 & 2), 15–19, 37–41.

Dyer-Bennett, J., Fuller, W. R., Seibert, W. E., & Shanks, M. E. (1958). Teaching calculus by closed-circuit television. *American Mathematical Monthly, 65*(6), 430–439.

Eisner, E. (1996). *Instructional and expressive objectives: Their formulation and use in curriculum* (AERA Monograph Series in Curriculum Evaluation, no. 3). Chicago: Rand McNally.

Ellis, W., Jr. (2000). Technology and calculus. In S. L. Ganter (Ed.), *Calculus renewal: Issues for undergraduate mathematics education in the next decade* (pp. 53–68). New York: Kluwer Academic/Plenum Publishers.

Farr, W. W. & VanGeel, M. (1994). *Maple labs and programs for calculus.* Unpublished manuscript.

Ferrini-Mundy, J. & Lauten, D. (1994). Learning about calculus learning. *Mathematics Teacher, 87*(2), 115–121.

Foley, G. & Ruch, D. (1995). Calculus reform at comprehensive universities and two-year colleges. *UME Trends, 6*(6), 8–9.

Francis, E. J. (1992). *The concept of limit in college calculus: Assessing student understanding and teacher beliefs.* Unpublished doctoral dissertation, University of Maryland, College Park.

Frid, S. (1994). Three approaches to undergraduate instruction: Their nature and potential impact on students' language use and sources of conviction. In E. Dubinsky, A. H. Schoenfeld, & J. Kaput (Eds.), *Research in collegiate mathematics education I* (pp. 69–100). Providence, RI: American Mathematical Society.

Ganter, S. L. (1997). Impact of calculus reform on student learning and attitudes. *Association for Women in Science Magazine, 26*(6), 10–15.

Ganter, S. L. (1999). An evaluation of calculus reform: A preliminary report of a national study. In B. Gold, S. Keith, & W. Marion (Eds.), *Assessment practices in undergraduate mathematics* (MAA Notes no. 49, pp. 233–236). Washington, DC: Mathematical Association of America.

Ganter, S. L., & Kinder, J. S. (2000). Targeting institutional change: Quality undergraduate science education for all students. In *Targeting curricular change: Reform in undergraduate education in science, math, engineering, and technology* (pp. 1–27). Washington, DC: American Association for Higher Education.

George, M. D. (2000). Calculus renewal in the context of undergraduate SMET education. In S. L. Ganter (Ed.), *Calculus renewal: Issues for undergraduate mathematics education in the next decade* (pp. 1–10). New York: Kluwer Academic/Plenum Publishers.

Gordon, S. P. (2000). Renewing the precursor courses: New challenges, opportunities, and connections. In S. L. Ganter (Ed.), *Calculus renewal: Issues for undergraduate mathematics education in the next decade* (pp. 69–90). New York: Kluwer Academic/Plenum Publishers.

Heid, M. K. (1988). Resequencing skills and concepts in applied calculus using the computer as a tool. *Journal for Research in Mathematics Education, 19*(4), 3–25.

Holden, L. S. (1967). *Motivation for certain theorems of the calculus.* Unpublished doctoral dissertation, The Ohio State University.

Hurley, J. F., Koehn, U., & Ganter, S. L. (1999). Effects of calculus reform: Local and national. *American Mathematical Monthly, 106*(9), 800–811.

Johnson, R., Johnson, R., & Stanne, M. (1986). Comparison of computer-assisted, cooperative, competitive, and individualistic learning. *American Educational Research Journal, 23*(3), 382–392.

Keith, S. Z. (1995). How do students feel about calculus reform, and how can we tell? *UME Trends, 6*(6), 6, 31.

Kroll, D. L. (1989). *Cooperative mathematical problem solving and metacognition: A case study of three pairs of women.* Unpublished doctoral dissertation, Indiana University.

Larsen, C. M. (1961). *The heuristic standpoint in the teaching of elementary calculus.* Unpublished doctoral dissertation, Stanford University.

Levi, H. (1963). An experimental course in analysis for college freshmen. *American Mathematical Monthly, 70*(8), 877–879.

Levine, J. L. (1968). *A comparative study of two methods of teaching mathematical analysis at the college level.* Unpublished doctoral dissertation, Columbia University.

Lightbourne, J. H., III (2000). Crossing the discipline boundaries to improve undergraduate mathematics education. In S. L. Ganter (Ed.), *Calculus renewal: Issues for undergraduate mathematics education in the next decade* (pp. 147–158). New York: Kluwer Academic/Plenum Publishers.

Mathews, D. M. (1995). Time to study: The C^4L Experience. *UME Trends, 7*(4), 14.

May, K. O. (1962). Small versus large classes. *American Mathematical Monthly, 69*(5), 433–434.

May, K. O. (1964). *Programmed learning and mathematical education.* San Francisco, CA: The Committee on Educational Media of the Mathematical Association of America.

McCallum, W. (2000). The goals of the calculus course. In S. L. Ganter (Ed.), *Calculus renewal: Issues for undergraduate mathematics education in the next decade* (pp. 11–22). New York: Kluwer Academic/Plenum Publishers.

McKeachie, W. J. (1954). Student-centered versus instructor-centered instruction. *Journal of Educational Psychology, 45*(3),143–150.

McKeen, R. L. (1970). *A model for curriculum construction through observations of students solving problems in small instructional groups.* Unpublished doctoral dissertation, University of Maryland, College Park.

Moise, E. E. (1965). Activity and motivation in mathematics. *American Mathematical Monthly, 72*(4), 407–412.

Monroe, H. L. (1966). *A study of the effects of integrating analytic geometry and calculus on the achievement of students in these courses.* Unpublished doctoral dissertation, University of Pittsburgh.

Mumford, D. (1997). Calculus reform for the millions. *Notices of the American Mathematical Society, 44*(5), 559–563.

National Council of Teachers of Mathematics. (1989). *Curriculum and evaluation standards for school mathematics.* Reston, VA: Author.

National Council of Teachers of Mathematics. (1995). *Assessment standards for school mathematics.* Reston, VA: Author.

National Science Foundation. (1987). *Undergraduate curriculum development in mathematics: Calculus* (Program announcement, Division of Mathematical Sciences). Washington, DC: Author.

National Science Foundation. (1991). *Undergraduate curriculum development: Calculus* (Report of the committee of visitors). Washington, DC: Author.

National Science Foundation. (1996). *Shaping the future: New expectations for undergraduate education in science, mathematics, engineering, and technology* (Report of the advisory committee for review of undergraduate education). Arlington, VA: Author.

Newell, J. C. (1994). *Student experiences in a first semester university calculus course: A study using ethnographic methods.* Unpublished master's thesis, Simon Fraser University.

Padgett, E. E., III. (1994). *Calculus I with a laboratory component.* Unpublished doctoral dissertation, Baylor University.

Roberts, W. (Ed.). (1996). *Calculus: The dynamics of change.* Washington, DC: Mathematical Association of America.

Rochowicz, J. A., Jr. (1993). *An analysis of the perceived impact of computing devices on calculus instruction in engineering curricula.* Unpublished doctoral dissertation, Lehigh University.

SAS Institute, Inc. (1990). *SAS/STAT user's guide* (version 6, 4th ed., vol. 2). Cary, NC: Author.

Schneider, E. M. (1995). *Testing the rule of three: A formative evaluation of the Harvard based calculus consortium curriculum.* Unpublished doctoral dissertation, The University of Texas at Austin.

Schoenfeld, A. H. (1996). *Student assessment in calculus* (A report of the NSF working group on assessment in calculus). Arlington, VA: National Science Foundation.

Shelton, R. M. (1965). *A comparison of achievement resulting from teaching the limit concept in calculus by two different methods.* Unpublished doctoral dissertation, University of Illinois.

SIAM News. (1995). *SIAM's MII project moves into implementation phase.* (January), 4–5.

Simonsen, L. M. (1995). *Teachers' perceptions of the concept of limit, the role of limits, and the teaching of limits in advanced placement calculus.* Unpublished doctoral dissertation, Oregon State University.

Slavin, R. E. (1983). When does cooperative learning increase student achievement? *Psychological Bulletin, 94,* 429–445.

Slavin, R. E. (1985). An introduction to cooperative learning research. In R. Slavin (Ed.), *Learning to cooperate, cooperating to learn* (pp. 5–15). New York: Plenum Press.

Slavin, R. E. (1999). A rejoinder: Yes, control groups are essential in program evaluation: A response to Pogrow. *Educational Researcher, 28*(3), 36–38.

Smith, D. A. (2000). Renewal in collegiate mathematics education: Learning from research. In S. L. Ganter (Ed.), *Calculus renewal: Issues for undergraduate mathematics education in the next decade* (pp. 23–40). New York: Kluwer Academic/Plenum Publishers.

Smith, D. A. (1996). Trends in calculus reform. In A. Solow (Ed.), *Preparing for a new calculus* (MAA Notes no. 36, pp. 3–13). Washington, DC: Mathematical Association of America.

Stannard, W. A. (1966). *The effect on final achievement in a beginning calculus course resulting from the use of programmed materials written to supplement regular classroom instruction.* Unpublished doctoral dissertation, Montana State University.

Stevens, F., Lawrenz, F., & Sharp, L. (1993). *User-friendly handbook for project evaluation: Science, mathematics, engineering and technology education.* Washington, DC: National Science Foundation.

Stockton, D. S. (1960). An experiment with a large calculus class. *American Mathematical Monthly, 67*(10), 1024–1025.

Tall, D., et al. (1984). The mathematics curriculum and the microcomputer. *Mathematics in School, 13*(4), 7–9.

Tall, D. (1990). Inconsistencies in the learning of calculus and analysis. *Focus on Learning Problems in Mathematics, 12*(3 & 4), 49–63.

Tucker, A. C. & Leitzel, J. R. C. (Eds.). (1995). *Assessing calculus reform efforts: A report to the community.* Washington, DC: Mathematical Association of America.

Turner, V. D., Alders, C. D., Hatfield, F., Croy, H., & Sigrist, C. (1966). A study of ways of handling large classes in freshman mathematics. *American Mathematical Monthly, 73*(7), 768–770.

Urion, D. K. & Davidson, N. A. (1992). Student achievement in small-group instruction versus teacher-centered instruction in mathematics. *Primus, 2*(3), 257–264

Webb, N., Ender, P., & Lewis, S. (1986). Problem solving strategies and group processes in small groups learning computer programming. *American Educational Research Journal, 23*(2), 243–262.

Wells, P. J. (1995). *Conceptual understanding of major topics in first semester calculus: A study of three types of calculus courses at the University of Kentucky.* Unpublished doctoral dissertation, University of Kentucky.

West, R. D. (1994). *Evaluating the effects of changing a mathematics core curriculum.* Presentation at the Joint Mathematics Meetings, Cincinnati, OH.

Williams, C. G. (1994). *Using concept maps to determine differences in the concept image of function held by students in reform and traditional calculus.* Unpublished doctoral dissertation, University of California, Santa Barbara.

Yackel, E., Cobb, P. & Wood, T. (1991). Small-group interactions as a source of learning opportunities in second-grade mathematics. *Journal for Research in Mathematics Education, 22*(5), 390–408.

Appendix. Final examination questions for D94

1. Suppose that $g(u,v) = \sin(u^2 - 3uv)$. Find $g_{uv} = \frac{\partial^2 g}{\partial v \partial u}$.
2. If z is a function of x and y, and x and y are functions of s and t, find $\frac{\partial z}{\partial t}$ if $\frac{\partial z}{\partial x} = 3$, $\frac{\partial z}{\partial y} = -2$, $\frac{\partial x}{\partial s} = 5$, $\frac{\partial x}{\partial t} = 6$, $\frac{\partial y}{\partial s} = 2$, and $\frac{\partial y}{\partial t} = 4$.
3. Given that $f(x,y) = x \sin y + e^{x \cos y}$,
   (a) find the gradient of f at the point $P(2, \frac{\pi}{2})$ and
   (b) find the directional derivative of f at $P(2, \frac{\pi}{2})$ in the direction of the vector $\vec{v} = \langle 3, 4 \rangle = 3\vec{\imath} + 4\vec{\jmath}$.
4. Find the equation of the plane passing though the points $A(1,2,3)$, $B(2,1,3)$, and $C(3,1,2)$.
5. Consider the vectors $\vec{a} = \langle 1, 2 \rangle, \vec{b} = \langle 3, 1 \rangle, \vec{c} = \langle -1, 2 \rangle$.
   (a) Find a number x such that $\vec{a} + x\vec{b}$ is perpendicular to $\vec{c}$.
   (b) Find a number y such that $\vec{a} + y\vec{b}$ is parallel to $\vec{c}$.
6. Let $f(x,y) = 2x^3 - 2xy + y$.
   (a) Find all critical points of the function f.
   (b) Determine the absolute maximum and minimum of f in the region D plotted below [graph is not shown in this article]. D is bounded by the lines $x = 0, y = 4$, and the curve $y = x^2$.

DEPARTMENT OF MATHEMATICAL SCIENCES, O-106 MARTIN HALL, BOX 340975, CLEMSON UNIVERSITY, CLEMSON, SC 29634-0975
E-mail address: sganter@clemson.edu

THE UNIVERSITY OF NORTH CAROLINA, SCHOOL OF PUBLIC HEALTH, DEPARTMENT OF BIO-STATISTICS, 3101 MCGAVRAN-GREENBERG HALL, CHAPEL HILL, NC 27599
E-mail address: mjiroute@bios.unc.edu

CBMS Issues in Mathematics Education
Volume **8**, 2000

A Longitudinal Study of the C^4L Calculus Reform Program: Comparisons of C^4L and Traditional Students

Keith E. Schwingendorf, George P. McCabe, and Jonathan Kuhn

This paper presents the results of a longitudinal statistical study in which comparisons are made between students who were taught introductory calculus courses using the Calculus, Concepts, Computers and Cooperative Learning (C^4L) pedagogical methodology and students taught in the traditional way (TRAD). Two basic, related questions considered in this study are:

- Which program, C^4L or TRAD, provides a student with a better understanding of the required calculus concepts?
- Which program, C^4L or TRAD, better inspires students to pursue further study in calculus or, more generally, mathematics?

The results of this study favor the C^4L program over the TRAD program. On the one hand, for example, it is found the C^4L students earn higher grades in calculus courses; in fact, almost half a grade higher, on average, than the TRAD students. On the other hand, it is found the C^4L students are as prepared as the TRAD students, but not more so, for mathematics courses beyond the calculus program. After some discussion of the C^4L program, the paper describes the statistical model used to perform the comparison between the two teaching methods and then presents the results of this statistical analysis. This paper reports on one phase of the evaluation of the C^4L calculus program, a quantitative evaluation based on a statistical model which is similar to that used in some health studies (e.g., studies on the relationship between second-hand smoke and cancer) and industrial studies (e.g., studies of the effects of the notion of Total Quality Management as related

This research has been partially supported by the National Science Foundation grants USE-9053432 and DUE-9450750. Any opinions, findings, and conclusions expressed in this paper are those of the authors and do not necessarily reflect the views of the National Science Foundation. The statistical model used in this study was designed by George P. McCabe, Professor of Statistics and Head of Statistical Consulting, Purdue University, West Lafayette, IN.

The authors are extremely grateful to each of the following people for their support, helpful discussions, comments and/or insightful suggestions for improving this paper: Linda McCabe who served as a statistical consultant and without whom this paper could not have been completed; Ed Dubinsky for his ideas regarding the analyses of the data; David M. Mathews, Jim Cottrill (Illinois State University), Lisa Schwingendorf, and Shonda Kuiper for their careful reading of the manuscript and suggestions for improvements of the paper. A special thanks goes to the 205 Purdue students reported on in this paper for their patience, hard work and dedication to the goals and principles of the C^4L Calculus Program.

©2000 American Mathematical Society

to productivity and employee satisfaction) which are observational in nature. The other phase of the C^4L evaluation involves qualitative research studies involving both C^4L and TRAD students.

The C^4L Program

The Calculus, Concepts, Computers and Cooperative Learning Calculus Reform Program is part of the National Calculus Reform Movement which began at the Tulane Conference (sponsored by the Sloan Foundation) in January 1986. The C^4L program is currently co-directed by Dubinsky, Schwingendorf, and David M. Mathews (Southwestern Michigan College, Dowagiac, MI). Numerous papers and reports have been written or presented at national conferences in connection with the C^4L program (Asiala et al., 1996; Asiala, Cottrill, Dubinsky, & Schwingendorf, 1997; Baker, Cooley, & Trigueros, in press; Czarnocha, Dubinsky, Loch, Prabhu, & Vidaković, in preparation; Clark et al., 1996; Cottrill et al., 1996; Cottrill, 1999; Donaldson, 1993; Dubinsky & Schwingendorf, 1990b; Dubinsky, 1992; Mathews, Narayan, & Schwingendorf, 1996; Mathews, 1995; Monteferrante, 1993; Schwingendorf & Dubinsky, 1990; Schwingendorf, Hawks-Hoover, & Beineke, 1992; Schwingendorf & Wimbish, 1994; Schwingendorf, Mathews, & Dubinsky, 1996; Schwingendorf, 1999; Silverberg, 1999, and McDonald, Mathews, & Strobel, this volume). The C^4L program texts (Dubinsky, Schwingendorf, & Mathews, 1995; Dubinsky & Schwingendorf, 1996) were written in an effort to support the C^4L pedagogical approach which is based on a constructivist perspective on how mathematics is learned (Asiala et al., 1996, 1997; Baker et al., in press; Czarnocha et al., in preparation; Clark et al., 1996; Cottrill et al., 1996; Dubinsky & Schwingendorf, 1990b; Dubinsky, 1992; Dubinsky et al., 1995; Dubinsky & Schwingendorf, 1996; Schwingendorf et al., 1996, and McDonald, Mathews, & Strobel, this volume). According to the Action-Process-Object-Schema (APOS) theory of learning (Asiala et al., 1996) on which the program courses are based, students need to construct their own understanding of each mathematical concept.

C^4L calculus courses differ from traditionally taught calculus courses and also from courses in other calculus reform programs in a number of ways. The traditional courses, delivered primarily via the lecture and recitation system, in general, attempt to "transfer" knowledge, emphasize rote skill and drill together with memorization of paper-and-pencil skills. In contrast, the primary emphasis of the C^4L program is to minimize lecturing, explaining, or otherwise attempting to "transfer" mathematical knowledge, but rather to create situations for students to foster their making the necessary mental constructions to learn mathematics concepts. The emphasis of the C^4L program is to develop student understanding of concepts together with the acquisition of necessary basic skills through active student involvement in what is termed the Activity-Class-Exercise (ACE) learning cycle.

Each unit of the ACE learning cycle, which generally lasts about a week, begins with students performing group computer activities in a laboratory setting in an effort to help students construct their own meaning of mathematical concepts and reflect on their experiences with their peers in a cooperative learning environment. Students use a mathematical programming language (MPL), such as *ISETL*, or that which is part of the *MapleV* system (Dubinsky & Schwingendorf, 1990b, 1990a; Dubinsky, 1992, 1995, n.d.; Mathews et al., 1996; Schwingendorf & Dubinsky, 1990; Schwingendorf et al., 1996) to write "programs" which model mathematical concepts, then investigate the concepts using various computer tools. Such tools include graphics and table facilities to help students reflect on their experiences in an effort to foster mental constructions of mathematical concepts. Many computer activities have students transfer information, sometimes requiring a reconstruction, back and forth between a formula in a symbolic computer system (SCS) such as *Derive*, *Mathematica*, or *Maple V*; the constructions in the MPL; and graphs or tables constructed on the computer. Laboratory periods are followed by class meetings during which carefully constructed "class tasks" and appropriate questions in conjunction with cooperative problem solving are used to help students to build upon their mathematical experiences in the computer laboratory. Finally, relatively traditional exercises are assigned to reinforce the knowledge students are expected to have constructed during the first two phases of the learning cycle.

The study reported on in this paper is one of several studies and reports that are beginning to emerge regarding the long-term effects of calculus reform (Bookman, 2000; Hurley, Koehn, & Ganter, 1999; Ganter, 1997, 2000; Schoenfeld, 1997). By long term, we mean studies done at least two to four years after students have completed their calculus courses. Susan L. Ganter reports (1997, 2000) the following trends concerning student performance:

> Evaluations conducted as part of the curriculum development projects revealed better conceptual understanding, higher retention rates, higher scores on common final exams, and higher confidence levels and involvement in mathematics for students in reform courses versus those in traditional courses; the effect on computational skills is uncertain.
>
> ... in general, regardless of the reform method used, the attitudes of students and faculty seem to be negative in the first year of implementation, with steady improvement in subsequent years if continuous revisions are made based on feedback. (Ganter, in press)

In her conclusions, Ganter (in press) also states that

> The success or failure of a reform effort is not necessarily dependent upon what is implemented but rather how, by whom, and in what setting. This raises important issues of design and interpretation of evaluation studies and points to just one of many factors that can complicate matters.

The C⁴L program qualitative research studies (Asiala et al., 1997; Baker et al., in press; Czarnocha et al., in preparation; Clark et al., 1996; Cottrill et al., 1996; Cottrill, 1999, and McDonald, Mathews, & Strobel, this volume), which included both C⁴L and TRAD students, generally suggest that C⁴L students have a deeper understanding of the conceptual nature of the calculus and computational skills as good as those of their TRAD counterparts (which was also suggested by the results on common final exams in 1988 and 1989). In addition, a recent analysis of questionnaire results concerning "study time" by calculus students that was completed by 38 faculty who have taught using the C⁴L program (Mathews, 1995) indicates a positive side-effect: C⁴L students' weekly study hours had mean 9.37 and standard deviation 4.27, while TRAD students' weekly study hours had mean 5.36 and standard deviation 2.41. Hence, C⁴L students generally appear to be spending more time studying calculus than do their TRAD counterparts which suggests that

C^{4}L students may reap the rewards for studying more than TRAD students in that
they often receive higher grades in calculus. Moreover, C^{4}L students appear not to
be adversely affected in their other courses, as had been conjectured by advisors,
students, and other faculty. No statistically significant differences in C^{4}L students'
grade point averages as compared to TRAD students were found in this study.

Statistical Method

We now describe the statistical model used to perform the comparison between
the two teaching methods. We first identify the population for this study. Then,
we state the specific questions asked that address the issues of whether or not
students do better in and/or take more subsequent courses that require calculus for
either the C^{4}L and TRAD programs. Next, we look at the problem of non-random
selection (or self-selection) of students which occurs in this study and how we dealt
with this problem. We note that the problem of self-selection is common to other
studies like this one as has been noted by other researchers, in particular, by Jack
Bookman in his evaluations of the Duke University PROJECT CALC (Bookman,
2000). Lastly, we summarize briefly the primary statistical model used.

Population. The students in this study were those enrolled at Purdue University, West Lafayette, IN from Fall 1988 through Spring 1991 in either the traditional
lecture and recitation three semester calculus sequence consisting primarily of engineering, mathematics and science students (TRAD); or the corresponding C^{4}L
reform calculus courses (C^{4}L). By "other" mathematics courses we mean mathematics courses beyond the basic three-semester calculus sequence.

We made several comparisons of the 205 C^{4}L students with 4431 TRAD students who took their first year of calculus in the academic years 1988, 1989 or 1990.
Thus, a total of 4636 students were included in these analyses.

Only data on those students enrolled for the first time in: (1) beginning calculus
during the fall semesters of 1988, 1989, and 1990; (2) second-semester calculus
during the spring semesters of 1989, 1990, and 1991; and (3) third-semester calculus
during the fall semesters of 1990 and 1991 were included in the study reported on
in this paper.

The sample of data in the study might be considered, in one sense, the *population* of all data since it is the complete set of data for all students who took the
indicated calculus courses over the given three-year period at Purdue. Of course,
we would like the results of this study to be used to infer conclusions about *all*
students who *might* take the indicated calculus courses at Purdue University or,
for that matter, more generally, a "typical U.S. university," say. To the extent
that the 4636 students in the study are somewhat representative of these more general populations, the results in this study can be extended to these more general
populations.

Comparisons made. We now consider, the first, more important, question
related to which program, C^{4}L or TRAD, provides a student with a better understanding of the required calculus concepts. One possible way of deciding this would
be to compare, say, the average final grades for the two programs. Because the final
grade reflects a student's knowledge and understanding of the course material, the
program whose students obtained the highest final average grade would be deemed
the better program.

Three statistics were used in the present study to compare a student's level of understanding of calculus concepts under the C^4L program with a student's level of understanding under the TRAD program and are given below.

- Average final grade.
- Average final grade for all mathematics courses taken after the completion of the calculus programs.
- Last available overall grade point average.

Instead of the average final *grade*, which includes both term component and a final mark component, it might have made more sense to compare the average *final grade* alone. Although common finals for first year calculus were given in 1988 and 1989 for the C^4L and TRAD programs, common finals were not given in 1990, and so a comparison using final marks was not feasible. We should note that the common finals given in 1988 and 1989 were those written by the TRAD faculty, and the C^4L average final exam scores were either about the same or higher than the TRAD averages (Dubinsky & Schwingendorf, 1990b). Common finals were not given in 1990 primarily due to a lack of agreement between TRAD faculty and the C^4L faculty on the nature and scope of questions that should be included on the final exams.

Comparing the average final grade for all mathematics courses taken by students after the completion of calculus in the two programs shed some light on whether students performed better in other mathematics courses after having taken the C^4L program or after having taken the TRAD program. Presumably, a student with a good understanding of calculus would perform better in a mathematics course that required (or did not require, for that matter) a knowledge of calculus than a student who had a poor understanding of calculus. Similarly, comparing the last available overall grade point average taken by students in the two programs gave some indication as to whether students performed better in *general*, for all courses, after having taken the C^4L program or after having taken the TRAD program. Moreover, the demands on C^4L students due to an additional workload (e.g., the laboratory component of the course) and the notion that this might interfere with C^4L students' performance in other course was a consideration we wished to try to measure using this third statistic.

There are other ways of deciding which of the two programs provides a student with a better understanding of the calculus concepts. Certainly, one possible way of measuring this would be to attempt to conduct extensive impartial qualitative interviews with students from each course and, on this basis, decide which of the two programs provided a student with a better understanding of the calculus concepts. This has been done in several studies involving both TRAD and C^4L students reported on in this paper (Asiala et al., 1997; Baker et al., in press; Czarnocha et al., in preparation; Clark et al., 1996; Cottrill et al., 1996; Cottrill, 1999, and McDonald, Mathews, & Strobel, this volume).

Consider, now, the second basic question considered in this study, related to which program, C^4L or TRAD, better inspires students to pursue further study in calculus and, more generally, mathematics. In this case, the interest a student showed in either calculus or mathematics, in general, was measured by counting the number of either calculus or mathematics courses this student took while pursuing his/her degree.

Two statistics considered in the study to compare a student's level of interest in calculus or mathematics courses under the C^4L program with a student's level of interest under the TRAD program and are given below.

- Average number of calculus courses.
- Average number of non-calculus mathematics courses.

The issue of non-random selection of students. Some students were encouraged to take the C^4L reform calculus course by academic advisors in the School of Science at Purdue. Academic advisors informed students about the nature of the C^4L course through a brief course description written by the original C^4L program directors. Every effort was made to enroll more female students than is usual in the first-semester C^4L Calculus course (MA 161A) and also to enroll as many prospective mathematics education teachers as possible (to provide a "role model" for teaching mathematics). The Director of the Women in Engineering program advised both female and male Freshman Engineering students and, in addition, coordinated and worked closely with other Freshman Engineering advisors in selecting students for the C^4L courses. Students were allowed to switch between the C^4L and TRAD calculus courses if they so desired. (However, few students made the switch from MA 161A to the traditional MA 161.) Hence, in effect, students either were directed or self-selected themselves, or some combination of the two, into the C^4L calculus courses.

Random assignment of students to the C^4L or TRAD programs would have been preferable to the non-random approach taken in the study. Random assignment would have offset any possible confounding factors which might have influenced the conclusions of the present study. Two factors which probably did influence the study and which were not taken into consideration were the grading policies and the amount of time spent by the students in the C^4L and TRAD programs.

The C^4L students were strongly encouraged to work in their permanent assigned groups (of three or four students) throughout the course whereas TRAD students worked as individuals. Four of the six course grade items for C^4L students were based on group achievement (where each individual was assigned the group grade except in the case when the individual did not fully participate) while two course grade items were based on individual achievement. All five course grade items for TRAD students were based on individual achievement. The C^4L students were assigned grades according to a predetermined benchmark or standard point total while TRAD students were assigned grades according to the rank of their point total. For example, C^4L students were each assigned a grade of A on a group test if the group score was 90% or higher, say, while TRAD students were assigned a grade of A if their point total on an exam was in the top 20 percent. In general, benchmark-grading C^4L instructors were allowed greater flexibility and responsibility than rank-grading TRAD instructors in assigning course grades. (Dubinsky 1999 provides a more complete discussion of the C^4L grading policy.)

The C^4L students spent seven in-class hours per week on course material (four in lab and three in class), whereas TRAD students spent five in-class hours on course material. The C^4L students engaged in a variety of thought-provoking group activities in addition to participating in class discussions and summaries of class concepts. The TRAD students attended three lectures and two recitation classes.

The issue of non-random assignment of students to programs is a limitation of the present study. However, it would seem there are no practical alternatives to this design. Random selection would almost certainly imply assigning at least some of these students, against their wishes, to either the C^4L program or the TRAD program. For most, if not all students, how well they perform academically is an overriding concern and so to place them in a course in which they have doubts about could possibly bias the results of the experiment. To randomly assign a student into one or the other program without this student's consent raises ethical issues. Furthermore, an approach which did not take into consideration the informed consent of each C^4L student would have ultimately lead to bias anyway because the two programs are so obviously different, a student would have quickly become aware of these differences.

Dealing with non-random assignment. A possible solution to the problem of the non-random assignment of students to either the C^4L or TRAD programs, is to "chip away at it" by making separate comparisons between smaller more homogeneous groups, defined by known extraneous factors, of C^4L students and TRAD students, or, in other words, to *control* for known extraneous factors. We note that Bookman and other researchers also use the notion of control variables to deal with non-random assignment. For example, say academic ability is identified as a possible confounding factor with academic performance of C^4L students and TRAD students. That is, suppose it is felt students are performing better academically not because they are taking the C^4L program but simply because they are better students. In this case, the effect of academic ability on C^4L students should be investigated and then, separately, the effect of academic ability on TRAD students would be investigated. Then, if there was shown to be a large positive association between academic ability and C^4L students and only a moderate positive association between academic ability and the TRAD students, say, this would indicate the students were performing better academically because of the C^4L program and not simply because they were better students.

Making comparisons between smaller more homogeneous groups of the C^4L or TRAD programs, to control for known extraneous factors, is not as effective as assigning students to programs at random. There could well be extraneous factors the authors were unaware of which are confounded with the type of teaching method program and so would not be dealt with by this procedure. Nonetheless, it appeared as though (at least) the following extraneous factors needed to be addressed as they appeared to be potentially confounding to the results of the study.

- Predicted grade point average (see below).
- Major course of study (the last available major).
- Gender.

A student's academic ability was measured using a modified grade point average called a *predicted grade point average*, or PGPA. This is a statistic which the registrar's office computes for entering freshman at Purdue to predict students' first semester grade point averages (Miller, May 1981; Suddarth & Wirt, 1974). Using the 1971 beginning first year students as the population (1970 and 1971 combined for those schools within Purdue with small enrollments), multiple linear regression analysis was performed to determine which combination of available predictor variables were most strongly associated with the obtained first semester index. The best predictors were found to be SAT-Verbal, SAT-Math, average high school grade (a weighted average of the number of semesters taken and the grades received for math, English, science, modern language and speech), and high school percentile rank cubed (cubed rank was used since it spreads the distribution and reduces the skew). Prediction equations involving these four variables were developed for each of the schools within Purdue, and in some cases, specific programs within schools. Virtually all students compared in this study were from the Freshman Engineering Program (in which all beginning engineering students must first complete the same first-year courses to apply for admission to the specific second-year engineering majors) or the School of Science (which includes prospective mathematics majors), and these two sets of students are comparable in quality and high school preparation. The equations provided the weights needed to combine a given student's scores on the four variables to arrive at a predicted grade point average.

Nonlinear effects of predicted grade point average (PGPA) were investigated by considering quadratic terms. It seemed reasonable to suppose a student would perform better in either of the two programs, C^4L or TRAD, the higher this student's PGPA score. However, in spite of the anticipated positive association between academic performance in either of the two programs, C^4L or TRAD, and PGPA, it was felt the association might not be linear. For instance, the highest PGPA students might perform better than the moderate PGPA students in either of the two studied programs, but by an amount less than the amount that the moderate PGPA students performed better than the modest PGPA students in either of the two studied programs. A student's major was grouped into the four categories of either engineering, math, science or other. To control for a student's major allowed a distinction to be made between a student's academic performance in calculus due to taking the C^4L or TRAD program from a student's academic performance in calculus due to how close this student's major was related to calculus. For instance, it might be the case that a student in mathematics who, because s/he had been accustomed to TRAD type mathematics courses, might actually perform worse than a student in engineering. Or, a freshman engineering student at Purdue may have been exposed to more mathematics courses and have done better in high school courses, traditional or otherwise, and would thus be more likely to perform better in the nontraditional C^4L program.

Gender was also accounted for in the study. To control for gender allowed a distinction to be made between a student's academic performance in calculus due to taking the C^4L or TRAD course from a student's academic performance in calculus due to his/her gender. Gender was controlled for because there was anecdotal evidence provided by the Director of Women in Engineering Program at Purdue that females tended to perform better in the group oriented C^4L program as opposed to females in the more individually oriented (competitive) TRAD program.

In addition to considering just the main effects of predicted grade point average (PGPA), major and gender on a student's academic performance in taking either the C^4L or TRAD program, the possible interaction effects of these three main effects were also considered. It was thought possible that a student's academic performance in calculus might be due to not just to a simple addition of the independent factors of PGPA and major, for instance, but to some dependent combination of these two factors, such as, say, a bright (high PGPA) engineering student performing as well as, but not better than, an average (moderate PGPA) mathematics student.

This study had to contend with missing data. In some cases, students started but did not finish one or the other of the C^4L or TRAD programs, for example.

This study included only those students who started and completed the programs. Important questions of why and how many students started, but did not finish, the two types of programs is left for a possible future study.

Although emphasis was placed on the influence of the three factors of PGPA, major and gender on academic performance, the study also looked at the influence of these three factors on which program, C⁴L or TRAD, better inspired students to pursue further study in calculus or, more generally, mathematics.

Statistical model. The main statistical model used in the comparison of the C⁴L and TRAD programs is an additive multivariate multiple regression model. The basic idea behind the use of this model is that the explanatory variables, such as predicted grade point average or major of a student, are treated as extraneous factors to be controlled for when assessing the type of program, C⁴L or TRAD, effect. The primary statistical model used in this study was of the form,

$$Y_{ij} = \alpha_i + \beta X_{ij} + e_{ij}$$

where,

- a type of program effect, α_i, is determined for both the C⁴L program, $i = 1$, and the TRAD program, $i = 2$, where this effect has accounted for, or has been adjusted for, the various explanatory (extraneous) variables.
- a response Y_{ij}, such as the final grade or number of calculus courses, is observed for each student j.
- an explanatory (extraneous) variable X_{ij}, such as predicted grade point average or major of a student j, is used in conjunction with an appropriately weighted regression coefficient, β, in an attempt to "explain" the various student responses.
- a residual, e_{ij}, measures the difference between the observed and modeled responses.

The statistical model described above is used to determine the estimated difference in the type of program effect, $\hat{\alpha}_{\text{C}^4\text{L}} - \hat{\alpha}_{\text{TRAD}}$, such as, for instance, the difference in average final grades for the C⁴L and TRAD students. Two kinds of estimated differences in the type of program effects are calculated. One kind of estimated difference in type of program is *adjusted* for the explanatory (extraneous) variables. The other kind of difference in type of program is *not* adjusted (or is unadjusted) by the explanatory (extraneous) variables (where β and X_{ij} are left out of the model here). Comparing the adjusted and unadjusted differences gives some idea of the amount of influence the extraneous variables have on the type of program. For example, if the adjusted difference in average final grades for the C⁴L and TRAD students was lower, say, than its unadjusted equivalent, then this would indicate the explanatory (extraneous) variables, such as predicted grade point average, is confounded with the type of program and that the adjustment was necessary.

Both 95% confidence intervals and tests of the estimated difference in the type of program effect, $\hat{\alpha}_{\text{C}^4\text{L}} - \hat{\alpha}_{\text{TRAD}}$, adjusted and unadjusted, were determined. For example, the adjusted 95% confidence interval for the difference in average final grades for the C⁴L and TRAD students was,

$$0.40 \pm 0.06.$$

where the grading scale for this study was given as: A = 4.00, B = 3.00, C = 2.00, D = 1.00 and F = 0.00. This means, with 95% confidence, the C^4L students attained, on average, 0.4 of a final grade more than the TRAD students, give or take a standard deviation of 0.6 of a grade. Similarly, a test revealed the chance, or p-value, of observing an adjusted difference in average final grades for the C^4L and TRAD students of 0.40, assuming the actual adjusted difference in average final grades is zero, is so small,

$$p\text{-value} = 0.0001,$$

in fact, that it would almost surely be concluded the average difference could not be zero. The complete set of 95% confidence intervals and tests are given in the next section.

The statistical model was, for the most part, analysed using the computer SAS procedure GLM (Neter, Kutner, Nachsheim, & Wasserman, 1996; SAS Institute, 1989). However, other parts of the SAS computer package were used, such as the logistic regression procedures CATMOD and LOGISTIC (Agresti, 1990; SAS Institute, 1989).

Results and Discussion

Both 95% confidence intervals and tests of the estimated difference in the type of program effect, $\hat{\alpha}_{C^4L} - \hat{\alpha}_{TRAD}$, adjusted and unadjusted, were determined and summarized in Table 1. An interpretation of this summary follows.

Table 1. C^4L versus TRAD programs: Unadjusted and adjusted results

	95% CI	Test: p-value
C^4L Versus TRAD Unadjusted Results		
Average final grade	0.40 ± 0.06	0.0001
Average final grade for all mathematics courses	no statistical significant differences	
Last available overall grade point average	0.13 ± 0.05	0.0074
Average number of calculus courses	0.16 ± 0.05	0.0038
Average number of noncalculus mathematics courses[a]	0.27 ± 0.07	0.0001
C^4L Versus TRAD Adjusted Results		
Average final grade	0.42 ± 0.06	0.0001
Average final grade for all mathematics courses	no statistical significant differences	
Last available overall grade point average	0.09 ± 0.05	0.05
Average number of calculus courses	0.24 ± 0.05	0.0001
Average number of noncalculus mathematics courses[a]	0.16 ± 0.07	0.025

[a]The last unadjusted and adjusted results are somewhat misleading, as is explained in greater detail below.

Average final grade. Both the adjusted and unadjusted 95% confidence intervals and p-value for zero difference in average final grades demonstrate C^4L students earn higher grades in calculus courses; in fact, almost half a grade higher, on average, than the TRAD students.

Average final grade for all mathematics courses. There is no evidence, using either the adjusted or unadjusted results, to conclude that there is a difference between the average grades of C^4L and TRAD students in mathematical courses beyond calculus. This suggests that C^4L students are as adequately prepared for mathematics courses beyond the calculus programs, as the TRAD students. We have had much anecdotal evidence from C^{4}L students who have communicated their displeasure with the traditional lecture, or lecture and recitation, teaching format of math (and other science) classes after taking C^{4}L calculus courses. This may explain why the two groups of students, C^{4}L and TRAD, performed at the same academic level in mathematical courses taken after the calculus programs.

Last available overall grade point average. The unadjusted, and particularly the adjusted, 95% confidence intervals and p-value for zero difference in average overall grade point average demonstrated the C^{4}L students performed only *slightly* better, on average, than the TRAD students. Again, this suggests that C^{4}L students are as adequately, possibly slightly better, prepared as the TRAD students for any academic courses beyond the calculus programs.

Number of calculus courses taken. Both the adjusted and unadjusted 95% confidence intervals and p-value for zero difference in average number of calculus courses demonstrate C^{4}L students take more calculus courses, on average, than the TRAD students. This suggests that the C^{4}L program better inspires students, rather than the TRAD program, to pursue further study in calculus. Indeed, the unadjusted 95% confidence interval suggests, that on average, out of 100 courses, the C^{4}L students take 24 more calculus courses, give or take a standard deviation of 5 courses, than their TRAD counterparts.

Number of noncalculus math courses taken beyond calculus. At first, based on both the adjusted and unadjusted 95% confidence intervals and p-value for zero difference in average number of noncalculus mathematics courses above, it appeared as though the C^{4}L students took more noncalculus mathematics courses, on average, than the TRAD students. This suggested that the C^{4}L program better inspires students, rather than the TRAD program, to not only pursue further study in calculus, but to pursue further study in mathematics courses, in general.

However, these results seemed to be at odds with the raw data given in Table 2 which seem to show the average number of noncalculus mathematics course taken by the C^{4}L and TRAD students to be about the same.

Table 2. Number of mathematics courses taken beyond calculus

	Overall	Male	Female
TRAD Students			
Zero courses	1763/4431 = 39.79%	1196/3265 = 36.63%	567/1166 = 48.63%
One course	2080/4431 = 46.94%	1607/3265 = 49.22%	473/1166 = 40.57%
Two courses	431/4431 = 9.73%	358/3265 = 10.96%	73/1166 = 6.26%
Three or more courses	157/4431 = 3.54%	104/3265 = 3.19%	53/1166 = 4.55%
C^{4}L Students			
Zero courses	81/205 = 39.51%	47/127 = 37.01%	34/78 = 43.59%
One course	81/205 = 39.51%	57/127 = 44.88%	24/78 = 20.41%
Two courses	18/205 = 8.78%	11/127 = 8.66%	7/78 = 8.97%
Three or more courses	25/205 = 12.20%	12/127 = 9.45%	13/78 = 16.67%

As a consequence of this discrepancy, we further examined this aspect of the data in two different ways.

In one analysis, we *excluded* the 1844 students who did not take any mathematics courses beyond calculus and ran the statistical model, described above, for the remaining students. The resulting unadjusted p-value of 0.0001 contrasted with the adjusted p-value of 0.38 for zero difference in average number of noncalculus mathematics courses taken after the calculus programs. This mixed result tends to support the idea that the C^4L students took about the same number of noncalculus mathematics courses after the calculus programs, on average, as the TRAD students.

In the other analysis, we again excluded those students who did not take any mathematics courses beyond calculus, but, this time, ran a different statistical model, a *logistic* model, for the remaining students. An unadjusted p-value of 0.078 results for a test of zero difference in average number of noncalculus mathematics courses. Again, this result tends to support the idea that the C^4L students took about the same number of noncalculus mathematics courses after the calculus programs, on average, as the TRAD students.

Summary and Conclusions

The results of this study favor the C^4L program over the TRAD program. In answer to the question of which program, C^4L or TRAD, provides a student with a better understanding of the required calculus concepts, we found that:

- The C^4L students earn higher grades in calculus courses; in fact, almost half a grade higher, on average, than the TRAD students.
- The C^4L students are as adequately prepared for mathematics courses beyond calculus as the TRAD students.
- The C^4L students are as adequately, possibly slightly better, prepared as the TRAD students for any academic courses beyond calculus.

In answer to the question of which program, C^4L or TRAD, better inspires students to pursue further study in calculus or, more generally, mathematics, we found,

- The C^4L program better inspires students, more than the TRAD program, to pursue further study in calculus.
- The C^4L students took about the same number of noncalculus mathematics courses after the calculus, on average, as the TRAD students.

References

Agresti, A. (1990). *Categorical data analysis.* New York, NY: Wiley.

Asiala, M., Brown, A., DeVries, D., Dubinsky, E., Mathews, D., & Thomas, K. (1996). A framework for research and curriculum development in undergraduate mathematics education. In A. Schoenfeld, E. Dubinsky, & J. Kaput (Eds.), *Research in collegiate mathematics education II* (pp. 1-32). Providence, RI: American Mathematical Society.

Asiala, M., Cottrill, J., Dubinsky, E., & Schwingendorf, K. (1997). The students' understanding of the derivative as slope. *Journal of Mathematical Behavior*, *16*, 399-431.

Baker, B., Cooley, L., & Trigueros, M. (in press). A schema triad—a calculus example. *Journal for Research in Mathematics Education.*

Bookman, J. (2000). Program evaluation and undergraduate mathematics renewal: the impact of calculus reform on student performance in subsequent courses. In S. L. Ganter (Ed.), *Calculus renewal: Issues for undergraduate mathematics education in the next decade.* New York, NY: Kluwer Academic/Plenum Publishers.

Clark, J. M., Cordero, F., Cottrill, J., Czarnocha, B., DeVries, D. J., St. John, D., Tolias, G., & Vidaković, D. (1996). Constructing a schema: The case of the chain rule. *Journal of Mathematical Behavior, 16*, 345–364.

Cottrill, J. (1999). *Students' understanding of the concept of chain rule in first year calculus and the relation to their understanding of composition of functions.* Unpublished doctoral dissertation, Purdue University, West Lafayette, IN.

Cottrill, J., Dubinsky, E., Nichols, D., Schwingendorf, K., Thomas, K., & Vidaković, D. (1996). Understanding the limit concept: Beginning with a coordinated process schema. *Journal of Mathematical Behavior, 15*, 167-192.

Czarnocha, B., Dubinsky, E., Loch, S., Prabhu, V., & Vidaković, D. (in preparation). *Students' intuition of area and the definite integral: Chopping up or sweeping out.*

Donaldson, J. A. (1993). *A report on Calculus I (Mathematics 015-156-01) 1993 spring semester.* Washington, DC: Howard University.

Dubinsky, E. (1992). A learning theory approach to calculus. In Z. Karian (Ed.), *Symbolic computation in undergraduate mathematics education* (MAA Notes no. 24, pp. 43-55). Washington, DC: Mathematical Association of America.

Dubinsky, E. (1995). *ISETL*: A programming language for learning mathematics. *Communications in Pure and Applied Mathematics, 48*, 1027-1051.

Dubinsky, E. (1999). Assessment in one learning theory based approach to teaching: A discussion. In B. Gold, S. Keith, & W. Marion (Eds.), *Assessment practices in undergraduate mathematics* (MAA Notes no. 49, pp. 229-232). Washington, DC: Mathematical Association of America.

Dubinsky, E. (n.d.). *Programming in calculus.* Unpublished manuscript.

Dubinsky, E., & Schwingendorf, K. (1990a). Calculus, concepts and computers: Some laboratory projects. In C. Leinbach (Ed.), *The laboratory approach to teaching calculus* (MAA Notes no. 20, pp. 197-212). Washington, DC: Mathematical Association of America.

Dubinsky, E., & Schwingendorf, K. (1990b). Constructing calculus concepts: Co-operation in a computer laboratory. In C. Leinbach (Ed.), *The laboratory approach to teaching calculus* (MAA Notes no. 20, pp. 47-70). Washington, DC: Mathematical Association of America.

Dubinsky, E., & Schwingendorf, K. E. (1996). *Calculus, concepts, and computers: Multivariable and vector calculus* (Preliminary version). Raleigh, NC: McGraw-Hill.

Dubinsky, E., Schwingendorf, K. E., & Mathews, D. M. (1995). *Calculus, concepts, and computers* (2nd ed.). Raleigh, NC: McGraw-Hill.

Ganter, S. L. (1997). The impact of ten years of calculus reform on student learning and attitudes. *Association for Women in Science Magazine, 26*(6), 10-15.

Ganter, S. L. (2000). *Calculus renewal: Issues for the next decade.* New York, NY: Kluwer Academic/Plenum Publishers.

Ganter, S. L. (in press). *Ten year of calculus reform: A report on evaluation efforts and national impact* (MAA Notes). Washington, DC: Mathematical Association of America.

Hurley, J. F., Koehn, U., & Ganter, S. L. (1999). Effects of calculus reform: Local and national. *American Mathematical Monthly, 106*(9), 800-811.

Mathews, D. M. (1995). Time to study: The C^4L experience. *UME Trends, 7*(4), 14.

Mathews, D. M., Narayan, J., & Schwingendorf, K. E. (1996). Using *Maple V* as a mathematical programming language in calculus. In *Proceedings of the seventh annual international conference on technology in collegiate mathematics* (pp. 309-313). Reading, MA: Addison-Wesley.

Miller, M. E. (May 1981). *Letter to Mr. John Krivacs.* (Available from K. E. Schwingendorf)

Monteferrante, S. (1993). Implementation of calculus, concepts and computers at dowling college. *Collegiate Microcomputer, 11*(2), 95-99.

Neter, J., Kutner, M. H., Nachtsheim, C. J., & Wasserman, W. (1996). *Applied linear models* (4th ed.). Homewood, IL: Irwin.

SAS Institute Incorporated (1989). *SAS/STAT user's guide* (Version 4, Vols. 1/2, 4th ed.). Cary, NC: SAS Institute Inc.

Schoenfeld, A. (Ed.) (1997). *Student assessment in calculus: A report of the NSF working group on assessment in calculus.* (MAA Notes no. 43). Washington, DC: Mathematical Association of America.

Schwingendorf, K. E. (1999). Assessing the effectiveness of innovative educational reform efforts. In B. Gold, S. Keith, & W. Marion (Eds.), *Assessment practices in undergraduate mathematics* (MAA Notes no. 49 pp. 249-252). Washington, DC: Mathematical Association of America.

Schwingendorf, K. E., & Dubinsky, E. (1990). Calculus, concepts and computers: Innovations for learning calculus. In T. Tucker (Ed.), *Priming the calculus pump: Innovations and resources* (MAA Notes no. 17, pp. 175-198). Washington, DC: Mathematical Association of America.

Schwingendorf, K. E., Hawks-Hoover, J., & Beineke, J. (1992). Horizontal and vertical growth of the students' conception of function. In G. Harel & E. Dubinsky (Eds.), *The concept of function: Aspects of epistemology and pedagogy* (MAA Notes no. 25). Washington, DC: Mathematical Association of America.

Schwingendorf, K. E., Mathews, D. M., & Dubinsky, E. (1996). Calculus, concepts, computers and cooperative learning, C^{4}L: The Purdue calculus reform project. In *Proceedings of the seventh annual international conference on technology in collegiate mathematics* (pp. 402-406). Reading, MA: Addison-Wesley.

Schwingendorf, K. E., & Wimbish, G. J. (1994). *Attitudinal changes of calculus students using computer enhanced cooperative learning.* Paper presented at the Annual Joint Winter Meetings of the AMS and MAA, Cincinnati, OH.

Silverberg, J. (1999). Does calculus reform work? In B. Gold, S. Keith, & W. Marion (Eds.), *Assessment practices in undergraduate mathematics* (MAA Notes no. 49, pp. 245-248). Washington, DC: Mathematical Association of America.

Suddarth, B. M., & Wirt, S. E. (1974). Predicting course placement using percentage information. *College and University*, Winter, 186-194.

PURDUE UNIVERSITY NORTH CENTRAL

PURDUE UNIVERSITY

PURDUE UNIVERSITY NORTH CENTRAL

CBMS Issues in Mathematics Education
Volume **8**, 2000

Understanding Sequences: A Tale of Two Objects

Michael A. McDonald, David M. Mathews, and Kevin H. Strobel

ABSTRACT. This paper is part of a continuing series of research studies by members of a collaborative group of mathematics education researchers (Research in Undergraduate Mathematics Education Community). Data for this study consisted of extensive interviews with students who had completed two semesters of calculus in either a traditional lecture/recitation calculus course or the learning theory based Calculus, Concepts, Computers and Cooperative Learning reform program. The Action-Process-Object-Schema theory was used to examine students' cognitive construction of the concept of sequence. We show that students tend to construct two distinct cognitive objects both of which they refer to as a sequence. One construction, SEQLIST, is what we might see as a listing representation of a sequence. The other, SEQFUNC, is what we see as a functional (expression with domain N) representation of a sequence. As the connections between these two entities become stronger, and the students reflect on these connections, they begin to view sequence as a single cognitive entity and SEQLIST and SEQFUNC as merely mathematical representations of this entity. In this paper we detail the construction of SEQLIST and SEQFUNC by the students, and characterize the connections between them through a model of schema development.

Much evidence (Carlson, 1998; Ferrini-Mundy & Graham, 1994; Selden, Selden, & Mason, 1994; Tall, 1992a) suggests that a conceptual understanding of important mathematical concepts such as function, derivative and integral, and a general ability to solve problems is lacking in calculus students. Many calculus instructors would agree that students also have difficulty with the concept of sequence and have difficulty applying their understanding of sequence to other mathematical concepts such as power series and uniform convergence of functions. This paper adds to the literature by reporting on a study of the nature of college students' understanding of the sequence concept after completing at least two semesters of calculus.

The goal of the current study is to use the Action-Process-Object-Schema (APOS) theory (Asiala et al., 1996) to increase our understanding of how learning the sequence concept might occur. The study was carried out using a research methodology that is being developed by members of the Research in Undergraduate Mathematics Education Community (RUMEC) for the purpose of studying

Work on this project was partially supported by National Science Foundation Grant No. USE 90-53432 and a grant from the ExxonMobil Education Foundation. Any conclusions or recommendations stated here are those of the authors and do not necessarily reflect official positions of the NSF or the ExxonMobil Education Foundation.

©2000 American Mathematical Society

how collegiate mathematics is learned. We find that APOS theory does appear to be useful in describing the possible cognitive development of the concept of sequence. And our analysis suggests that it might be helpful for students to study the concept of sequence in courses which implement pedagogy specifically aimed at helping the students foster the mental constructions identified by this theoretical analysis. We give preliminary evidence that students who took calculus using the Calculus, Concepts, Computers, and Cooperative Learning (C^4L) approach may have developed a deeper understanding of the multi-faceted concept of sequence as seen through the lens of APOS theory than students who took traditionally taught calculus courses.

We begin our discussion by outlining APOS theory. Following this, we give a brief survey of the literature as it relates to our study. Next, the research methodology employed in the study will be described. A detailed analysis of the data will then be given.

Theoretical Framework

In our analysis of the interview transcripts, we view students' cognitive constructions through the lens of the Action-Process-Object-Schema theory. This theoretical framework has developed through attempts to understand Piaget's ideas on reflective abstraction when extended to collegiate mathematics learning. We present a brief introduction to APOS theory below; a complete discussion of APOS analysis can be found in Asiala et al. (1996).

In general, research using APOS theory begins with an initial *genetic decomposition* of the mathematical concept of interest in the study. This genetic decomposition is simply a description of how students may come to understand a mathematical concept in terms of the components of the theory as described below. If no previous examination of students' understanding of this concept has been undertaken using APOS theory, as is the case in the present study, the researchers may begin with a genetic decomposition based on their own mathematical knowledge and their experience helping students learn this concept. Mathematicians have multiple, context-sensitive ways to think about sequences and operations on sequences. However, because a sequence can be thought of as a function on a discrete domain, the initial genetic decomposition proposed for sequence was essentially the same as the genetic decomposition of function. Thus, in addition to the general discussion of the components of APOS theory, we give examples of these components for the concept of function. A detailed APO analysis for the function concept was introduced and refined in Breidenbach, Dubinsky, Hawks, and Nichols (1992) and summarized in Dubinsky and Harel (1992). A function schema has yet to be fully investigated.

Action. An action is any transformation of objects to obtain other objects. We say that an individual has an *action conception* of a given concept[1] if her or his depth of understanding is limited to performing actions relative to that concept. The transformation is perceived by such an individual as being externally directed, as it has the characteristic that at each step, the next step is triggered by what has come before. Clearly much of mathematics can be described as the performance

---

[1]We distinguish between conception and concept as the first is intrapersonal (i.e., the individual's idea or understanding) and the latter is communal (i.e., a concept as agreed upon by mathematicians).

of actions. Someone with a deeper understanding of a concept may well perform actions when appropriate but may not be limited to performing actions. This person has moved beyond an action conception of the concept.

For the function concept, an individual performs an action when he or she plugs a given number into a given expression. If a student is limited to an action conception, then he or she may require an explicit formula for calculating an output value given an explicit input value. In addition, a student who is limited to an action conception of function may be able to consider the composition of two functions by replacing the variable in the outer function by the expression for the inner function and simplifying. However, he or she may not be able to work with composition of functions in more general settings, such as when given piecewise defined functions or functions defined graphically or recursively, rather than by an expression.

Process. When an action is repeated and reflected upon, it may be *interiorized* into a process. Thus, in contrast to actions, processes are perceived as being internal to, and under the control of, the individual. An important characteristic of a process is that the individual is able to describe, or reflect upon, the steps of the transformation without actually performing those steps. We say that an individual has a *process conception* of a given concept if the individual's depth of understanding is limited to thinking about the idea as a process. Such an individual is not able to perform an action on the process itself.

Someone with a process conception of function will be able to think about transforming input objects in a specific and repeatable way into output objects. They think of this transformation as a complete activity and often do not need to be given or even refer to specific expressions or inputs to discuss this process. They will be able to combine this process with other processes, or reverse the process. However, someone limited to a process conception will not be able to perform actions on this process and transform it. As a student's process conception strengthens, notions such as one-to-one or onto become more accessible.

Object. When an individual reflects on operations applied to a particular process, becomes aware of the process as a totality, realizes that transformations (whether they be actions or processes) can act on it, and is able to actually construct such transformations, then we say the individual is thinking of this process as an object. In this case we say that the process has been *encapsulated* into an object. We say that an individual has an *object conception* of a concept when that individual's depth of understanding includes treating that idea or concept as an object. Such an individual is also able to *de-encapsulate* an object back into the process from which it came when necessary.

A student with an object conception of function has encapsulated the underlying process as described above into an object on which he or she will be able to perform actions. Such actions may include differentiating a function to obtain another function. When necessary, a student is able to de-encapsulate the function into its underlying process.

Schema. A schema for a certain piece of mathematics is an individual's collection of actions, processes, objects and other schemas which are linked consciously or subconsciously in a coherent framework in the individual's mind and may be brought to bear upon a problem situation involving that area of mathematics. An important function and characteristic of coherence is its use in deciding what is in the scope of the schema and what is not.

Cottrill et al. (1997) saw the schema structure as a useful tool to characterize the development of students' understanding of limit. This theory has been further refined to include a triad development of a schema when examining students' understanding of the chain rule (Clark et al., 1997). We found this triad development to be useful in characterizing the connections that students must make among processes and objects as they work toward constructing a more complete understanding of sequence. This triad concept was introduced by Piaget and Garcia (1989) in their discussion of the development of knowledge and it provides us with the language to analyze further the development of students' cognitive construction of the sequence concept. The three stages of schema development are summarized below, with examples at each stage given from a study of the understanding of students' development of a chain rule schema (Clark et al., 1997).

Intra Stage. This stage is characterized by a focus on an individual cognitive item in isolation from other actions, processes, and objects of a similar nature. One is not aware of any cognitive relationships between these similar items.

While developing chain rule schema, a student may use distinct, special rules such as the general power rule and not recognize them as related in any way or as special cases of the chain rule.

Inter Stage. This stage is characterized by the recognition of relationships between actions, processes and objects. For example, one may see several examples as special cases of a more general concept. At this level one begins to group together items of a similar nature and perhaps even call them by the same name.

In terms of the development of the chain rule, students may now see several rules as linked together. The student does not yet understand why they are special cases of the chain rule, but perhaps thinks in terms of similar procedures involving inside and outside functions.

Trans Stage. The trans stage is characterized by the individual having constructed an underlying structure through which the relationships discovered in the inter stage are understood and which gives the schema coherence. As stated earlier, the importance of coherence is its use in deciding what is in the scope of the schema and what is not.

One is at the trans stage of development of the chain rule schema if one understands various special cases as applications of the chain rule by identifying certain functions which are composed. The student is also able to apply the chain rule in some new situations by looking for a composition of functions.

Viewing the development of a complete understanding of sequence as largely the act of building relationships is consistent with the following comment by Hiebert and Lefevre (1986):

> Some relationships are constructed at a higher, more abstract level than the pieces of information they connect. We call this the reflective level. Relationships at this level are less tied to specific contexts. They often are created by recognizing similar core features in pieces of information that are superficially different. The relationships transcend the level at which the knowledge currently is represented, pull out the common features of different-looking pieces of knowledge, and tie them together. (p. 5)

Summary of Relevant Literature

Throughout the past decade, numerous educational researchers have conducted investigations into college students' difficulties with and understandings of various topics in first year calculus. The relevant topics to the present study are the concepts of function and, of course, sequence. Although some researchers have examined issues related to sequences, such as limits (Monaghan, 1991; Tall, 1992b; Williams, 1991) and the role of infinity (e.g., Tall, 1992b), these issues are not central to the present paper.

Research on calculus students' understanding has highlighted the importance of a rich understanding of function as a prerequisite to understanding many of the fundamental ideas in calculus. We believe this understanding of function is necessary for students to have a solid understanding of the sequence concept. Although it is often assumed by college mathematics teachers that beginning calculus students bring with them a sufficiently strong concept of function, our study supports numerous researchers' observations that this is not the case for many students. In particular, Eisenberg (1992) presents 10 problems that he claims most calculus course graduates will fail, partially due, in his view, to the fact that students do not have a strong visual understanding of the function concept. Carlson (1998) investigates various aspects of function understanding previously studied separately. These aspects include students' ability to interpret graphs, their cognitive understanding of the function concept, and their ability to move between various representations of functions. She studied college algebra, calculus, and mathematics graduate students and found that a full development of the concept of function appears to evolve over a long period of time and is enhanced by having students participate in activities where they must construct their understanding and reflect upon these constructions. In addition, she finds that the most successful calculus students still have a great many difficulties including: understanding of covariant aspects of functional situations, defining piecewise continuous functions, and bringing newly acquired knowledge from calculus to bear on unfamiliar function-related problems. Other key resources of research on students' understanding of function are a volume of research papers edited by Harel and Dubinsky (1992) and a survey article by Thompson (1994).

Several studies (Breidenbach, Dubinsky, Hawks, & Nichols, 1992; Dubinsky & Harel, 1992) have utilized APOS theory in examining students' understanding of the function concept. Breidenbach et al. (1992) propose action, process, and object conceptions of function and then examine student responses to mathematical situations to refine these definitions. Their conclusions of action, process, and object conceptions for the function concept are detailed in the previous section. Dubinsky and Harel (1992) study undergraduate students to see how far beyond an action conception students can get when given specific instruction. This instruction consists of having students write *ISETL* computer code to implement function-related procedures and having them discuss and reflect on these activities. The authors find that many students do progress to the process conception, but the nature of their process conception was determined by several factors. The ability to perform manipulations or not was important as to whether students could describe given situations as functions. Students needed to feel autonomy to construct a functional process for given situations where the process was not explicit. Students needed to overcome the restriction many had previously held that only relations with continuous graphs were considered functions.

As stated above, many researchers have examined issues which are closely tied to the concept of sequence, such as limit and infinity, but none has looked specifically at the conceptual nature of infinite sequence in and of itself. However, Dubinsky and Harel (1992) elaborate further on a set of "function situations" using finite sequences first reported by Breidenbach et al. (1992). After a period of general experience using the computer language *ISETL*, but before specific instruction on functions, students were asked to consider whether three finite sequence situations could be seen as functions. The three sequences were given in *ISETL* code:

- $[2n + n^3 \ : \ n \text{ in } [1 \ldots 100]]$;
- $[2^n > n^2 + 3n \ : \ n \text{ in } [1 \ldots 100]]$; and
- $[2n + \text{random}(9) \ : \ n \text{ in } [1 \ldots 50]]$.

Based on simple yes or no responses, students saw functions in these situations 55.2%, 54.1%, and 61.6% of the time respectively (Breidenbach et al., 1992). Dubinsky and Harel conjecture that because the notational structure of these situations suggests the construction necessary for the function concept, most of the success students had can not be ascribed to any more than an action conception of function. However, the fact that the output of the second sequence are boolean values and the third sequence involved random numbers suggests that successful students had more than an action conception of function.

Research Methodology

Participants. Twenty-one students at a large midwestern university were selected from students completing at least two semesters of calculus—six completed the traditional calculus track, and fifteen completed the Calculus, Concepts, Computers, and Cooperative Learning (C^4L) track, an NSF-sponsored reform calculus curriculum development project. The traditional courses consisted of larger lectures and smaller group problem solving sessions (recitations). The C^4L courses were activity-based courses where students worked in cooperative groups on computer programming activities using *ISETL* to solve mathematical tasks. The underlying structure and the justification for the specific activities in which the students were engaged was a preliminary learning-theoretical analysis of what might be required to understand the concepts of calculus. The goal of many of these activities is to help students develop an object conception of various calculus concepts. In particular, because the C^4L text authors believe that an object conception of function is prerequisite to a mature understanding of most calculus topics, students are expected to engage in numerous activities to help them develop an object conception of function. For example, C^4L activities often require students to pass functions as parameters within an *ISETL* program. For more detailed information, see Schwingendorf, Mathews, and Dubinsky (1994). We note that the positive effect that the computer programming activities can have has been reported in the literature by multiple authors, including Tufte (1989) and Asiala, Cottrill, Dubinsky, and Schwingendorf (1997). All students in the study reportedly completed second semester calculus with a grade of A or B, except for one C^4L student who received a C and one C^4L student who did not report her grade.

Many of the students from both calculus tracks were computer science or engineering majors, thus had completed subsequent courses in mathematics. The purpose of our research was to characterize the cogitive constructions the students had made and perhaps to understand how students in general tend to construct the concept of sequence. Thus we were not looking for immediate recall on problems and there was no need to differentiate between students who just completed calculus with those who were in a more advanced mathematics course.

Data. In-depth interviews with these twenty-one students were audiotaped and transcribed. The interviews consisted of fourteen questions, mainly on sequences. The initial genetic decomposition of the sequence concept, which was essentially the same as the genetic decomposition of function, informed the construction of these interview questions. In summary, the interview questions asked students to:

- define, compare, contrast, and give examples of sequences and series;
- discuss the sequence $a_n = 1 + (-1)^n$;
- discuss the sequence where each term is the average of the previous two terms;
- explain the difference between sequences and functions;
- define and give examples of sequences which are and are not monotone, bounded, or convergent;
- draw a Venn diagram relating sequences which are, or are not, monotone increasing, bounded above, or convergent;
- find $\lim_{n \to \infty} \dfrac{2n^2 - 1}{1 + 5n^2}$ and then rigorously discuss what it means to say that $\lim_{n \to \infty} \dfrac{2n^2 - 1}{1 + 5n^2} = \dfrac{2}{5}$; and
- discuss the behavior of sequences of numbers such as $\{0.9, 0.99, 0.999, \ldots\}$ or $\{2.7, 2.71, 2.718, 2.7182, 2.71828, 2.718281, \ldots\}$.

Interviewers followed the interview guide closely, but based on student responses, asked additional questions for clarification and to further probe understanding. A copy of the students' written work was attached to the transcript of their interview.

Analysis of the data began with each member of the research team reading all of the interview data for every student. Next, the complete research team met and discussed the major themes that seemed to appear in the transcripts. As a result of this preliminary look at the data, a grid for organizing the data using an APOS analysis of various concepts was developed and tested. Each interview was then individually reread by a pair of researchers who each catalogued the students' performance in a formal way on the grid. Finally, the pairs of researchers met to develop a consensus view of each student's observed behavior and possible cognitive constructions. Any differences in findings were negotiated.

Results: SEQLIST and its Origins

A major finding of this investigation was that many students were limited to a concept image (Tall & Vinner, 1981) of sequence as a list of numbers. When confronted with the term "sequence" all that comes to such a student's mind is a list; we term the corresponding cognitive construction "SEQLIST." Although some students were limited to this list conception of sequence, this conception was surprisingly useful for most of the typical calculus problems the students encountered.

Although we would like to give a genetic decomposition of SEQLIST based on the data, this is difficult as we found all students at what we would consider an object conception. We summarize briefly what we found in the data and what we conjecture based on these data. All students in this study appeared to have an object conception of SEQLIST because they could refer to these lists of objects (usually numbers) separated by commas as single entities and they could perform actions on these lists such as comparing two lists. Some students exhibit this object conception by putting set braces or brackets around the list or by refering to the list with a singular noun. With an object conception, students could also de-encapsulate the object into an underlying process when necessary. We saw evidence of the underlying process when students spoke dynamically about "listing the values" as they focused on the values which made up the list rather than on the list as a single entity. We hypothesize that an action conception of SEQLIST might include the ability to write a list of integers (or another specified list) when requested to do so, and not include interiorizing the writing of such a list in general. In the rest of this section we give further details on this SEQLIST conception of sequence.

Central to this particular image of sequence was the appearance of commas separating the numbers. The importance of the commas as the identifying feature is exemplified by the student below who was contrasting sequences and series.

> Fran:[2] Well, the sequence actually has the commas and the series actually has little plusses in between 'em and you add 'em all together.

In addition, many students expressed the belief that the numbers in the list had to follow some pattern.

> I: OK, what's a sequence?
> Beth: (Pause) uh, (mumbles) a set that has a definite pattern that repeats.

Or take Larry's response as to whether the list of numbers approaching e could be considered a sequence.

> I: ...would you say this list is a sequence of numbers?
> Larry: I don't think so. Um, I may be wrong with my definition of what the sequence is ...
> I: OK.
> Larry: ...but I think that it's a, has to have, um, some kind of, um, formula to it, as far as like a pattern.
> I: OK.
> Larry: And since that doesn't have a pattern, I, I may be wrong with my definition but that's what I, that's what I think the definition is.

We see that pattern is essential in this student's conception of sequence.

The characteristics of commas and patterns are not important only for the weaker students. Even one of the top performing students in our study had these characteristics as an important part of his SEQLIST object. His focus on both commas and patterns can be seen throughout his transcript as in the example below.

---

[2]The names of the students have been changed. We also use the notation I: for the interviewer in all excerpts even though the actual interviewing was done by several different researchers.

James: An infinite sequence is one which continues infinitely. I mean, really, every sequence, as far as I know, is an infinite sequence because, I mean, it's going to keep going on if you have certain pattern. Like if you have a sequence where it's one, comma, one, comma, one, comma, one, it's going to continue indefinitely. Uhm, although, I guess, if you restricted your sequence then it would no longer be an infinite sequence and that would be the case. If you have something where your sequence is two to the n, then it's going to be 2, 4, 8, 16, 32. It's going to keep going indefinitely.

In addition to the importance of commas and patterns, the actual objects within a sequence seem to be important to many students. The importance of the elements of the sequence being numbers is underscored by the next exchange between Beth and the interviewer. When asked to compare and contrast sequences and series, she produces the list "$1, 2, 3, 4, 5, 6, \ldots$" as an example of sequence, and "$a^0 + a^1 x + a^2 x^2 + \ldots a^n x^n$" for an example of series.

I: Using these two examples, compare and contrast sequences and series.
Beth: Um. Well my sequence doesn't depend on any variables. Whereas my series does. Um, there are no unknowns in my sequence, but I'm sure we could come up with a sequence if there were. Um, series (pause) there is a definite relation between the numbers in my sequence because I've just added one, whereas (pause) these depend on the coefficients. I don't know.

Part of this response seems to indicate that she does not think of sequence in a functional way. The other part might indicate otherwise. Thus the interviewer probed more deeply by suggesting that she consider series of numbers. To this, Beth produces the series "$1 + 2 + 3 + 4 + 5 \ldots$" When asked again to compare and contrast series and sequences after having produced this series of numbers, Beth replies only that the series "uses variables" and the sequence doesn't. The importance of numbers in sequences is still apparent even after the interviewer prodded Beth to examine series of numbers.

SEQLIST is not a worthless or dispensible mathematical object. Indeed, as we will argue it is one necessary component of a fully developed sequence schema. Students with an object conception of SEQLIST can refer to this list and can act upon the list. They think of this list as one entity rather than a listing of many entities. A few students exhibit this notion of one entity by putting set braces or brackets around the list, or by refering to it with a singular noun.

There seem to be few obstacles to developing an object conception of SEQLIST.[3] We conjecture that the ease of developing an object conception of SEQLIST is because:

i. it is easy to come by since (a) number is such a strong cognitive object for most students, and (b) a strong ability to enumerate is usually developed very early; and

---

[3] An Appendix is provided with a summary of specific student progress in developing the various cognitive entities discussed throughout the paper. For instance, you can see in the Appendix that all students in this study exhibited an object conception of SEQLIST.

    ii. it is productive since students seem to be able to solve most of the prob-
lems they are typically faced with in calculus using SEQLIST.

As mathematicians, we realize that a sequence can also be thought of as a
function. But for many students, the image of sequence as merely a list is not
necessarily just a preference for the list over some functional representation. This
is evident in the following excerpt.

> I: Now for a sequence, uh, does it make sense to talk about the domain?
> Anne: No.
> I: OK.
> Anne: I, I would not understand that.
> I: OK.
> Anne: I wouldn't understand that at all.

Although Anne does not offer much in the way of a response, she does show a
complete denial of the possibility of a sequence having a domain. We see that for
some students it is not the case that this listing conception of sequence is simply a
mathematical representation of a more sophisticated cognitive construct.

When students have an object conception of SEQLIST, they can de-encapsulate
this object into an underlying process. Although all students in this study exhibited
an object conception of SEQLIST, in the interviews we can see evidence of the
process when students speak about the dynamic listing out of the values. They
may state this fact in general as Vince does.

> Vince: ...a sequence is just where you are listing the values....

Or they may exhibit this process through an example.

> James: ...Another example would be one, comma, one, comma, one, comma,
> one....

There is a dynamic sense to the listing process for both Vince and James rather
than the static sense of an object, the list. Stated another way, it is the process
of "write a number—write a comma—write a number—write a comma—write a
number—..." which is encapsulated into the cognitive object SEQLIST.

As in the case of students learning to compute the mean of a data set (Clark
& Mathews, in preparation), the action of "computation" is apparently readily
interiorized and thus no one was observed with merely an action conception. We
hypothesize that an action conception might include the ability to write a list of
the integers (or another specified list) when requested to do so, and not include
interiorizing the writing of such a list in general.

Results: SEQFUNC and its Origins

A second cognitive entity, distinct from SEQLIST, that students tend to construct in dealing with problem situations involving sequences is that of sequence
as a function. We will refer to this object as SEQFUNC. We saw in the previous
section that some students do not have a fully developed functional notion of sequence. An object conception of SEQFUNC appeared to be much more difficult
for the students in this study to develop than an object conception of SEQLIST.

Recall that all students appeared to have an object conception of SEQLIST. Only about two-thirds of the students had an object conception of SEQFUNC.

Object Conception of SEQFUNC. An object conception of SEQFUNC may be indicated by students who are comfortable giving examples of sequences in closed form. For example, when asked for an example of a convergent sequence and of a sequence which does not converge, Nathan writes "$a_n = \frac{1}{n^2}$" and "$a_n = n^2$", respectively. In addition this student displays a good understanding of sequence as function when he is comparing his original example of a sequence and a series. His examples were "$1, 2, 3, \ldots, n$" and "$\displaystyle\sum_{n=1}^{\infty} \frac{n+1}{n}$".

> I: How are they [sequences and series] alike?
> Nathan: How are they alike?
> I: Yeah.
> Nathan: Um, they both still have some sort of correspondence, I mean, in the sequence you know the first element is 1, the second element is 2, the third is 3. In the series, the terms have a correspondence as well.
> I: OK.
> Nathan: I mean, you know when n is 1, you've got 2. When n is 2, you've got $\frac{3}{2}$ and so on and so forth.

Most conclusively, when asked about the difference between sequence and function, students with an object conception of SEQFUNC usually state that a sequence is a function, and that functions may be sequences if the input or domain is appropriately restricted.

> I: What's the difference between a sequence and a function?
> Khalid: The difference between a sequence and a function ... well ... a function has a unique output for every input ... or ... yeah, or unique input, well (mumbling) ... yeah, well for every input you have a unique output ... or is it vice-versa? Well, the function has to pass the vertical line test, so we talking like ...
> I: OK.
> Khalid: ... the real, like everything has to be within that definition of a function?
> I: Um ...
> Khalid: Like a real, or just like, are you just saying like an expression?
> I: Well function in whatever sense you're comfortable with.
> Khalid: OK. I just think of a function as a ... something that'll pass the vertical line test and every, every input has one output.
> I: OK.
> Khalid: Yeah, that's what I meant to say. Every input has one output. I mean the output can be the same for a different input but one for each particular input.
> I: Mm-hmm.
> Khalid: That's the way I think of a function. Variable ... and an independent and dependent variable.
> I: OK.

Khalid: Um ... but for a sequence ... well a sequence has ... an independent variable which is n ...

I: OK.

Khalid: ... and it has basically a dependent variable too ... but I don't know if, well, but every ... I wonder if every independent variable has to have one dependent in a sequence ... I'm trying to think ... well so far the one's we've done have ... well, but wait, oh I know, that's still the same, like I was thinking of the, the 0, 2, 0, 2 ...

I: Mm-hmm.

Khalid: ... that still has ... boy, you know I can't really think of a real ... difference between a function and a sequence.

I: OK.

Khalid: I wonder if a sequence is a function.

I: What do you think?

Khalid: Oh, I think it makes sense because you put in an n and you get back a number and your n goes to infinity, the only difference is a sequence you can only plug in integers.

I: OK.

Khalid: So your n has to be a 1, 2, 3, 4, blah blah blah, so you, on your, if you were to make a graph you would have points ...

I: OK.

Khalid: ... you wouldn't necessarily have a smooth line so you would have maybe a, you know ...

I: OK.

Khalid: ... and as your n's increased ... I guess if you think of it that way, well this could still be a function, because it still passes the vertical line test, but it's, it's different because unless you like interpolate between this you really, you can't plug in any number, you know ...

I: OK.

Khalid: ... so you're basically restricted toward, to integers and that ... is really the only difference I can think of between them.

I: OK. So you'd say a sequence is a function, and it just has integers as its input?

Khalid: Right.

This lengthy exchange between the interviewer and Khalid indicates that he now understands that there is a strong functional component to the sequence concept. Even thought he often refered to specific examples, he could discuss both sequence and function as general processes of inputing values and obtaining unique outputs. Although he seems to be developing (or redeveloping) this conception of sequence as a function, his ability to give closed form expressions for sequences and to give properties for and manipulate or act on these closed form expressions throughout the interview shows he has an object conception of SEQFUNC.

Although Khalid seemed to have a solid understanding of the SEQFUNC object, we saw evidence that students can have a SEQFUNC object that is different from our understanding of sequence as represented by a function. The difference could be related to a lack of understanding of the discrete nature of the domain or an inconsistent use of discrete domains. For example, when asked for an example of a convergent sequence, Helen draws two graphs: one which has the same qualitative shape as the graph of $y = \frac{1}{x^2}$ (the domain appears to be $x \neq 0$) and another which has the same shape as $y = \frac{1}{x}$ (with domain $x > 0$). One might say that she did this because it is easier to draw unbroken curves rather than a collection of dots. However, throughout the interview, her predominant choices for sequences appear to have as domains the real numbers or non-discrete subsets of the real numbers. Her choices are things like: $x_n = \ln n$, $x_n = \sin n$, or $x_n = \frac{1}{n}$; her accompanying figures are mostly unbroken curves.

After careful examination of Helen's complete transcript, there is no question that there is *an* object present for this student. We argue that it must be the object SEQFUNC due to the context in which this object is developed and the problems to which this object is applied. The examples above admittedly do not display a fully functional conception of SEQFUNC because she does not seem to include the discrete nature of the domain in her definition. This conception of sequence, with the inconsistent use of discrete domains, will be a major obstacle in Helen's ability to understand more sophisticated mathematical ideas such as series. The important pedagogical lesson here is that students such as Helen should be faced with new problem situations where their current object conception of SEQFUNC will be inconsistent with the problem presented. This will serve to help them revise their original process and encapsulate the new process into a more consistently useful SEQFUNC object.

The development of an object conception of SEQFUNC is obviously tied to the students' function schema. As Thompson (1994) observed, "If we have learned anything in mathematics education research it is that a person's thinking does not respect topical boundaries." Depending upon its richness, a function schema may be quite helpful in the development of SEQFUNC as an object, or it may be a hindrance. For example, in response to a question on the properties of sequences, many students produced graphical examples. Some of these were properly constructed as points in the plane. Others repeatedly produced continuous graphs. Still others either produced points and then connected them, or produced continuous curves and then superimposed points along the curve.

Not surprisingly, students who appeared to have a poor function schema were seen to be particularly disadvantaged in their ability to work with SEQFUNC. An example of a student without a well-developed function schema is Emily, who when asked to find the limit of the sequence given by $a_n = \frac{2n^2-1}{1+5n^2}$, immediately searched for the zeros of the denominator.

> I: Does this sequence have a limit? (Pause.) What are you trying to establish? What are you trying to look for here?
>
> Emily: To make sure that doesn't go to zero.
>
> I: Can it ever go to zero?
>
> Emily: If n squared equals one, if n squared equals negative one fifth ...

Students like Helen who have difficulty dealing with a discrete domain, or like Emily who have not created a sense of coherence among the various actions, processes, and objects related to functions, are going to have difficulty with the many mathematical ideas which build upon the foundation of functions.

Many college mathematics faculty will agree with the National Research Council's (1989) statement in *Everybody Counts* that if "it does nothing else, undergraduate mathematics should help students develop function sense...." Nonetheless, much of the research discussed in the Literature section of this paper has documented that a strong function sense is, for many students, not an outcome of the undergraduate mathematics curriculum.

Thompson (1994) observed that "a predominant image evoked in students by the word 'function' is of two written expressions separated by an equal sign." We observed precisely this concept image held by Ilene.

> I: ...Is there any relationship between them?
>
> Ilene: Between a sequence and a function?
>
> I: Uh-huh.
>
> Ilene: Well you can, I mean you can turn a sequence into a function by setting it equal to something else.
>
> I: OK. So a sequence can be a function. Can a function be a sequence?
>
> Ilene: I would say no, because ...(pause).
>
> I: Because what?
>
> Ilene: Mm ...because one, because you can just take a, a side of a, a equation and just, you have to have it equal to whatever ...it's equal to, you can't set it equal to something that it's not equal to.

A limited understanding of function for Ilene leads her to have a limited view of SEQFUNC as well. We should note that despite this limited view of function, Ilene received an A in traditional Calculus II. This student will most likely have difficulty building a viable object conception of SEQFUNC because functions are not objects for her as defined in our theoretical framework. She has not encapsulated the input-output process into an object conception of function. Rather she sees the function as an expression with the appropriate variables and an equal sign. What is noteworthy in Ilene's case is that even with such an inadequate conception of SEQFUNC, she is able to produce examples of increasing, decreasing, not monotone, bounded, not bounded, convergent, and not convergent sequences and settles on a correct drawing of the Venn diagram relating these properties. But will she be able to understand more sophisticated mathematical concepts such as series, which require a fuller understanding of sequences?

Process Conception of SEQFUNC. In examining our data, we saw that, in general, a process conception of SEQFUNC fits nicely within the the framework that has been developed for a process conception of function (Breidenbach et al., 1992; Dubinsky & Harel, 1992). However, the development of SEQFUNC is not completely determined by a student's function schema. Vince is an example of a student who, through close examination of his entire transcript, exhibited an object conception of function. But he, like many students, is limited to a process conception of SEQFUNC. These students had not yet encapsulated this process into an object. This development may be indicated by the use of plural descriptors for the sequence and singular descriptions of function. The following excerpt from Vince's interview occurs just after the interviewer asks what the difference between a sequence and a function is.

Vince: A function would be something like one over two to the x, and a sequence would be the terms that come out of that function. It would be a listing of these terms. Like, well, we'll use the example. We'll use x^2 plus one. Cause that's what I've been using. That's the function. And the sequence is the terms that come out when you supply the values for x into the function.

I: OK. So, say that again.

Vince: OK.

I: I'm not sure I quite caught that.

Vince: Alright. The function would be x^2 plus one. OK? And then the sequence would be, say you are given a range for x, you know, that goes from zero to infinity, for x. And then, that's, when you supply, when you put each of the terms that's in that range into the function, the values that you get out form the sequence.

I: OK.

Vince: OK? So you'd start off with zero and you'd get the one. And then you'd put in the one, you get a two. And you keep going like that until you get the series that we had from before.

I: So then, how are the function and the sequence different?

Vince: Are they different? I think ... the func ... well, I don't know how to say this. The function is, like, the equation and the sequence are the values that you get from the equation. And you plug in the values.

And later when asked for specific examples of monotone, bounded, and convergent sequences, Vince asks the following question.

Vince: Do you have to see, do you want to see the function that goes with it, too, or just the sequence of numbers?

Here, and throughout the transcript, Vince indicates an object conception of function. And he clearly exhibits a process conception of SEQFUNC. But he has not yet encapsulated this process in the context of sequences.

Action Conception of SEQFUNC. We do not characterize an action conception of SEQFUNC, as all students had at least a process conception and none were limited to an action conception. Although in general, interiorizing actions to processes can be quite difficult, we see that students made more progress in the specific case of interiorizing the action of evaluating an expression with integer inputs. We attribute this to the fact that students are provided with the additional external cue of the nature of the output to this action, the elements of SEQLIST. And SEQLIST is already a strong cognitive object for them.

We have explained above that the SEQFUNC process generates as output the SEQLIST object. This is a mathematical link, albeit a very simple one for most students to make, between the two constructs. We discuss this in more detail presently as we discuss students' schema development for sequence.

Results: Development of a Sequence Schema

To understand the concept of sequence in a way that is compatible with the mathematician's view requires not only the construction of sequence as list and sequence as function, but also ultimately the realization that these cognitive constructions can be used as mathematical representations of one *single* mathematical concept, that of sequence. Moving beyond two individual constructions to an understanding that lists and functions can be used to represent a multi-faceted idea, sequence, requires the development of cognitive links between SEQLIST and SEQFUNC. The development of these links implies the construction of a more mature schema. In the Theoretical Framework section, we described the intra, inter, and trans stages of schema development in a general way along with the specific example of the chain rule concept. We now characterize these stages of schema development for the particular case of the sequence schema.

Intra Stage of Sequence Schema Development. The initial level of development in coordinating the constructions in the developmental chains of SEQFUNC and SEQLIST is that of the intra stage. Recall that in the case of sequences, there is a natural mathematical link between SEQLIST and SEQFUNC—that is, the SEQLIST object is the natural output of the SEQFUNC process. Thus, given that all students had at least a process conception of SEQFUNC, they could generate a sequence list from a closed-form expression. But this does not necessarily imply that there are explicit cognitive links between SEQLIST and SEQFUNC for these students.

For a student at the intra stage, the cognitive links between these two constructs are typically non-existent. In other words, once a student produces a list from a closed-form expression, she then turns her attention to the list and, in essence, ignores the expression from which it came. It is as if someone had covered up the expression once the list was produced.

Our data do not clearly show anyone in this stage of sequence schema development. But two students, Emily and Ilene, may be at this stage or may be just beginning the transition to the inter stage. Within the context of her entire interview, one example of why we might classify Ilene at (or newly transitioning from) the intra stage occurs when she is asked whether the sequence $a_n = \frac{2n^2-1}{1+5n^2}$ has a limit. Ilene's first reaction is to substitute $n = 0, 1, 2, 3$ and find the first four terms in the sequence. Not seeing a discernible pattern, she pauses and returns to the closed form for the sequence and then produces the correct limit (by looking at the coefficients of the dominant terms). She doesn't ever explicitly relate the list back to the function, and when trying to explain her method for computing the limit and in discussing the idea of a limit in general does not ever refer back to her sample computations, i.e. the list.

For students at the intra stage, the two constructs of SEQLIST (object) and SEQFUNC (process or object) are distinct and cognitively unrelated. They are not representations of some higher-order construct. Students may use both constructs in attacking a particular mathematical problem, as Ilene did with the limit problem above. But they see those as two constructs, and these two methods for attacking the problem as unrelated.

Inter Stage of Sequence Schema Development. The next level of development in terms of the existence of cognitive links and the awareness of these links is the inter stage. As was the case with students at the intra stage, a student at the inter stage still may not have constructed a cognitive object of SEQFUNC. Also, such a student may not have developed an awareness of all the intimate connections among the constructions in the developmental chain for SEQLIST and for

SEQFUNC. And certainly there is no underlying coherence to these connections. However, a student at the inter stage may demonstrate the existence of links between sequence as function and as list not only by being able to generate a list from a functional expression, but also by showing a certain level of comfort in examining patterns in a list and writing an algebraic expression for it. Students at this level of schema development preferentially use sequence as list or sequence as expression depending on the context or problem situation, without the explicit awareness of the equivalence between what we see as two representations of the same thing. In contrast with students at the intra stage whose focus of attention was on one construct at a time in relative isolation from the others, students at the inter stage may simultaneously be aware of the two constructions but not aware of their equivalence.

Half of the students at the inter stage of development of a sequence schema were also limited to a process conception of SEQFUNC. An example is Vince who seemed limited to a process conception of SEQFUNC, but in a previous excerpt of his interview does show an object conception of SEQLIST and some awareness of the links between the SEQFUNC process and SEQLIST.

However, it is possible for students to have constructed SEQLIST as object and SEQFUNC as object, and yet still not have the level and awareness of connections and the underlying coherence to take them beyond the inter stage. Khalid is an example of such a student who has an object conception of SEQFUNC and is still at the inter stage. The reader is referred back to the lengthy excerpt from his interview presented in the section on the development of SEQFUNC. Khalid is at the early stages of the inter level because he has yet to fully establish the links between SEQLIST object and SEQFUNC object. In fact, in his transcript excerpt, we see that some of these links seem to be rapidly developing in the interview setting. He has yet to develop all the necessary links between SEQLIST and SEQFUNC, to reflect upon these links, and to build an understanding of the equivalence of these two constructs. Only then will he be able to move into the trans stage.

Trans Stage of Sequence Schema Development. Students at the trans stage of schema development for sequence have constructed the individual cognitive objects SEQLIST and SEQFUNC. They have also constructed strong and numerous connections between sequence as function and sequence as list, they are aware of these connections, and they can now subordinate each of these objects to an overall concept of sequence. This coherent framework of related concepts will allow students, even when faced with new situations, to determine whether or not this sequence schema is applicable to these situations. For such a student, SEQLIST and SEQFUNC can be viewed as representations, both cognitive and mathematical, of the concept of sequence.

The totality of David's transcript convinces us he is at the trans stage. Note his certainty in discussing the multiple ways of viewing a sequence.

> David: That's uh ... I don't know, a sequence could be, uh, n over $n + 1$, that'd be a sequence.
> I: OK.
> David: And, uh, whatever the terms of this would be, you know, n starting at 1 would be $\frac{1}{2}$, $\frac{2}{3}$, $\frac{3}{4}$, and you could also represent it graphically. But you'd have to put dots instead of, you know, a line because it's

just a sequence and it's only at n's of real, real numbers, you know, you can't have, you know, n, an n of .5, it can only be whole numbers.

And later in his interview we find the following exchange.

I: Do all sequences have some kind of a closed form?

David: Uh, yeah.

I: OK.

David: They, they do but some of them are harder to find than others, if they're not given. Because sequences don't have to be given in a closed form, you can just represent them graphically, or represent them with numbers like I did here.

I: Mm-hmmm.

David: See this is two representations of the same thing.

Note that an object conception of SEQLIST and SEQFUNC are necessary, but not sufficient, for a student to progress to the trans stage. Recall that in the case of Khalid, these constructions seemed to be present, but an explicit awareness of their equivalence was not. In fact Khalid was somewhat surprised when he began seeing some of the links between SEQLIST and SEQFUNC.

It is also important to note that students must have a SEQFUNC object which allows them to make the reasonable links between the functional input for SEQFUNC and the position in SEQLIST. Thus, recall Helen who we claim had a limited object conception of SEQFUNC because of problems restricting the domain. We claim that she would not be able to fully develop her sequence schema until she reconstructs her SEQFUNC object. The following statement out of context might make one think she is at the trans stage.

I: ... why don't you write down an example of a sequence for me.

Helen: OK, um what I remember is that are lots of different ways to write them but you can always write them as numbers.

Yet the totality of her interview shows that there are still some missing connections, mainly related to her lack of understanding of domain issues, and that she is not yet cognitively aware of many of the connections between SEQLIST and SEQFUNC.

It is clear that one of the implicit goals of sequence instruction is for our students to develop a trans level sequence schema. Consider, for example, Wendy who says the following when writing down an example of a sequence.

Wendy: OK. Um, my example of a sequence, um, you could represent it as a sub n, and just say that the nth term is n, and then that would be the same as 1, 2, 3, ... all the way out for all integers.

Here she shows an awareness of the multiple representations of sequence. Later when dealing with the sequence $1 + (-1)^n$, she shows clear connections between representations of sequence as function and sequence as list.

Wendy: OK. Um, when I look at a sequence, I usually write out the first few terms, instead of just looking at the formula ...

But unlike Ilene, who also wrote down the first few terms in the context of finding the limit of the rational function, Wendy does not isolate one representation from the other. She understands the equivalence of examining the formula or looking at the list.

For students who have developed to the trans stage, the word sequence can evoke a powerful concept image which includes the ability to use lists and functions as representations of what for them is now the single concept of sequence. The coherent framework underlying their schema will help them utilize this schema in new situations, such as with sequences of functions in real analysis or sequences of integral equations used in Picard iteration in differential equations.

Results: Student Progress in Understanding Sequence

We now present the number of students who were able to progress to each of the levels of development for SEQLIST and SEQFUNC and how many were able to achieve the trans level of schema development.[4] Additionally, we will report the number of students in each of the instructional methods who appeared to have attained these levels. Of course, because the groups were very small, because a variable number of mathematics courses had been taken by the students after calculus, because of self-selection between the instructional treatments and because of self-selection into the interview process, attempts at statistical inference would be fruitless. Still, our experience teaching calculus leads us to believe that these students are typical of many who have taken two semesters of calculus. Additionally, the difference between the percentages of students who were successful in developing a rich conception of sequence as a result of the different instructional treatments, though not conclusive, is suggestive. These comparative data, although weak on their own, do complement a collection of other studies which have compared the performance of C^4L students and traditionally taught students. In totality, this collection may shed light on possible performance differences between C^4L students and traditionally taught students. Thus, even though a serious statistical comparison of the limited data in this study would be seriously flawed, we include the numbers so as to add to the broader set of comparative data. We give an overview of the progress made by all students, then consider progress made toward developing the individual cognitive objects of SEQLIST and SEQFUNC, and progress in schema development for the two instructional treatments.

Every student of the 21 we interviewed had an object conception of the very primitive construct, SEQLIST. In contrast, only 12 students clearly demonstrated an object conception of SEQFUNC. In addition, two students were apparently in transition between the SEQFUNC process and SEQFUNC object conception. All students in the study had attained at least a process conception of SEQFUNC. Finally, only seven of the students interviewed were at the trans level of sequence schema development. The rest were at the inter stage, with two possibly still in transition from the intra stage. These numbers highlight the importance of function in the development of this concept. And the different emphasis on function, very strong in the C^4L approach, is probably a key factor in the different performance levels between the two classes as highlighted below.

---

[4]Recall, a table listing all students at each of these levels of cognitive development is provided in an Appendix.

Obviously there were no differences between the results of the two groups in terms of SEQLIST development. There were some notable differences in the development of SEQFUNC. Ten of the 15 students who had participated in the C^4L reformed calculus course had achieved an object conception of SEQFUNC, while only two of the six traditionally taught students attained this level. The additional two students who appeared to be making the transition between SEQFUNC process and SEQFUNC object were both from the C^4L course. Alternatively, we see that four of the six traditionally taught students were limited to a process conception of SEQFUNC, while only three of the 15 C^4L students were clearly at the process level. Given the emphasis in C^4L on the notion of function, it is not surprising that as a whole the C^4L students attained a substantially higher level of understanding of sequence as a function than did their traditionally taught counterparts.

There is an important observation relating the mathematical performance of students to the constructions SEQLIST and SEQFUNC. Some students who had only developed an object conception of SEQLIST, and not an object conception of SEQFUNC, and who had not made many of the crucial links between these two constructs, performed quite well on interview questions which required them to discuss and relate the properties of sequences by working entirely with lists. An example is Anne, who earned a B in Calculus II, and had completed three further mathematics courses, Multivariable Calculus, Linear Algebra with Differential Equations, and Elements of Vector Calculus. Moreover, at the time of the interview she was enrolled in three mathematics courses including Elementary Linear Algebra, Fundamental Concepts of Geometry, and Topics in Mathematics. Despite this mathematical exposure, Anne did not appear to have an object conception of SEQFUNC. However, she still performed extremely well when asked to produce examples of sequences with certain properties and to draw the Venn diagram relating these properties. In generating examples related to these questions, she wrote a number of appropriate lists and several suitable closed form expressions. Without the benefit of the clinical interviews, we might incorrectly conclude from her written work that she had attained an object conception of SEQFUNC.

We now turn to the question of progress in developing a sequence schema. It was in two cases difficult to classify individual students as to intra or inter stages. In fact, it is also possible for students to exhibit different levels of schema development in different problem contexts. It has been frequently observed that what a student knows and is capable of doing may not be available to her or him at a given moment and in a given situation. Indeed, Asiala et al. (1996) characterize an individual's mathematical knowledge as

> ... her or his tendency to respond to perceived mathematical problem situations by reflecting on problems and their solutions in a social context and by constructing or reconstructing mathematical actions, processes and objects and organizing these in schemas to use in dealing with the situations. (p. 7)

For this reason, except for the isolated cases like David who stated "See, this is two representations of the same thing," who can be recognized as being at the trans stage through a single statement they make, the determination of level of schema development was more often the result of considering the interview transcript as a whole. That is we considered the totality of responses to the approximately hour long interview. As we noted previously, Helen made several statements that might lead us to believe she is at the trans stage. However, the totality of her interview showed us that she had yet to make many of the important connections and had yet to become fully aware of these connections and place them into a coherent framework.

Recall that seven of the 21 students were at the trans level of schema development. Of these, six were students from the C^4L course, and one was from the traditional course. Because an object conception of SEQFUNC was a prerequisite for this development, it is not surprising that a higher percentage of C^4L students than traditionally taught students were able to construct a trans level sequence schema.

Conclusion

The Action-Process-Object-Schema theory and the triad development of schema are useful tools in understanding the interview data relative to students' cognitive construction of sequence. We found that all students studied had a concept image which included sequence as a list of numbers. This SEQLIST object was relatively easy for students to develop. Students also tended to develop another cognitive concept, SEQFUNC, while coming to understand sequence. This SEQFUNC object was more difficult for students to develop partly because it is tied to their function schema, a schema which is relatively weak for many students. We found that only about two-thirds of the students studied had an object conception of SEQFUNC.

Students then must build connections between their construction of SEQLIST and of SEQFUNC. For students to develop a rich sequence schema, they must build multiple connections and must be aware of these connections. When students reach the trans stage, they have the ability to use lists and functions as representations of what for them is now a single concept for sequence. Unfortunately only one-third of the students studied had developed such a robust sequence schema.

Our findings support the observations made in other contexts that the language we use for the cognitive constructions we have will not necessarily evoke the same cognitive constructions in the students. We saw a recurrence of what Thompson (1994) saw with functions. In general we may use the term "sequence" to mean a rich connection between sequence as function and sequence as list. To many students it is heard only as sequence as list.

So the term "sequence" evokes a limited (and yet often powerful) image in many students' minds, despite the rich connotations that an instructor might attach to this term. Given this, it is naive to believe that carefully crafted explanations will be sufficient to induce trans level development of schema for sequence. Indeed the students involved in this study were by national standards above average, and represented a variety of instructors. Schoenfeld (1994) notes that this carefully crafted lecture

> ... conception of teaching assumes, implicitly, that students are recipients of knowledge. It fails to take into account the fact that what the students already understand—whether right or wrong—will shape the ways in which they interpret new information that they encounter. (p. 10)

We must understand the constructions that students will need to make in order to reach our conception of sequence. Our curricular choices and our teaching should help guide students to make these constructions and not simply allow for them to "make do," as they seem to be able to now, with only a SEQLIST conception of sequence.

We may not always consciously think about getting students to the trans stage of development of their sequence schema. But it seems that we implictly assume that they, and not just our A students, do have this level of understanding. We should want to get even our C students to this level, as it is prerequisite knowledge for their future work. In the same course where sequences are introduced, students will need a rich sequence schema to be able to fully understand series as the limit of a sequence of partial sums. And in this course, or later, they will need to understand sequences to work with Taylor polynomials, Taylor series, series solutions of differential equations, and to understand the equivalence of sequence and ϵ–δ definitions of continuity. We must help our students make the appropriate constructions related to sequence, and not simply by giving clear lectures on our understanding of sequence.

This study raises several other interesting questions intimately related to students' understanding of sequence such as their understanding of limits, monotonicity, and bound. There are several studies of students' conception of limit. In our data we noted that students' discussion of the sequence $\{0.9, 0.99, 0.999, \dots\}$ were consistent with the results presented by Tall (1992b). Cottrill et al. (1996) have examined the concept of limits of continuous functions using APOS theory. Further research may shed additional light on limits of sequences.

Two characteristics of sequences which our data addressed to a limited extent and which may be worth further examination are that of monotonicity and bound. The word monotone (or monotonic) did not seem to mean anything in reference to sequences to most students in our study. There seems to be confusion between everday usage of this word (e.g., he spoke in a monotone voice) and its specialized use in the context of sequences. Clearly students could benefit from more deliberate discussion of the meaning of monotone sequences.

Although most students understood the basic notion of bound, there were two very typical problems associated with their understanding. The first is that most students saw bound as a unique number. For example, many students would state that 2 is *the* upper bound for the sequences $a_n = 1 + (-1)^n$ or 1 is *the* upper bound for $a_n = 1 - \frac{1}{n}$ and no other numbers could be called upper bounds. Some evidence in our data show that they appear to be envisioning a process of drawing a horizontal line above all points in their graph (or possibly their continuous curve) and then lowering this line until it just "touches" a point (or the curve). This probably leads to the second problem, that the bound is optimal. This often leads students to connect the bound with the limit of a sequence, even if the supremum or infimum is not the same value as the limit. Clearly we need to do a better job in emphasizing through examples that bounds are not unique and are often not equal to the limit. The issue of bound might be examined as part of research on students' understanding of limits of sequences.

In addition to these conceptual issues, the student progress data leave open the question whether the trends suggested in these small samples of traditionally-taught students and reform program students are general trends. There are other such studies comparing small samples of C^4L students with traditionally-taught students. A close examination of all of these studies as well as a larger, more formal study of these two cohorts of students may be worthwhile.

Possibilities for future studies which might build directly on our results include generating an interview that quickly illuminates whether students exhibit an object conception of SEQFUNC and are at the trans stage of schema development. Beacuse a few students in our study used lists or recursively defined sequences almost exclusively, it may also be worthwhile to examine students' conceptions of recursively defined sequences and the relationship between their conceptions and what we have termed SEQLIST and SEQFUNC.

References

Asiala, M., Brown, A., DeVries, D. J., Dubinsky, E., Mathews, D., & Thomas, K. (1996). A framework for research and development in undergraduate mathematics education. In J. Kaput, E. Dubinsky, & A. H. Schoenfeld (Eds.), *Research in collegiate mathematics education II* (pp. 1–32). Providence, RI: American Mathematical Society.

Asiala, M., Cottrill, J., Dubinsky, E., & Schwingendorf, K. (1997). The development of students' graphical understanding of the derivative. *Journal of Mathematical Behavior, 16*(4), 399–431.

Breidenbach, D., Dubinsky, E., Hawks, J., & Nichols, D. (1992). Development of the process conception of function. *Educational Studies in Mathematics, 23*, 247–285.

Carlson, M. P. (1998). A cross-sectional investigation of the development of the function concept. In E. Dubinsky, A. H. Schoenfeld, & J. Kaput (Eds.), *Research in collegiate mathematics education III* (pp. 114–162). Providence, RI: American Mathematical Society.

Clark, J., Cordero, F., Cottrill, J., Czarnocha, B., DeVries, D., St. John, D., Tolias, G., & Vidakovic, D. (1997). Constructing a schema: The case of the chain rule. *Journal of Mathematical Behavior, 23*(4), 345–364.

Clark, J., & Mathews, D. (in preparation). *Successful students' conceptions of mean, standard deviation, and the central limit theorem.*

Cottrill, J., Dubinsky, E., Nichols, D., Schwingendorf, K., Thomas, K., & Vidakovic, D. (1996). Understanding the limit concept: Beginning with a coordinated process schema. *Journal of Mathematical Behavior, 15*(2), 167–192.

Dubinsky, E., & Harel, G. (1992). The nature of the process conception of function. In G. Harel & E. Dubinsky (Eds.), *The concept of function: Aspects of epistemology and pedagogy* (MAA Notes no. 25). Washington, DC: Mathematical Association of America.

Eisenberg, T. (1992). On the development of a sense for functions. In G. Harel & E. Dubinsky (Eds.), *The concept of function: Aspects of epistemology and pedagogy* (MAA Notes no. 25, pp. 153–174). Washington, DC: Mathematical Association of America.

Ferrini-Mundy, J., & Graham, K. (1994). Research in calculus learning: Understanding of limits, derivatives, and integrals. In J. Kaput & E. Dubinsky (Eds.), *Research issues in undergraduate mathematics learning* (MAA Notes no 33, pp. 31–45). Washington, DC: Mathematical Association of America.

Harel, G. & Dubinsky, E. (Eds.). (1992). *The concept of function: Aspects of epistemology and pedagogy* (MAA Notes no. 25). Washington, DC: Mathematical Association of America.

Hiebert, J., & Lefevre, P. (1986). Conceptual and procedural knowledge in mathematics: An introductory analysis. In J. Hiebert (Ed.), *Conceptual and procedural knowledge: The case of mathematics* (pp. 1–27). Hillside, NJ: Erlbaum.

Monaghan, J. (1991). Problems with the language of limits. *For the Learning of Mathematics, 11*(3), 20–24.

National Research Council (1989). *Everybody counts.* Washington, DC: National Academy Press.

Piaget, J., & Garcia, R. (1989). *Psychogenesis and the history of science.* New York, NY: Columbia University Press. (Original work published 1983.)

Schoenfeld, A. (1994). Some notes on the enterprise (research in collegiate mathematics education, that is). In E. Dubinsky, A. H. Schoenfeld, & J. Kaput (Eds.), *Research in collegiate mathematics education I* (pp. 1–19). Providence, RI: American Mathematical Society.

Schwingendorf, K., Mathews, D., & Dubinsky, E. (1994). Calculus, concepts, computers and cooperative learning. In *Proceedings of ICTCM-VII* (pp. 402–406). Reading, MA: Addison-Wesley.

Selden, J., Selden, A., & Mason, A. (1994). Even good calculus students can't solve nonroutine problems. In *Research issues in undergraduate mathematics learning* (pp. 19–26). Washington, DC: Mathematical Association of America.

Tall, D. (1992a). *Students' difficulties in calculus.* (Plenary presentation at the International Congress of Mathematics Education, Quebec, Canada.)

Tall, D. (1992b). The transition to advanced mathematical thinking: functions, limits, infinity, and proof. In D. Grouws (Ed.), (pp. 495–511). New York, NY: MacMillan.

Tall, D., & Vinner, S. (1981). Concept image and concept definition with particular reference to limits and continuity. *Educational Studies in Mathematics, 12,* 151–169.

Thompson, P. (1994). Students, functions, and the undergraduate curriculum. In E. Dubinsky, A. H. Schoenfeld, & J. Kaput (Eds.), *Research in collegiate mathematics education I* (pp. 21–44). Providence, RI: American Mathematical Society.

Tufte, F. (1989). Revision in calculus instruction: Suggestions from cognitive science. In J. Malone, H. Burkhardt, & C. Keitel (Eds.), *The mathematics curriculum: Towards the year 2000. Content, technology, teachers, dynamics* (pp. 155–161). Perth, Western Australia: The Science and Mathematics Education Center, Curtin University of Technology.

Williams, S. (1991). Models of limit held by college calculus students. *Journal of Research in Mathematics Education, 23*(3), 219–236.

Appendix: Student Progress Summary

The following table presents student progress in constructing SEQLIST and SEQFUNC. Students from the traditional course are noted in boldface.

TABLE 1. Student Progress in Constructing SEQLIST and SEQFUNC.

	PROCESS	OBJECT
SEQLIST	None	All students
SEQFUNC	**Anne**	Curt
	Beth	David
	Fran	Emily
	Greg	Helen [5]
	Ilene	James
	Sam	Khalid
	Vince	Madelyn
		Nathan
		Peter
		Rita
		Thomas
		Wendy

[5]Helen had a limited SEQFUNC object.

Note: Larry and Olivia (not listed) clearly had achieved a process conception of SEQFUNC. However, it was not clear whether they had achieved an object conception.

The following table presents student progress on developing a schema for the sequence concept. Again, students from the traditional course are noted in boldface.

TABLE 2. Student Progress in Constructing a Sequence Schema.

INTER	TRANS
Anne	Curt
Beth	David
Fran	James
Greg	Madelyn
Helen	**Rita**
Khalid	Thomas
Larry	Wendy
Nathan	
Olivia	
Peter	
Sam	
Vince	

Note: Emily and **Ilene** (not listed) appeared to be in transition from Intra to Inter stages of sequence schema development.

The last table gives a listing of students by course, with their course grade in parentheses.

TABLE 3. Students Listed by Course with Course Grade.

Traditional	C^{4}L
Anne (B)	Beth (A)
Ilene (A)	Curt (B)
Peter (A)	David (A)
Rita (A)	Emily (B)
Sam (B)	Fran (A)
Vince (B)	Greg (A)
	Helen (B)
	James (A)
	Khalid (A)
	Larry (C)
	Madelyn (?)
	Nathan (B)
	Olivia (B)
	Thomas (A)
	Wendy (A)

OCCIDENTAL COLLEGE, LOS ANGELES, CALIFORNIA
E-mail address: `mickey@oxy.edu`

SOUTHWESTERN MICHIGAN COLLEGE
E-mail address: `dmathews@smc.cc.mi.us`

AEGON USA INSURANCE GROUP

CBMS Issues in Mathematics Education
Volume **8**, 2000

A Theoretical Framework for Analyzing Student Understanding of the Concept of Derivative

Michelle J. Zandieh

This paper develops a theoretical framework for exploring student understanding of the concept of derivative. The framework is an attempt to clarify, describe and organize the facets that we as a mathematical community consider to be part of the understanding of the concept of derivative to allow us to more systematically analyze and discuss issues such as:

- Individual understanding. How is a particular student's understanding of the concept similar to and different from the concept as seen by the mathematical community?
- Comparative individual student understandings. How are individual student understandings of the concept similar or different?
- Individual learning. Do students tend to learn the various aspects of the concept in a particular developmental order? Are some aspects of the concept usually learned before others?
- Teaching strategies. Would students learn more efficiently if they were introduced to the various aspects of the concept in a particular order?
- Effectiveness of pedagogical practices or curricula. How are the understandings of a class of students using a particular pedagogy or curriculum similar or different from those in a class using different pedagogical strategies or curricula?
- Evaluation of curriculum materials. What understandings of a concept are emphasized or de-emphasized by a set of curriculum materials, a calculus text, or a set of homework problems?

The framework is not meant to explain how or why students learn as they do, nor to predict a learning trajectory. Rather the framework is a "map of the territory," a tool of a certain grain size that we, as teachers, researchers and curriculum developers, can wield as we organize our thinking about teaching and learning the concept of derivative and try to answer questions like those listed above.

Obviously, this paper will not demonstrate all the ways this framework may be used. The purpose of this paper is to highlight the development of the framework and to demonstrate its use in a few examples, in particular, examples that debunk the seemingly hierarchical nature of the framework.

©2000 American Mathematical Society

The first four sections of this paper lay out the framework and its roots in other work. The framework is then used to evaluate data collected from several case studies of high school advanced placement calculus students. We will see that these gifted students have surprising weaknesses in their understanding of the concept of derivative. The students also show a broad diversity in the order in which they build their understanding of the concept of derivative.

Roots of the Framework

The initial premise of my investigation into understanding is that for a concept as multifaceted as derivative it is not appropriate to ask simply whether or not a student understands the concept. Rather one should ask for a description of a student's understanding of the concept of derivative—what aspects of the concept a student knows and the relationships a student sees between these aspects.

The work of Tall and Vinner (1981), Hiebert and Carpenter (1992) and Lakoff (1987) sets the stage for defining understanding of a concept in terms of a structured framework. Tall and Vinner (1981) describe a person's *concept image* for a particular concept as "the total cognitive structure that is associated with the concept, which includes all the mental pictures and associated properties and processes" (p. 152). This notion is broad enough to describe what I will mean by a student's understanding of derivative. Other notions developed by Vinner and Dreyfus (1989), including compartmentalization and concept definition, suggest that the concept image is structured, but these notions are not adequate to provide a detailed description of a student's understanding of derivative.

Hiebert and Carpenter (1992) say that a "mathematical idea or procedure or fact is understood if it is part of an internal network. . . . The degree of understanding is determined by the number and strength of the connections" (p. 67). This description provides more detail, but one must still analyze the structures specific to the concept of derivative. Lakoff's (1987) book, *Women, Fire and Dangerous Things*, provides the notion that we can analyze language to determine what a community of people includes in the understanding of a concept and do this in a structured way.

This paper first lays out a description of the mathematical community's notion of understanding the concept of derivative at the first-year calculus level. An individual's understanding of derivative is then defined in terms of its overlap or lack of overlap with the mathematical community's notion of that concept. Similarly, a passage in a mathematics text or an exercise, problem or activity in a curriculum may be examined for the aspects of the concept of derivative that it asks students to explore.

My framework was developed by observing the way the concept of derivative is described in textbooks and by listening to the way that mathematics education researchers, mathematicians, mathematics graduate students, and students in calculus classes discuss the concept of derivative. The textbooks and curriculum materials examined (as well as those used by the students, teachers, and researchers) ranged from traditional (e.g., Thomas & Finney, 1992) to "reform" (e.g., Dick & Patton, 1992; Hughes-Hallett & Gleason, 1994) to "experimental" (e.g., Monk & Nemirovsky, 1995; Tall, 1986, 1991). As Kaput (1997) makes clear, differences in these curricula may make an incredible difference in the contexts for learning that students experience and thus the types of understandings that emerge in those students.

It is my belief that my framework is a tool which is flexible enough to be used in any of these settings since it does not predict which understandings will emerge nor in what order, but rather organizes and structures a wide range of possible understandings. However, this paper will illustrate the framework primarily with examples of student understandings of calculus from nine high school seniors who had experienced both a traditional (Swokowski, 1988) and a reform text (Dick & Patton, 1992) and had some experience with using graphing calculator technology (HP-48) and computer software (*Mathematica*) to view functions and derivatives in multiple representations. The collection of these student data and the latter stages of the development of the framework occurred in tandem, each influencing the other.

An Overview of the Framework

My framework for describing the concept of derivative has two main components: multiple representations or contexts and layers of process-object pairs. I will discuss these two ideas briefly here, referencing the work that most influenced my ideas in these directions. This will be followed by a comparison of the process-object pairs to other research, primarily on student understanding of function, that uses related structures.

Multiple representations. Researchers using the notion of concept image often find that a person's concept image includes a number of different representations of the concept (e.g., Hart, 1991; Dreyfus & Vinner, 1989; Krussel, 1995; Tall & Vinner, 1981). For functions these include analytic or symbolic, graphic, numeric, verbal, and physical representations. Many calculus reform texts have emphasized the use of multiple representations as a way to develop student understanding (e.g., Dick & Patton, 1992, 1994; Hughes-Hallett & Gleason, 1994; Ostebee & Zorn, 1997). Concepts that involve functions, such as the limit of a function or the derivative of a function, may be described in terms of these representations.

The concept of derivative can be represented (a) graphically as the slope of the tangent line to a curve at a point or as the slope of the line a curve seems to approach under magnification, (b) verbally as the instantaneous rate of change, (c) physically as speed or velocity, and (d) symbolically as the limit of the difference quotient.

Many other physical examples are possible, and there are variations possible in the graphical, verbal, and symbolic descriptions. What do each of these descriptions of derivative have in common that cause us to call them by the same name? What are the relationships between the different representations or contexts? The next section on the process-object framework describes the similar structure of the concept of derivative in each context. The term "representation" tends to be used to mean that there is an underlying structure that each of the contexts "represents." However, it may well be that the "representations" are all there is, that it is our "subjective sense of invariance" (Thompson, 1995b, p. 39) that causes us to call each by the same name. In addition, I will sometimes say "contexts" instead of "representations" to indicate more broadly ways in which we think of the concept of derivative. For example, a context not already listed would be the context of measuring the outdoor temperature, T, for a specific hour, t, past noon.

Contexts

	Graphical	Verbal	Paradigmatic Physical	Symbolic	Other
Process-object layer	Slope	Rate	Velocity	Difference Quotient	

Ratio

Limit

Function

FIGURE 1. Outline of the framework for the concept of derivative.

The derivative resides in this physical context as "the instantaneous rate at which the temperature changes per hour."

Layers. Consider the formal symbolic definition of the derivative:

$$f'(x) = \lim_{h \to 0} \frac{f(x+h) - f(x)}{h}.$$

The derivative of f, f', is a <u>function</u> whose value at any point is defined as the <u>limit</u> of a <u>ratio</u>. I will call these underlined aspects of the concept of derivative (ratio, limit, function) the three "layers" of the framework. At this point we may combine the layers and the contexts or representations to form a matrix outlining the structure of the concept of derivative (Figure 1). Although only four columns are named specifically, there can be many other columns, one for each situation in which there is a functional relationship for which one may discuss the concept of derivative, e.g., temperature as a function of time. Velocity is a paradigmatic example (an exemplar) of this possibility.

Velocity is the physical example most commonly used in the teaching of beginning calculus. It is labeled paradigmatic because it is the context that is most representative of the group of such contexts. Velocity is an exemplar because it is an extremely familiar phenomenon for which we have additional natural language structure (e.g., increasing velocity is called acceleration) and because it is metaphorically related to most other derivative contexts (e.g., the phrases, "he enjoyed a speedy recover," or "the economy slowed in the fourth quarter").

Each empty box in the matrix represents an aspect of the concept of derivative. For example, the box in the ratio row and the graphical column represents the slope of a secant line on a graph of the function whose derivative we are concerned about. More details about the layers in terms of slope, velocity and the difference quotient will be discussed in the section below, Understanding the Concept of Derivative. Then the section Using the Framework to Describe Individual Conceptions of Derivative will provide examples of specific student quotes that further illuminate the content of some of the boxes in the matrix.

Process-object pairs. Mathematically it is clear that the concept of derivative involves a ratio, a limit and a function. However, thinking of these as "layers" comes out of my thinking about Sfard's (1992) framework of processes acting on previously established objects. In addition, her notions of operational and structural conceptions add a third dimension to my framework.

Sfard's (1992) research on the historical and psychological evolution of mathematical concepts suggests a transition from a process or *operational* conception to a static *structural* conception. According to Sfard's processes are operations on previously established objects. Each process is reified into an object to be acted on by other processes. This forms a chain of what I will call *process-object* pairs.

Each of the layers of the concept of derivative—ratio, limit and function— can be viewed both as dynamic *processes* and as static *objects*. For example, a ratio or a rational number may be thought of operationally as division, but also structurally as a pair of integers within a multiplicative structure. Limits may be viewed dynamically as a process of approaching the limiting value or statically through the epsilon–delta definition. Functions may be seen as the process of taking an element in the domain and acting on it to produce an element in the range. Functions may also be viewed statically as a set of ordered pairs.

The concept of derivative involves three process-object pairs which are linked in a chain. The ratio process takes two objects (two differences, two lengths, a distance and a time, etc.) and acts by division. The reified object (the ratio, slope, velocity, etc.) is used by the next process, that of taking a limit. The limiting process "passes through" infinitely many of the ratios approaching a particular value (the limiting value, the slope at a point on a curve, the instantaneous velocity). The reified object, the limit, is used to define each value of the derivative function. The derivative function acts as a process of passing through (possibly) infinitely many input values and for each determining an output value given by the limit of the difference quotient at that point. The derivative function may also be viewed as a reified object, just as any function may. (The derivative function may be thought of as an object that is the output of another process, the derivative operator.)

Pseudo-objects. Suppose a student has not developed a structural conception of one of the layers. How can that student consider the next process in the derivative structure without an object to operate on? One simple solution is to use what Sfard (1992) calls a *pseudostructural* conception. A pseudostructural conception may be thought of as an object with no internal structure. In fact, even for a person who can conceptualize each layer as both a process and an object, it is often simpler to describe a process by having it operate on a pseudostructural "object." For brevity, I will refer to a pseudostructural object as a *pseudo-object*. Note that pseudo-object is not meant to have a negative connotation, rather it merely denotes that the object a person is using does not refer to the underlying process (nor underlying static structure) of the true object. Pseudostructural examples for the concept of function include viewing a graph or symbolic expression as an object to be manipulated without recognizing the domain, range and relationship (either dynamic or static) between the input and output values. A pseudostructural conception of limit refers exclusively to the value of the limit, the end result of the process, without recognition of the process that leads to that result or the epsilon-delta criterion that requires that result. A pseudostructural conception of a ratio would be to see a common fraction (e.g., $\frac{1}{2}$) as a single value or a place on the number line without recognizing that the ratio can also represent a division process or a pair of numbers in a multiplicative structure.

For an example of a process operating on a pseudostructural object consider the derivative function as a process that gives us the speed at each point, like a car's speedometer. For this description a student can concentrate on the function process and its output without, for the time being, working with the complications of the underlying limit or ratio processes.

Pseudostructural conceptions usually have the form of a gestalt. By this I mean that the conception is thought of as a whole without parts, a single entity without any underlying structure. Sometimes, as in the speedometer example above, a gestalt may be used to simplify a thought process. The details underlying the gestalt may be known, but not emphasized in that context. In other cases a student may not be aware of any underlying structure or may have compartmentalized any knowledge about the underlying structure so that the student does not evoke this information in an appropriate context. The recognition of underlying structure is the transition from a pseudostructural conception to an operational or process conception. This transition will be examined here as closely as the other transitions emphasized by Sfard.

Comparison with Other Frameworks

This section makes comparisons between several theoretical frameworks for advanced mathematical thinking that include notions related to process-object pairs. The concept of function is used for this comparison because that is the concept around which most of these frameworks initially developed and because it allows me to make a distinction between my way of viewing the notion of function as two process-object pairs in contrast to the more customary way of thinking of the function concept as involving only one such transition. Further discussion of the details of my derivative framework and student results will follow this section.

As discussed above, my thinking about process-object pairs developed most notably from the work of Sfard, but several other researchers have done similar work. Monk (1992) distinguishes students as having either a *pointwise* or *across-time* view of functions. The students with a pointwise view of function concentrate on particular input and output pairs and are reluctant to discuss the qualitative change in the function over time. Students who do not have this difficulty are said to have an across-time view of function.

Breidenbach et al. (1992) use the words *process* and *object* to describe a similar dual nature of functions. Following Piaget, processes and objects are said to be constructed by *reflective abstraction* (see also Dubinsky, 1991). Breidenbach et al. describe this theoretical view for any arbitrary mathematical object and then relate it specifically to functions. Their framework is strikingly similar to that of Sfard's, but there are differences in language that the reader should be careful to note in the following description.

Breidenbach et al. describe an *action* as "any repeatable physical or mental manipulation that transforms objects (e.g., numbers, geometric figures, sets) to obtain objects" (p. 249). When the action can be thought of as a whole without running through all of the steps, the action is said to be *interiorized* to become a process. The subject can then use the process to obtain new processes. When the person can imagine the process as transformed by some action, then it has become *encapsulated* as an *object*.

Sfard describes three stages in the transition from an operational to a structural conception—*interiorization, condensation*, and *reification*. Interiorization occurs when a person can step through the relevant process. Condensation occurs when the person can view that process as a whole and use it as a subprocess in other processes. Reification occurs when the process may be viewed structurally as an object. (Note that the interiorization of Breidenbach et al. is most similar to Sfard's description of condensation, but is different from her use of the term interiorization.)

Breidenbach et al. apply their framework to functions by considering three ways of thinking about functions (see also Dubinsky & Harel, 1992). The *prefunction* stage is used to describe a student's understanding which involves a minimal conception of function. Additionally, that student is characterized as being unable to perform basic tasks involving functions. Thinking of a function as an *action* includes being able to input a number into an algebraic expression for a function and calculate the corresponding output. However, a person with this understanding will think of this for only one value at a time. A *process conception* of function involves a dynamic transformation understood for all values at once. The subject can therefore imagine combining this function process with other function processes and can also imagine reversing the process. The subject is said to have an *object* conception of function if it is possible for him or her to perform actions on it, in particular actions that transform it.

Monk's description of pointwise understanding of function is similar to the action conception of function described by Breidenbach et al. and the interiorization stage presented by Sfard. It is simply the ability to take an input value for a function and find its associated output value, and is therefore less than a full process conception. In the latter, the action is condensed so that a person may imagine the evaluation of one or more output values without having to actually go through the process. Monk's across-time view of function is different than the process conception of function. Monk's interest in across-time questions and articulation questions involving rate of change suggests the importance of understanding function in terms of the covariation of two quantities. The importance of covariation in student understanding of function is also reported by Confrey and Smith (1994) and Thompson (1995b).

Thompson (1995b) relates the idea of covariation to the process-object duality in the following way. Once a person has a process conception of function, then that person "can begin to imagine 'running through' a continuum of numbers, letting an expression evaluate itself (very rapidly!) at each number" (p. 26). "Once students are adept at imagining expressions being evaluated continually as they 'run rapidly' over a continuum, the groundwork has been laid for them to reflect on a *set* of possible inputs in relation to the *set* of corresponding outputs" (p. 27). The correspondence between the sets would then be a view of function as object. Thus, Thompson's work suggests that the view of function as covariation depends on an understanding of function as process, but is less comprehensive than a view of function as object.

The process-object pairs that I emphasize in my framework are a single dichotomy unlike the three-part structure of action-process-object or Sfard's three transitions: interiorization, condensation and reification. Therefore I see a generic function as consisting of not one, but two, process-object pairs. Examining what I have done for derivative, we see that it takes two of my layers, the ratio and limit layers, to take an input and get an output, i.e. to calculate a single value of a derivative function. The third layer, the function layer, emphasizes the notion of covariation. So for any function there would be at minimum two process-object pairs. The pairs would be:

1) A process of taking an input of a function to its output paired with an object that is the value of the function at a point (noting that although the value itself is named as the object, the corresponding input is referenced).

2) A process of covariation, i.e. imagining "running through" a continuum of domain values while noting each corresponding range value, paired with an object that is the function itself, the set of ordered pairs.

If, as in the case of derivative, the researcher wants to emphasize other processes involved in the calculation of the function value at a point, then more layers are added. Note that for the concept of derivative we might have added an additional layer prior to the ratio layer to denote the calculation of the differences, or increments, involved in the numerator and denominator of the derivative ratio (Jacobs & Zandieh, 2000).

Understanding the Concept of Derivative

In preparation for examining specific student quotes, this section describes the three process-object layers—ratio, limit, and function—of the framework in several representations in more detail than was done previously. As an introduction we will note the framework in terms of rate of change and Leibniz notation. Then we will follow with a more detailed symbolic notation as well as the details of the framework in terms of slope and velocity.

Rate of change. Consider that a difference quotient can be used to measure the average rate of change of the dependent variable with respect to change in the independent variable. The calculation of this ratio of differences is a process. We might represent this ratio in Leibniz notation as $\frac{\Delta y}{\Delta x}$.

The consolidated process, the average rate, may be used as an object in the second process, the limiting process. The limiting process consists of analyzing a sequence of average rates of change as the difference in the denominator of the ratios goes to zero. We can represent this in Leibniz notation by $\lim_{\Delta x \to 0} \frac{\Delta y}{\Delta x}$. The limiting process is consolidated to an instantaneous rate of change, represented by $\frac{dy}{dx}$.

This consolidated process, the instantaneous rate of change at each input value, is used as an object in the construction of the derivative function. The function value at each point has already been described by the limiting process. The function process we will stress here is the covariation of the input values with the output or instantaneous rate of change values. The function as a process and object is not easily represented by the Leibniz notation.

Symbolic. The first layer, the symbolic difference quotient, is often written as

$$\frac{f(x) - f(x_0)}{x - x_0} \quad \text{or} \quad \frac{f(x_0 + h) - f(x_0)}{h}$$

where x and x_0 are values in the domain of the function and h is the distance between x and x_0. This quotient may be thought of as a process or an object. As an object it is acted on in the second layer by a limiting process:

$$f'(x_0) = \lim_{x \to x_0} \frac{f(x) - f(x_0)}{x - x_0} \quad \text{or} \quad f'(x_0) = \lim_{h \to 0} \frac{f(x_0 + h) - f(x_0)}{h}.$$

These expressions give the value of the derivative function at x_0. This limiting process must be thought of as consolidated and repeated for every value in the domain of f' to progress to the third layer, the derivative function. The formula

$$f'(x) = \lim_{h \to 0} \frac{f(x + h) - f(x)}{h}$$

is now considered as one that applies to a domain of possibly infinitely many values of x.

The object in the second layer, the derivative at a point, and the object in the third layer, the derivative function, are conceptually very different. However, symbolically the difference is extremely subtle (the use of a subscript) and easily glossed over by students.

Slope: A graphic interpretation of derivative. Now we examine these three layers in a graphic representation. The first layer is the slope, specifically the rise over run, of a line connecting two points on the curve described by the graph of the function in question. The line connecting these two points is often referred to as a secant line. The process is the calculation of the rise over run. The object is the slope itself. For the purpose of building the layers we will think of one of the two points as being fixed; however, this layer may exist without reference to further layers as the slope between any two points.

For the second layer we look at the limit of a sequence of slope values. These slope values may be thought of as the slopes of secant lines all going through one common point. We see that as the second point on the curve which determines the secant line approaches the common point, the secant lines approach the tangent line at the common point. This is the limiting process. The object is the slope of the tangent line at that point.

Another way to view this limiting process graphically is to think of zooming in on the curve at a point of interest. At each or any step in the process of zooming in one may find the slope of a line between two points close to the point being zoomed in on. As one zooms in, one finds the slope determined by two points closer and closer to the point of interest. The object here is again the slope of the tangent line at the point of interest, but in this case we think of the tangent line as becoming a better and better local approximation to the curve itself under magnification as opposed to a sequence of secant lines rotating to the position of the tangent.

In either case the third layer is the same. The limiting process, viewed through magnification or rotating secant lines, becomes consolidated so that this process may be thought to have occurred for every point on the curve of the original function's graph. The function process is the notion of running through every point on the original curve and extracting the instantaneous slope value. The object is the derivative function, whose graph can also be thought of as a curve itself.

Motion: Interpreting the derivative function as a "speedometer." The context of motion actually gives us two models for derivative: velocity if the function is displacement and acceleration if the function is velocity. I will concentrate on the former as a physical representation in which to examine the three layers.

The first layer process is the ratio of the change in distance (displacement) to the change in time. The first layer object is this average velocity. The second layer process consists of looking at average velocities over shorter and shorter intervals of time. This limiting process culminates in an instantaneous velocity. The third layer process consists of imagining the consolidated limiting process occurring for every moment in time so that the final result is a function that has associated with each moment in time an instantaneous velocity. The derivative function serves as a speedometer (or more accurately, a velocimeter) in this context.

Other interpretations or contexts are, of course, possible. In each case a parallel three-layered structure could be described. The totality of the three-layered structure paralleled in different environments and contexts and the links between these environments and contexts will be what we refer to as the structure of the concept of derivative.

Methods and Data

Much of the data for this discussion comes from the work of Zandieh (1997). This study followed nine high school calculus students. The evolution of each student's understanding of the concept of derivative was treated as an individual case study. Any aggregate data is meant to show trends and is not statistically significant. However, these students provide counterexamples to some hypotheses one might make about the uniform development of understanding in students.

The nine-member Advanced Placement calculus class included six National Merit Finalists. All students had taken and were taking other Advanced Placement courses. Considering the make-up of the class, difficulties these students had are likely to be shared by other less well-qualified students. All but one of the students had been in the same math and science classes for ninth through eleventh grade. Hence, differences in the development of understanding of these students are likely due to their individual personalities since their classroom experiences had been so similar.

The data collected included field notes from the researcher written during and after each class session attended (75 classes over the course of nine months), student written work on exams, and a series of five interviews spread across the nine months of the calculus course. This paper will concentrate on data from the interviews.

Each of the interviews allowed the student to provide information about her or his understanding of derivative by answering open-ended questions and solving problems involving the concept of derivative. The first, second, and fifth interviews were the most comprehensive. Each of these interviews asked students a wide variety of questions to try to elicit each part of their understanding of the concept of derivative.

The first interview was conducted in the beginning of the school year after the students had reviewed functions and limits but had not yet reviewed the theoretical and applied notions of derivative that they had studied in the last few months of their junior year precalculus course. Student understanding of derivative at this point was assumed to be based on what they had learned the previous year and any reviewing the students had done over the summer. (Students were asked to solve some calculus problems over the summer, but not all did.) The second interview occurred after the students had just finished studying derivative in their senior year class. The fifth interview occurred at the end of the school year after the students had taken the AP Calculus exam.

Using the Framework for Describing Conceptions of Derivative

This section begins by describing how one may think of a student's understanding of the concept of derivative using the framework of this paper. The section continues by revealing a diagrammatic way to illustrate, in terms of the framework, the understandings that a student displays. Several examples will be given consisting of quotes taken from the calculus students of this study and corresponding grids (enhanced versions of Figure 1) that allow us to graphically contrast various student responses in terms of the derivative framework. This will give the reader a taste of how one may use the framework to aid in studies of individual understanding and comparative individual student understandings (the first and second items on the list at the beginning of this paper).

The derivative framework of this paper is meant to describe what the mathematical community means by the concept of derivative at the first-year calculus level. The same structures may also be used to describe the parts of an individual's concept image that coincide with the mathematical community's concept of derivative. Missing in either case are the rules for taking derivatives symbolically such as the power, product, and chain rules. These rules by-pass the limit of the difference quotient definition of derivative or any other description of the layers of the concept of derivative. We note that an individual's knowledge of these rules can be gained independently of the understanding of derivative as a three-layered structure that may be characterized in many different contexts or representations.

Each individual's understanding may be described in relation to the three-layered concept of derivative. Such a description will highlight the following:

1. What layers of the structure are available to the person?
2. In what representations or contexts are these layers available?
3. Does the person understand both the process and object nature of each layer?
4. Can the person coordinate all three layers simultaneously?
5. Does the person recognize the parallel nature of each of the layers in the symbolic, graphic, kinematic, and other settings?
6. Does the person prefer to use a particular representation or context as a model or prototype for the derivative concept when no representation or context is specified?
7. Does the person's understanding of derivative include ideas that do not fall into the three-layered structure of the concept of derivative? Does the student's concept image include understandings considered incorrect by the mathematical community?

It is not necessary for a person to think of the layers as "layers" or to name the dual nature of the layers with the terms "process" and "object." This terminology is that of the researcher. However, we suggest that a mathematician (or any person with a robust understanding of derivative) is aware of each of the layers even without naming them as such and is able to recreate that structure in any relevant context.

Part of an individual's understanding may be noted within a matrix or grid like Figure 1. The grids illustrate only whether a student mentions a context or layer or a process or pseudo-object aspect of the concept of derivative in answer to one or more interview questions. The grids do not illustrate the connections a student may or may not make, nor do they document misstatements or understandings outside of those considered correct by the mathematical community. Also, it is taken as given that the student may have other understandings that are not revealed by the particular interview questions asked.

In each grid, a circle in a box denotes that a student has demonstrated (at least) a pseudo-object understanding of the row and column which intersect in that box. If the circle is shaded, then a student has also demonstrated an understanding of the process involved in the layer and context which intersect in that box.

The first quotes will contrast two students' answers to the question, "What is a derivative?" This question was asked near the beginning of each interview immediately after a few warm-up questions about the student' thoughts on the book or graphing calculators or the difficulty of a recent test. Frances (all student names are pseudonyms) had the following response during the first interview:

MZ: What is a derivative?
Frances: I don't know. Well, I know it's the slope of the tangent line,
          you know to the... Like if you're taking the derivative of x^2
          so it would be $2x$.
MZ: OK.
Frances: I mean, I know how to take the derivative. The slope of the
          tangent line at a certain point.

Figure 2 illustrates that Frances describes derivative as slope by placing an open circle in the slope column and the ratio row. This circle would have been closed if Frances had described a ratio for slope such as rise over run. The open circle in the slope column and the limit row in Figure 2, indicates that Frances said the derivative was at a "certain point." Her specificity indicates that she does not mean an average slope or a slope between two points. This circle would be closed if Frances described the limiting processes of average slopes approaching the slope at one point. Note that Frances's mention of an example of the power rule is not indicated in the chart and would need to be noted separately

Now consider Brad's response at the beginning of the first interview.

MZ: Do you remember what a derivative is?
Brad: [laughs] Isn't it like the opposite of a—no, what's the opposite
       of a limit or something. I don't know. I have no idea how to
       say it. It's been a while. [pause] Isn't it like all those formulas
       for velocity and acceleration and something like that?

Note that Brad mentions velocity but does not specify the ratio associated with velocity. Therefore Figure 2 notes an open circle in the velocity column and the ratio row. A closed circle would have indicated that Brad described a ratio for velocity such as distance over time. Figure 2 also indicates an open circle in the symbolic column and the limit row. This is because Brad mentioned that a limit was involved. The circle would have been closed if Brad had indicated the symbolic details of the limiting process such as $h \rightarrow 0$.

Figure 2 and Figure 3 graphically demonstrate the contrast in Frances and Brad's answers to the "What is a derivative?" question. They also demonstrate an illustration of pseudo-object statements for four of the boxes.

Contexts

Process-object layer	Graphical	Verbal	Paradigmatic Physical	Symbolic	Other
	Slope	Rate	Velocity	Difference Quotient	
Ratio	◯				
Limit	◯				
Function					

FIGURE 2. Frances, Interview 1, answer to "What is a derivative?"

Contexts

Process-object layer	Graphical	Verbal	Paradigmatic Physical	Symbolic	Other
	Slope	Rate	Velocity	Difference Quotient	
Ratio			◯		
Limit				◯	
Function					

FIGURE 3. Brad, Interview 1, answer to "What is a derivative?"

The next example, from Ingrid's second interview, demonstrates how the symbolic column may fill up without students having an understanding of the symbols beyond the context of the symbolic calculations (see also Figure 4). (Note that Ingrid's responses in this interview segment are not meant as a direct contrast to Brad's and Frances's transcripts above since Ingrid is not only asked "What is a derivative?" but also whether anything else comes to mind.)

MZ: What is a derivative?
Ingrid: The slope of a line, the slope of a tangent line, the slope of a line tangent to a function at a certain point.
MZ: Does anything else come to your mind for what a derivative is?
Ingrid: The limit—that little picture that [Mr. Forrest] wants us to have in our mind of the graph. [pause] The limit as the change in x approaches 0—you know, that one equation.

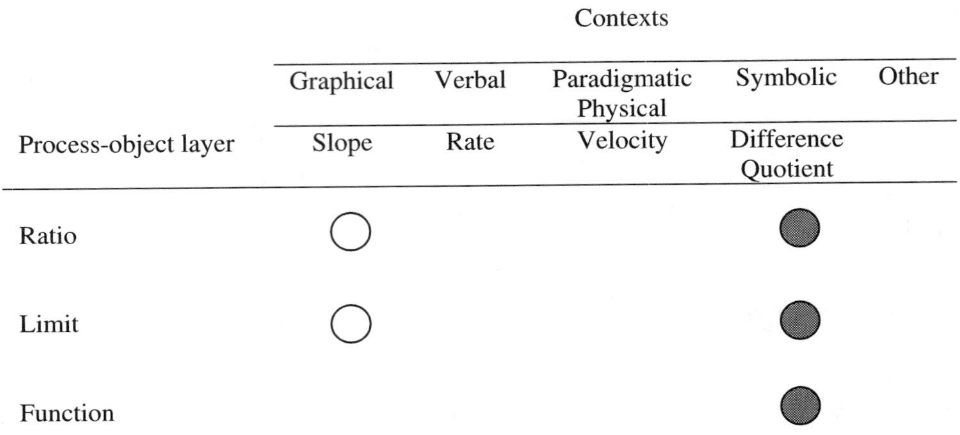

Contexts

Process-object layer	Graphical	Verbal	Paradigmatic Physical	Symbolic	Other
	Slope	Rate	Velocity	Difference Quotient	
Ratio	○			●	
Limit	○			●	
Function				●	

FIGURE 4. Ingrid, Interview 2, quote illustration.

Ingrid: [writes: $\lim_{\Delta x \to 0} \frac{f(x+\Delta x)-f(x)}{\Delta x}$]

MZ: And do you happen to remember that picture that goes with it?

Ingrid: I'm thinking about it. I don't know. Maybe it'll come to me later.

MZ: Do you happen to remember how these parts fit in, the f of—

Ingrid: Well, you have a function and this is like—whatever's over here would be x plus change in x and then $f(x)$, subtract those and divide by—I don't know.

Note that Ingrid describes the calculations of subtraction and division involved in the symbolic expression she has written but cannot remember how these relate to a graphical interpretation of derivative, a connection that has been described by her teacher, Mr. Forrest, several times in class. The closed circles in Figure 4 in the symbolic column indicate Ingrid's complete statement of the symbolic definition for derivative and her understanding that this symbolic expression involves the process of taking the limit of a division expression involving a subtraction of functions. In other words, she understands the symbolic procedures underlying those symbols even though she does not understand what those symbols represent in the graphical context.

The open circles in Figure 4 in the slope column indicate her statement of slope at a point as they did for Frances.

Now that we have seen a few examples of quotes illustrated in the diagrams we can consider how we might compare two students' responses across the totality of one interview. Figure 5 is a summary of Frances's first interview. In answer to a variety of questions about derivative in the first interview Frances discusses slope at a point, instantaneous rate of change, and velocity without explaining the details of any of these layers. She is also unable to state the formal definition of derivative even though she knows that there is a ratio and a limit involved. She guesses that the derivative is a function because of its symbolic form, i.e., "because when you take the derivative you're just like decreasing the exponent for it."

Contexts

Process-object layer	Graphical	Verbal	Paradigmatic Physical	Symbolic	Other
	Slope	Rate	Velocity	Difference Quotient	
Ratio	◯	◯	◯	◯	
Limit	◯	◯		◯	
Function				◯	

FIGURE 5. Frances, Summary of chart. Interview 1.

Contexts

Process-object layer	Graphical	Verbal	Paradigmatic Physical	Symbolic	Other
	Slope	Rate	Velocity	Difference Quotient	
Ratio	●	◯	◯	●	
Limit	●	◯	◯	●	
Function	●			●	

FIGURE 6. Alex, Summary of chart, Interview 1.

We can contrast Frances's first interview to Alex's first interview. Figure 6 is a summary chart for Alex's first interview which illustrates Alex's relatively strong understanding. Note that, unlike Frances, Alex has several circles filled in. It is beyond the scope of this paper to diagram each of Alex's statements, but we can take the following quote as an example of how Alex filled in the limit circle in the graphic column, as well as open circles for a mention of slope and average and instantaneous rate of change.

> MZ: OK. How would you explain what a derivative is to someone who's like an AB student or a precalc student that hasn't studied it yet?
>
> Alex: When you have two points on any graph and you get a line through them, and the slope of that line would be the average rate of change between those two points. If the derivative is

Contexts

Process-object layer	Graphical	Verbal	Paradigmatic Physical	Symbolic	Other
	Slope	Rate	Velocity	Difference Quotient	
Ratio	●	●	●	●	
Limit	●	○	○	●	
Function	●			●	

FIGURE 7. Frances, Summary of chart, Interview 5.

instantaneous rate of change because those two points are really close to each other. I mean they are so close that you can even have just one point. So what you do is like—OK, you can leave one of these points and you can move the other one closer and closer to the other one, the one you left. And that will give you more accurate rate of change between—if you take two points far apart that will give you just average rate of change, throughout the whole function. But as you start moving closer to that point you're going to get more accurate average rate of change between those two points because they are really close. And when you get those points to be so close together that you could even consider them to be one point, that's going to be instantaneous rate of change of that function at that point.

Alex: So that's just the slope of the function between the two points.

By the end of the school year, Frances's understanding had deepened and it showed in her responses to the fifth interview (see Figure 7).

As an example of this deeper understanding, consider that Frances is able to describe the relationship between the ratio in the formal definition and the notion that derivative is the slope.

MZ: OK. This is what I'm kind of more interested in, how this statement, instantaneous slope at a point, relates to this formal definition.

Frances: Because if you say, as h gets really small, then it's kind of like you're taking the slope of the secant line, but the slope of the secant line—the end points of the secant line are getting closer and closer together. So it's almost like you're taking the slope at that exact point.

MZ: OK. So how do the different little symbols here fit into that thing you just described?

Frances: This is like $f(b)$—if this was x, then that would be $x + h$, if that was a really small distance [sketches a pair of axes and marks x and $x + h$ on the horizontal axis].

MZ: OK.

Frances: So then this would be $f(b) - f(a)$ over $b - a$. But $x + h - x$, you just get h [writes: $x + h - x$ and then crosses out the x's].

MZ: OK. And then that describes the slope right?

Frances: Yeah.

MZ: Then the limit does what you were saying about—

Frances: Because you're making it a tiny distance between x and $x + h$.

MZ: OK. What does this formal definition have to do with this statement that you just gave me, how fast the function is changing?

Frances: [short pause] I don't know. 'Cause if the derivative is 0, then— If you have a line like—if you have a function like that [draws a flat curve], the slope of that is small because it's not changing very fast. But if you have a function like that [draws a steep curve], then the derivative is big because it's changing a lot.

These quotes and illustrative grids are meant to serve as preliminary examples of ways to use the framework to compare individual student comments. The next section considers the issue of a hierarchy in the framework.

Is There a Hierarchy Involved in Understanding the Concept of Derivative?

Since we are considering understanding the derivative in terms of a matrix where each layer or process-object pair may be noted in many different contexts or representations, we must consider both whether the representations need to be learned in a specific order and then whether the layers of each representation are learned in a specific order.

Representation hierarchy. Is there a hierarchy in learning representations for the concept of derivative? Are there representations or contexts for the concept of derivative that are more easily understood or preferred by students? Does this preference change over time?

Initial indications from the case studies of nine calculus students suggest that beginning students' preferences are not uniform, but that they become more similar as students' knowledge of the concept increases.

Using the information about the order and frequency with which a student mentions the various representations, we may hypothesize the preferred representation for each student at the time of the interview. The preferred representation for three interviews is given in Table 1.

We can see from the table that the nine students held a wide variety of representational preferences during the first interview, although slope is the most frequently mentioned interpretation with six students mentioning it most often. Only three students have rate as a preferred interpretation, and in each case it is second to slope. By the fifth interview, all but one student has rate as her or his first or second most prominent interpretation. Slope and rate (or rate then slope) become the most mentioned interpretations for all students except Helen who still mentions the formal definition as her second most prominent interpretation.

TABLE 1. Student Preferences in Interpreting the Concept of Derivative

Student	Interview 1	Interview 2	Interview 5
Alex	SR	SR	SR
Brad	V	RS	SR
Carl	TV	S	RS
Derick	SR	SR	RS
Ernest	SV	SV	SR
Frances	ST	SR	SR
Grace	SR	S	RS
Helen	F	SF	SF
Ingrid	S	SF	RS

S = slope; R = rate or rate of change; V = velocity or acceleration;
T = taking the derivative symbolically using rules
F = formal definition or (limit of) the difference quotient

From these data it appears that it is possible for students to have very different understandings of derivative even when this understanding has developed through the same class environment and homework assignments. From each of the starting points demonstrated in the first interview, the students were then able to build a more comprehensive understanding. This understanding included most of the primary representations listed in the table with a focus on using rate and slope as descriptors.

Although mathematicians may think of the formal definition as a primary means to interpret derivative in any relevant context, this was not the case for the students in this study. By the fifth interview, each of the best students in the class (four of the nine) could recreate the formal definition or a similar ratio such as that used in the Mean Value Theorem from his or her knowledge of derivative in terms of slope. The notion of slope was central and the formalization was peripheral to slope for these students. For each of the other students the formal definition was even less well integrated into his or her understanding. Three students had the formal definition memorized but could not accurately explain what it represented in any other context. Two students refused to learn the formal definition, specifically citing its irrelevance to their understanding (Zandieh, 1998).

Another notion not in the preferred list for students in the fifth interview was velocity. However, velocity still played an important role as a paradigmatic example for some students. For example, even though rate of change was a primary descriptor for both Carl and Derick, each needed to revert to the use of the more specific velocity model when interpreting $f'(3) = 4$ where f is a function that gives the temperature at a time given in hours since noon. Before comparing with velocity, both students thought that the temperature changed by 4 degrees instantaneously at 3 p.m. After a comparison with velocity each of them realized that the temperature was changing at a rate of 4 degrees per hour at 3 p.m.

Process-object hierarchy. Is there a hierarchy in learning process-object pairs? Must certain process-object pairs or layers of derivative understanding be learned before other layers? Must a process understanding of a layer come before an object understanding?

As discussed earlier in this paper, the process-object duality developed by Sfard assumes that the process conception occurs first and then is followed by a structural conception which is a consolidation or even a reification of the process layer. This would seem to assume that there is a hierarchy in learning processes since certain processes must be consolidated before they may be used by "higher" level processes. However, as Sfard states, this ordering does not eliminate the possibility that one may have a pseudostructural conception, a conception of an "object" for which the underlying process is unknown. This pseudostructural conception allows for an "object," which I call a pseudo-object, to be learned before a process, and for this pseudo-object to be used in other processes. In this way the three layers of the concept of derivative need not be thought of as hierarchical. Examples for each of the three layers show this at work.

The pseudo-object for ratio. For the ratio layer, a pseudo-object understanding would be to think of slope simply as steepness or velocity as a speed, a magnitude for how fast something is moving, without thinking about a ratio of rise over run or distance over time. Although the calculus students in this study by Zandieh (1997) were aware of the ratios underlying slope and velocity, they sometimes used only information about steepness or speed. More telling is the fact that younger students often, perhaps always, are aware of steepness or speed or rate before they construct an understanding of these notions in terms of ratios. Two studies with younger students (Confrey & Smith, 1994; Thompson, 1995a) demonstrate this idea.

Confrey and Smith (1994) note that children can distinguish a variation across rates of change. "A child recognizes a change of speed in an automobile; a gust of wind is heard 'blowing harder,' both implicit rate concepts" (p. 156). Confrey and Smith also suggest, "that students form a direct connection between slope and rate of change which is not mediated by numerical analysis. This . . . rate concept is more holistic than the analytic 'unit per unit,' and connects the experiential basis of slope with rate variation in contextual problems" (p. 157).

Thompson (1995a) examined the concept of rate held by a fifth-grader called JJ. Thompson met with JJ eight times for approximately 55 minutes each over a three week period. JJ initially thought of speed as a distance. When asked to determine the amount of time it would take to go 100 feet at 30 ft/sec, she reasoned based on the number of "speed-lengths" in 100 feet to find the time $3\frac{1}{3}$. JJ explained, "because there are three 30's in 100, and 10 is $\frac{1}{3}$ of 30, so $3\frac{1}{3}$ seconds." Thompson (1995a) noted that, "for JJ it is the case that *speed is a distance* (how far in one second) and *time is a ratio* (how many speed-lengths in some distance)" (Thompson's emphasis, p. 198). JJ appeared to measure the distance in speed-lengths to obtain an associated time. When given the distance and the time and asked for the speed, JJ tried to guess the correct speed and then checked to see if it gave the desired time. For JJ time was a ratio of the distance over the speed. It was not until later in the teaching experiment that JJ was able to construct speed as a ratio of distance over time.

When JJ was able to think of speed as a constant ratio, something invariant over changes in distance and time, Thompson says she understood speed as a rate. "A rate as a reflectively abstracted constant ratio symbolizes that structure as a whole, but gives prominence to the constancy of the result of the multiplicative comparison" (Thompson, 1995a, p. 192). Expressing this in the language of Piaget so that it can be easily compared with the work of Breidenbach et al. (1992) on the understanding of function, Thompson calls rate an *interiorized ratio*. Even though Thompson speaks of reflective abstraction and interiorization, he does not use the terms "process" or "object." Based on what he has said, it is useful to describe the ratio as the process, the rate as the reflectively abstracted object, and the speed (as in speed-lengths) as a pseudostructural construct, an object with no internal process structure.

Relationships between the ratio and limit layers. For the limit layer, a pseudostructural conception consists of an understanding that the derivative is a value at a point, for example, knowledge that the derivative is the slope at a point or the velocity at a specific moment in time. For this type of understanding it is not necessary to know how to calculate a limit or think about a limiting process. It is also not necessary to understand the process involved in the ratio layer underlying this layer. For example, a student might read from a speedometer the velocity at a specific moment in time or estimate the steepness of a curve at a particular point on the graph without thinking about any processes involving limits or ratios. The student may be aware of the underlying processes and just not bother to evoke those understandings at the moment or the student might truly be unaware of the underlying processes. The students in this study often mentioned the slope at a point or instantaneous rate of change as descriptors for the concept of derivative without describing any other details. When asked to elaborate, the answers they gave usually still failed to describe the underlying processes involved in the ratio and limit layers. Here is an example from the beginning of the fourth interview.

MZ: What do people usually mean if they say [instantaneous rate of change]?

Grace: Like, if you're measuring something that's always changing, but you catch it like at an exact moment.

. . .

MZ: How about what the rate of change part means, more specifically?

Grace: The rate that something is changing. [laughs] Like how fast the thing is changing, and the derivative is just when you catch it at one point.

Note that Grace emphasizes the instantaneous, "at one point," nature of rate of change without describing a limiting process. She also describes a pseudo-object notion of rate as a feeling of "changing" without describing a ratio or division process.

An overview of the nine students' answers to the first interview questions indicates that only four of the students (Alex, Carl, Derick, and Ernest) mentioned a ratio outside of the context of stating the formal definition. Derick's explanation was in a detailed answer of how to explain derivative to another person. Alex mentioned a limiting process for slope (or steepness) values in his explanation for another person, but did not mention the ratio involved until he needed to recreate for the interviewer the formal definition that he did not have memorized. Carl stated the ratio in explaining that rate of change refers to examples like "dx over dt, the change in x over time." Unfortunately he goes on to claim that this expression is equivalent to Δx. Ernest mentions the ratio in explaining why a graph with a vertical tangent line does not have a derivative at that point. "It would be undefined cause you go up one and over none. 1 over 0 would be the slope."

In terms of a limiting process, Alex and Derick's explanations were nicely given as described above. Two other students described limiting processes that not only did not use the ratio involved in the first layer, but did not completely correctly describe a pseudo-object for the first layer either. Ernest explains at the beginning of the first interview that the derivative is "the slope of the tangent lines leading up to the point, leading up to the limit." Grace's mention of a limiting process comes in trying to explain how slope or rate of change is related to the formal definition. "If you have a graph and you pick two points and you want to find the change between the two of them, the closer these two points move together the more accurate it is. . . . All I'm going to say is that you have two points and the closer they move together the more accurate your slope is going to measure and the rate of change."

All students except Ingrid eventually stated the ratio for the first layer of the derivative concept. Ernest became less communicative and never explained the ratio after the first interview. By the final interview, the other six students had explained the ratio in at least two contexts. However, eliciting this information usually took the asking of specific questions such as a request to explain how other contexts are related to the formal definition, to explain the equation involved in the Mean Value Theorem, or to estimate the derivative from a table of function values.

Similarly, information about the limiting process was difficult to elicit from the students. Carl never described a limiting process; nor did Ernest beyond the small quote from the first interview above. Perhaps coincidentally these were the two students who never memorized or learned the formal definition of derivative. Ingrid mentioned limiting processes, but never with appropriate objects or pseudo-objects being acted on.

It is interesting how difficult it is to elicit student explanations of the ratio and limit processes. In part, this may indicate a lack of knowledge or appreciation of these processes on the part of the students. On the other hand, it is clear that considerable understanding of calculus and ability to solve problems in calculus may be accomplished without thinking about some of these processes. If students can reasonably correctly discuss the derivative concept and solve problems in calculus classes without calling on a knowledge of these processes, then it brings into question the role of these processes in a student's understanding of calculus. Students should understand the processes involved in the derivative concept and value their importance. Such understanding is certainly necessary for an appreciation of the formal definition of derivative. However, the fact that much of a calculus course can be taught without this level of detail, should allow us to teach some concepts of calculus, the intuitive concepts, to younger students.

The function layer. For the function layer, a pseudostructural conception is to think of a function as an expression or a graph without an awareness of the input-output pairs involved. Thompson (1994) found that upper-division college-level students had a primary image of function as "a short expression on the left and a long expression on the right, separated by an equals sign" (p. 268). This was not the only image of function available to the students, but it seems to be the one most easily elicited.

In my study the primary data on student images of function comes from asking the students whether a derivative is a function. In the first interview this was an impromptu follow up to asking, in the long list of terms, whether function was related to derivative. By the second and fifth interview the question was a standard part of the protocol. In some ways this is a trick question since the word derivative may be used to refer to both the derivative function and its value at a point. Only one student (Ingrid) had trouble with this aspect of the question. Her answer in the second interview was that the derivative is not a function because "you can't graph a slope. . . . Or you can't graph a limit. But then . . . like on a test, 'This is the graph of the derivative.' I guess it has to be. It could be a function." Ingrid's answer is interesting because of her focus on a function as something that can be graphed.

A more typical response was Frances's statement in the second interview that the derivative probably is a function because "if you do something to a function, I guess it stays one." Students who used this kind of reasoning often gave examples such as the fact that the derivative of a polynomial is a polynomial. However, the reasons for stating that a polynomial is a function were not clear. The students may have been noting something beyond the fact that the derivative was still a symbolic expression, since some students mentioned continuity or places where the original function would not be differentiable. Ingrid, in her fifth interview, mentioned the vertical line test.

Only two students (Alex, Derick) gave responses that referred to anything specific to a derivative function such as the nature of the ratio or limit layers. In the fifth interview Derick explains that "If you have a function, at a particular point there will be only one rate of change of that function at that point. You can't have two values for that." Similarly, Alex explains in the fifth interview that "you can't have two slope values for one x value." Note that each of these explanations requires only knowledge of the pseudo-objects for the first two layers and a process knowledge for the function layer.

Based on the examples from each of the three layers of the concept of derivative, if pseudostructural conceptions are allowed, students may learn the various layers and process or object parts of the concept of derivative in any order, without regard for a hierarchy.

Conclusion

This paper introduces the notion of describing student understanding of the concept of derivative in terms of three process-object pairs, each viewed in a variety of representations or contexts. A pseudo-object is defined as an intuitive understanding that does not involve an understanding of the process underlying the object. A person may use a pseudo-object because he or she does not know the underlying process or simply because there is no need to refer to the underlying process in the context in which the person is working. The use of pseudo-objects and multiple representations allows individuals to build initial understanding of the concept of derivative that are very different from each other. These understandings are like the understandings of the blind men who examine the elephant, each taking a different part of the elephant (tail, ear, side) as his interpretation of the whole. As each student develops a more complete understanding, their understandings become more similar. In this way, a student's understanding of derivative develops from a partial understanding, each student with a different collection of pieces of the puzzle, to a more complete understanding with perhaps only a few pieces remaining unplaced. There seems to be a wide variability in terms of which pieces of the puzzle that a student initially places in the puzzle frame. The study by Zandieh

(1997) shows considerable variability in the understandings of nine students who had studied mathematics and science together. More variability might be observed in a study of students using different textbooks and pedagogical approaches.

The challenge for educators is to give students experiences that will enable each of them to come to a complete understanding of all three process-object layers involved in the derivative concept. One difficulty is that many problems involving the concept of derivative can be solved using only pseudo-object understandings. This allows weaker students to survive a calculus course without focusing too much attention on underlying processes. A second issue is that students do not automatically connect an understanding of a process in one context with the same process in another context. Extra care must be taken to develop in students a recognition of the parallel nature of the process in each context. A student will not have a complete understanding of the concept of derivative if he or she cannot recognize and construct each of the three processes involved in understanding the derivative concept in any relevant context.

The process-object framework may be used not only to note whether a student has developed each of these aspects of understanding the concept of derivative, but it also may be used to note whether a textbook or a set of curriculum materials provides students opportunities to learn the underlying processes in multiple contexts. The framework may also be used to analyze classroom discourse for which aspects of the concept of derivative are involved in student reasoning and teacher questions and explanations. In these ways the framework is broader than a tool for describing individual understanding. The framework may allow us to form analyses that seek to determine the relationship between student learning experiences and student understanding.

The framework need not be restricted to use in the setting of a first year calculus course. In any setting that requires knowledge of the concept of derivative, for example in studying differential equations, the framework may be used to note which aspects of understanding the concept of derivative a student uses in his or her reasoning. In addition, process-object pairs may be used to describe student understanding of other concepts that have process-object structures. Likely candidates are the concepts of integral and differential equations.

References

Breidenbach, D., Dubinsky, E., Hawkes, J., & Nichols, D. (1992). Development of the process conception of function. *Educational Studies in Mathematics*, *23*, 247–285.

Confrey, J. & Smith, E. (1994). Exponential functions, rates of change, and the multiplicative unit. *Educational Studies in Mathematics*, *26*, 135–164.

Dick, T. P. & Patton, C. M. (1992). *Calculus*. Boston: PWS.

Dick, T. P. & Patton, C. M. (1994). *Calculus of a single variable*. Boston: PWS.

Dreyfus, T. & Vinner, S. (1989). Images and definitions for the concept of function. *Journal for Research in Mathematics Education, 20*, 4, 356–366.

Dubinsky E. (1991) The constructive aspects of reflective abstraction in advanced mathematics. In L. P. Steffe (Ed.), *Epistemological foundations of mathematical experiences*. New York: Springer-Verlag.

Dubinsky, E. & Harel, G. (1992). The nature of the process conception of function. In G. Harel & E. Dubinsky (Eds.), *The concept of function: Aspects of epistemology and pedagogy*, MAA notes no. 25, pp. 85-106). Washington, DC: Mathematical Association of America.

Hart, D. K. (1991). Building concept images: Supercalculators and students' use of multiple representations in calculus. Doctoral dissertation, Oregon State University. *Dissertation Abstracts International, 52A*, 4254.

Hiebert, J. & Carpenter, T. (1992). Learning and teaching with understanding. In D. Grouws (Ed.), *Handbook of research on mathematics teaching and learning* (pp. 65-97). New York: Macmillan.

Hughes-Hallett, D. & Gleason, A. M. (1994). *Calculus*. New York: John Wiley & Sons, Inc.

Jacobs, A. & Zandieh, M. (2000). *The Case of Brad: Clarity about rate of change; confusion about change.* Unpublished manuscript, Arizona State University.

Kaput, J. J. (1997). Rethinking calculus: Learning and thinking. *American Mathematical Monthly, 104*(8), 731–737.

Krussel, C. E. (1995). Visualization and reification of concepts in advanced mathematical thinking Doctoral dissertation, Oregon State University, 1995. *Dissertation Abstracts International, 56A*, 127.

Lakoff, G. (1987). *Women, fire and dangerous things*. Chicago: The University of Chicago Press.

Monk, G. (1992). Students' understanding of a function given by a physical model. In G. Harel & E. Dubinsky (Eds.), *The concept of function: Aspects of epistemology and pedagogy* (MAA Notes no. 25, pp. 175–193). Washington, DC: Mathematical Association of America.

Monk, G. & Nemirovsky, R. (1995). The case of Dan: Construction of a functional situation through visual attributes. In E. Dubinsky, A. H. Schoenfeld, & J. Kaput (Eds.), *Research in collegiate mathematics education I* (pp. 139–168). Providence, RI: American Mathematical Society.

Ostebee, A. & Zorn, P. (1997). *Calculus: From graphical, numerical, and symbolic points of view.* Fort Worth: Saunders College Publishing.

Sfard, A. (1991). On the dual nature of mathematical conceptions: Reflections on processes and objects as different sides of the same coin. *Educational Studies in Mathematics, 22*, 1–36.

Sfard, A. (1992). Operational origins of mathematical objects and the quandary of reification—The case of function. In G. Harel & E. Dubinsky (Eds.), *The concept of function: Aspects of epistemology and pedagogy* (MAA notes, no. 25, pp. 59–84) Washington, DC: Mathematical Association of America.

Swokowski, E. W. (1988). *Calculus with analytic geometry* (4th ed.). Boston, MA: PWS-Kent.

Tall, D. O. (1986). *Building and testing a cognitive approach to the calculus using interactive computer graphics.* Unpublished doctoral dissertation, University of Warwick, Coventry, England.

Tall, D. O. (1991). *A graphic approach to calculus* [computer program]. Pleasantville, NY: SunBurst Communications, Inc.

Tall, D. O. & Vinner, S. (1981). Concept image and concept definition in mathematics with particular reference to limits and continuity. *Educational Studies in Mathematics, 12*, 151–169.

Thomas, G. B. & Finney, R. L. (1992). *Calculus and analytic geometry* (8th ed.). Reading, MA: Addison-Wesley.

Thompson, P. W. (1994). Images of rate and operational understanding of the fundamental theorem of calculus. *Educational Studies in Mathematics, 26*, 229–274.

Thompson, P. W. (1995a). The development of the concept of speed and its relationship to concepts of rate. In G. Harel & J. Confrey (Eds.), *The development of multiplicative reasoning in the learning of mathematics* (pp. 179–234). Albany, NY: SUNY Press.

Thompson, P. W. (1995b). Students, functions, and the undergraduate curriculum. In E. Dubinsky, A. H. Schoenfeld, & J. Kaput (Eds.), *Research in collegiate mathematics education I* (pp. 21–44). Providence, RI: American Mathematical Society.

Vinner, S. & Dreyfus, T. (1989). Images and definitions for the concept of function. *Journal for Research in Mathematics Education, 20*, 356–366.

Zandieh, M. (1997). *The evolution of student understanding of the concept of derivative.* Unpublished doctoral dissertation, Oregon State University.

Zandieh, M. (1998). The role of a formal definition in nine students' concept image of derivative. In S. Berenson, K. Dawkins, M. Blanton, W. Coulombe, J. Klob, K. Norwood, & L. Stiff (Eds.), *Proceedings of the 20th Annual Meeting of the North American Chapter of the International Group for the Psychology of Mathematics Education* (pp. 136–141). Columbus OH: ERIC Clearinghouse for Science, Mathematics, and Environmental Education.

DEPARTMENT OF MATHEMATICS, ARIZONA STATE UNIVERSITY, TEMPE, AZ 85287-1804
E-mail address: zandieh@asu.edu

CBMS Issues in Mathematics Education
Volume **8**, 2000

Why Can't Calculus Students Access their Knowledge to Solve Non-routine Problems?

Annie Selden, John Selden, Shandy Hauk, and Alice Mason

ABSTRACT. In two previous studies we investigated the abilities of students just finishing their first year of a traditionally taught calculus sequence to solve non-routine differential calculus problems. This paper reports on a similar study, using the same non-routine calculus problems, with students who had completed one and one-half years of traditional calculus and were in the midst of an ordinary differential equations course. More than half of these students were unable to solve even one problem and more than a third made no substantial progress toward any solution. A test of associated algebra and calculus skills indicated that many of the students were familiar with the key calculus concepts for solving the non-routine problems; nonetheless, students often used sophisticated algebraic methods rather than calculus in approaching the non-routine problems. We suggest a possible explanation. These students may have had too few *tentative solution starts* in their *problem situation images* to help prime recall of the associated factual knowledge. We also discuss the importance of this for teaching.

1. Introduction

1.1. Background. Two previous studies demonstrated that students with C's as well as those with A's or B's in a traditional first calculus course had very limited success in solving non-routine problems (Selden, Mason, & Selden, 1989; Selden, Selden, & Mason, 1994). Further, the second study showed that many of these students were unable to solve non-routine problems for which they appeared to have an adequate knowledge base. This raised the question of whether more experienced students, those towards the end of a traditional calculus/differential equations sequence, would have more success; in particular, would they be better able to access and use their knowledge in solving non-routine problems? Folklore has it that one only really learns material from a mathematics course in subsequent courses. The results reported here in part support and in part controvert this notion. As will be discussed, the differential equations students in this study often used algebraic methods (first introduced to them several years before their participation in the study) in preference to those of calculus courses taken more recently. These

---

We would like to acknowledge partial support from Chapman University, the ExxonMobil Education Foundation, Tennessee Technological University, and the National Science Foundation (under Grant #DGE-9906517).

©2000 American Mathematical Society

students, who had more experience with calculus than those in the first two studies, appealed to sophisticated arithmetic and algebraic arguments more frequently than students in the earlier studies. Although somewhat more accomplished in their problem-solving ability, slightly more than half of them still failed to solve a single non-routine problem, despite many having an apparently adequate knowledge base.

As in the previous two studies, what we are calling a non-routine or novel problem is simply called a problem, as opposed to an exercise, in problem-solving studies (Schoenfeld, 1985). A *problem* can be seen as comprised of two parts: a task and a solver. The solver comes equipped with information and skills and is confronted with a cognitively non-trivial task; that is, the solver does not already know a method of solution. Seen from this perspective, a problem cannot be solved twice by the same person, nor is a problem independent of the solver's background. In traditional calculus courses most tasks fall more readily into the category of exercise than problem. However, experienced teachers can often predict that particular tasks will be problems for most students in a particular course, and tasks that appear to differ only slightly from traditional textbook exercises can become problems in this sense.

1.2. Overview of the Paper and Related Literature. In Section 2 we describe the setting and subjects—differential equations students who had taken a traditional calculus course obtaining grades of A, B, or C, at least one fourth of whom went on to obtain master's degrees and one a Ph.D. We present the two tests—one with five moderately non-routine differential calculus problems, administered first, and a subsequent ten-question routine test of corresponding algebra and calculus skills. Section 3 contains a comparison of these students' performance on the two tests and introduces the notions of *full, substantial,* and *insubstantial factual knowledge.* In Section 4 we provide detailed information on the students' favored solution methods and compare these with what was observed in our previous two studies (Selden et al., 1989, 1994). Although the differential equations students were slightly more inclined to use calculus than students in the previous studies, they did so on only 39% of their solution attempts, preferring a combination of guessing, trial-and-error, arithmetic techniques, and algebra.

In Section 5 we analyze our results and suggest that students only slowly come to use their factual knowledge of calculus, or other mathematics, flexibly. Somewhat similar observations have also been made by Carlson (1998), Stacey and MacGregor (1997), and Dorier, Pian, Robert, and Rogalski (1998). While our differential equations students were quite ready to employ algebraic techniques, they were much less inclined, or able, to use calculus effectively. We introduce a non-routineness scale for problems. Although non-routineness is only one aspect of problems, the ability to solve moderately non-routine problems is often seen as a hallmark of deep understanding of a the material in a course. We ask why our students could not solve our non-routine problems and conjecture that they lacked a kind of knowledge, which when brought to mind produces what Mason and Spence (1999) have referred to as "knowing-to act in the moment." In describing this additional kind of knowledge we build on Tall and Vinner's (1981) idea of concept image. We introduce the notion of *problem situation image,* a mental structure associated with problem situations which may contain, amongst other things, *tentative solution starts,* i.e., various ways of beginning to solve a problem.

Finally some teaching implications of this conjecture are discussed in Section 6. We restrict our attention to moderately non-routine problem-solving, referring readers to the work of Schoenfeld and others (Arcavi, Kessel, Meira, & Smith, 1998; Santos-Trigo, 1998; Schoenfeld, 1985) for a general treatment of problem-solving. We suggest that the construction of a problem situation, and its image, depends on student activities (experiences) analogous to the construction of a concept, as described by Breidenbach, Dubinsky, Hawks, and Nichols (1992) and Sfard (1991).

The central question we ask is: Why could our students not access their knowledge of calculus when needed? This question, of how one comes to know to act in a given situation, has been largely neglected in the mathematics education research literature. We offer a conjecture: what is missing is a kind of knowledge—tentative solution starts, ways of beginning, that are part of an individual's problem situation image. Such knowledge arises from a habit of mind, that of reflecting on various possible starting points. Yet, how does such knowledge come to mind in the moment? This is a question of how one brings information from long-term memory (one's knowledge base) into short-term memory (see Baddeley, 1995) and makes it conscious (see, for example, James, 1910 or Mangan, 1993). We suggest that recognizing a problem situation partly activates the information in its image which then primes the recall of factual knowledge.

2. The Course, the Students, and the Tests

2.1. The Calculus/Differential Equations Sequence. The setting is a southeastern comprehensive state university having an engineering emphasis and enrolling about 7500 students—the same university of the earlier studies of C and A/B first-term calculus students (Selden, Mason, & Selden, 1989; Selden, Selden, & Mason, 1994). The annual average ACT composite score of entering freshman is slightly above the national average for high school graduates, e.g., in the year the data were collected the university average was 21.1, compared to the national average of 20.6.

A large majority of students who take the calculus/differential equations sequence at this university are engineering majors. The rest are usually science or mathematics majors. A separate, less rigorous, calculus course is offered for students majoring in other disciplines.

Until Fall Semester 1989, the calculus/differential equations sequence was offered as a five-quarter sequence of five-hour courses. Since then, it has been offered as a four-semester sequence. Under both the quarter and the semester systems it has been taught, with very few exceptions, by traditional methods with limited, if any, use of technology and with standard texts (Swokowski, 1983; Berkey, 1988; or Stewart, 1987 for calculus and Zill, 1986 for differential equations). Class size was usually limited to 35–40 students, but some sections were considerably smaller and a few considerably larger. All but three sections were taught by regular, full-time faculty of all ranks; the three exceptions were taught by a part-time associate professor who held a Ph.D. in mathematics. All instructors taught according to their usual methods and handled their own examinations and grades.

2.2. The Students. The pool of 128 differential equations students considered in this study came from all five sections of differential equations taught in

Spring 1991, omitting only a few students who had taken an experimental calculator-enhanced calculus course or who were participants in the two previous studies. All students in this pool had a grade of at least C in first-term calculus.

In the middle of the Spring semester, all of the 128 beginning differential equations students were contacted by mail and invited to participate in the study. As with the previous study of A/B calculus students (Selden et al., 1994), each student was offered $15 for taking the two tests and told he or she need not, in fact should not, study for them. The students were told that three groups of ten students would be randomly selected according to their first calculus grades and in each group there would be four prizes of $20, $15, $10, and $5. The latter was an incentive to ensure that all students would be motivated to do their best. Altogether 11 A, 14 B, and 12 C students volunteered and ten were randomly selected from each group. Of those, 28 students (nine A, ten B, and nine C) actually took the tests: three mathematics majors and 25 engineering majors (nine mechanical, five chemical, four civil, four electrical, two industrial, and one undeclared engineering concentration). These majors reflected the usual clientele for the calculus/differential equations sequence.

TABLE 1. Fall 1990 Calculus III grades for study participants and for all students enrolled in the course.

Grade	Participants		All Students	
A	4	(17%)	8	(5%)
B	6	(27%)	24	(15%)
C	7	(30%)	49	(30%)
D	4	(17%)	41	(25%)
F	2	(9%)	33	(20%)
W	0	(0%)	10	(6%)
Total	23	(100%)	165	(100%)

At the time of the study, all but one of the 28 students tested had taken the third semester of calculus at this university; the one exception was enrolled in Calculus III and Differential Equations simultaneously. Their grades in Calculus III were 5 A, 8 B, 8 C, 4 D, and 2 F. Of these, one D student and one F student were repeating Calculus III while taking Differential Equations. Twenty-three of the students had taken Calculus III in the immediately preceding semester (Fall 1990). Their grades and the grades of all students who took Calculus III that semester are given in Table 1, which indicates that the better mathematics students are over-represented in this study.

In Table 2 we give the mean ACT scores and the mean cumulative grade point averages (GPA) at the time of the test for the 28 students in this study and compare this with the same information for the students in our two earlier studies (Selden et al., 1989, 1994). The numbers for the Differential Equations (DE) students are quite close to those of the A/B calculus students but considerably above those of the C calculus students.

Eleven of the 28 differential equations students in this study had already taken additional mathematics courses. Of the three mathematics majors, two had completed, and the third was then currently taking, a "bridge to proof" course, and the

TABLE 2. Mean ACT and GPA of students in all three studies.

Study	Mean ACT	Mean Math ACT	Cumulative GPA
DE	26.26	27.74	3.145
(A/B) Calculus	27.12	28.00	3.264
(C) Calculus	24.18	25.65	2.539

third had also taken Discrete Structures. In addition, five students had taken Complex Variables, and another was enrolled in that course at the time of the study. One of these five had also taken an introductory matrix algebra course, as had two other students. Except for one C, all grades for these students in these additional mathematics courses were at least B. In our analysis of the results we will compare the students who had studied mathematics beyond the calculus/differential equations sequence with those who had not.

Of the 28 students in this study, 22 (79%) graduated within six years of their admission to the university as first-year students. In comparison, for the university as a whole over the same time period, the average graduation rate within six years of admission was 41%.

As of May 1999, it was known that all but two of the 28 students tested had earned bachelor's degrees at this university, three in mathematics and the others in engineering. In addition, five students had earned master's degrees in engineering and one had earned an MBA, all at this university. One student had earned a master's degree in mathematics at this university and a Ph.D. in mathematics at another university. There may be additional accomplishments of these kinds among the 28 students, but they could not all be traced.

The students in this study represented 33% of the A's, 26% of the B's, and 6% of the C's in Differential Equations that semester, and none of the 30 D's, F's and W's. In addition, after a minimum of three semesters at this university, these 28 students had a mean GPA of 3.145 for all courses taken. Their graduation rate was almost double that of the university as a whole, and at least 25% of them went on to complete a graduate degree. By all these indicators, at the time of the study and subsequently, these students were among the most successful at the university.

2.3. The Tests. The items for the non-routine test were originally chosen for the study of average calculus students' problem solving (Selden et al., 1989). Problems were chosen which could be solved using material covered in the first term of differential calculus. That the five non-routine problems were, in fact, novel for those students was determined by inviting department faculty to an informal seminar where possible problems were presented and the faculty were asked for suggestions. To the best of our knowledge, these problem types had not been taught or assigned in any of the classes that year. Solutions to the problems are no more complex than those traditionally covered in the university's calculus courses. However, in order to make progress towards a solution, students must access and combine ideas in ways that are new.

Students were allowed one hour to take the five-problem non-routine test, followed by half an hour for a ten-part routine test comprised of associated algebra and calculus exercises. Prior to the non-routine test, the students were told they might find some of the problems a bit unusual. No calculators were allowed. They were asked to write down as many of their ideas as possible because this would be helpful to us and to their advantage. They were told that A students (in first calculus) would only be competing against other A students for prizes, and similarly, for B and C students. They were assured all prizes would be awarded and partial credit would be given.

Each non-routine problem was printed on its own page, on which student work was to be done. All students appeared to work diligently for the entire hour.

As soon as the non-routine tests were collected, the students were given the two-page routine test. They were told answers without explanations would be acceptable, but they could show their work if they wished. Most students worked quickly, taking from 12 to 17.5 minutes to complete the routine exercises. None stayed the allotted half hour.

2.3.1. The Non-routine Test.

1. Find values of a and b so that the line $2x + 3y = a$ is tangent to the graph of $f(x) = bx^2$ at the point where $x = 3$.

2. Does $x^{21} + x^{19} - x^{-1} + 2 = 0$ have any roots between -1 and 0? Why or why not?

3. Let $f(x) = \begin{cases} ax, & x \le 1 \\ bx^2 + x + 1, & x > 1 \end{cases}$. Find a and b so that f is differentiable at 1.

4. Find at least one solution to the equation $4x^3 - x^4 = 30$ or explain why no such solution exists.

5. Is there an a such that
$$\lim_{x \to 3} \frac{2x^2 - 2ax + x - a - 1}{x^2 - 2x - 3}$$
exists? Explain your answer.

2.3.2. The Routine Test.

1. (a) What is the slope of the line tangent to $y = x^2$ at $x = 1$?
   (b) At what point does the tangent line touch the graph of $y = x^2$?
2. Find the slope of the line $x + 3y = 5$.
3. If $f(x) = x^5 + x$, where is f increasing?
4. If $f(x) = x^{-1}$, find $f'(x)$.
5. (a) Suppose f is a differentiable function. Does f have to be continuous?
   (b) Is $f(x) = \begin{cases} x, & x > 0 \\ 2, & x \le 0 \end{cases}$ continuous?
6. Find the maximum value of $f(x) = -2 + 2x - x^2$.
7. Find $\lim_{x \to 1} \dfrac{x^2 - 1}{x - 1}$.
8. Do the indicated division: $x - 1 \overline{)\ x^3 - x^2 + x - 1}$.
9. If 5 is a root of $f(x) = 0$, at what point (if any) does the graph of $y = f(x)$ cross the x-axis?
10. Consider $f(x) = \begin{cases} x^2, & x \le 1 \\ x + 3, & x > .1 \end{cases}$.
    (a) Find $\lim_{x \to 1^+} f(x)$.
    (b) What is the derivative of $f(x)$ from the left at $x = 1$ (sometimes called the left-hand derivative)?

As in the previous studies (Selden et al., 1989, 1994), each non-routine test problem was assigned 20 points and graded by one of the authors and checked by the others. If the student's work showed substantial progress toward a correct solution then at least 10 points were awarded. Arithmetic errors reduced scores by 1 point. The mean score on the non-routine test was 21.3, as compared with 20.4 for the A/B and 10.2 for the C calculus students. The lowest non-routine score in all three studies was zero.

On the routine test each question was assigned 10 points. Again, one point might be lost for an arithmetic or representational error.[1] The highest routine test score was 100 (out of 100) and the lowest score was 50, as compared to a high score of 90 and a low score of 53 for the A/B calculus students (Selden et al., 1994). The mean score on the routine test was 75.3.

Each problem on the non-routine test could be solved using a combination of basic calculus and algebra skills. The correspondence between routine questions and non-routine problems is given in Table 3.

TABLE 3. Correspondence between routine questions and non-routine problems.

Non-routine problem	1	2	3	4	5
Corresponding routine questions	1, 2	3, 4, 9	5, 10	6, 9	7, 8

3. The Results

We present the test results from several different perspectives. We examine the students' ability to solve non-routine problems, their knowledge base (of associated basic calculus and algebra skills), and whether they were able to access and use their resources effectively.

3.1. Ability to Solve Non-routine Problems. Slightly more than half (57%) the differential equations students (16 of 28) failed to solve a single non-routine problem. This is somewhat better than the two-thirds of A/B, and all of C, first-year calculus students who failed to solve a single non-routine problem in the previous studies (Selden et al., 1989, 1994).

In order to analyze the non-routine test results, we make a distinction between a solution *attempt,* a page containing written work submitted as a solution to a non-routine problem, and a solution *try,* any one of several distinct approaches to solving the problem contained within a single solution attempt. In only four instances did a student not attempt a non-routine problem; thus there were 136 attempts by the 28 students on the five non-routine problems. On these attempts there were a total of 243 solution tries. Of the 136 attempts, 20 were judged *completely correct* (except possibly for a minor computational error). Twelve other solution attempts were found to be *substantially correct* because they exhibited substantial progress toward a solution, that is, the proposed solution could have been altered or completed to arrive at a correct solution.

Of the 20 completely correct solutions, five were for Problem 1, three for Problem 2, five for Problem 3, none for Problem 4, and seven for Problem 5. These completely correct attempts came from 12 of the 28 students; the 12 substantially

---

[1]An answer of 5 on Routine Problem 9 received full credit as most students did not seem to distinguish between "meet" and "cross."

correct attempts came from 11 students. Altogether, 18 of the 28 students provided at least one substantially or completely correct solution attempt. That is, 36% of the differential equations students (10 of 28) were unable to make substantial progress on any non-routine problem; this is lower than the 42% and 71% reported previously for A/B and C first-year calculus students (Selden et al., 1989, 1994). Thus, the differential equations students did perform somewhat better than the first-year calculus students.

3.2. Comparison of Non-routine and Routine Test Results: Did Students Have Adequate Resources and Use Them?

The routine test was designed to determine whether the students' inability to do the non-routine problems was related to an inadequate knowledge base of calculus and algebra skills (i.e., Schoenfeld's (1985, 1992) "resources"). Did the students lack the necessary factual knowledge or did they have it without being able to access it effectively? Scores on the corresponding routine questions (shown in Table 3) were taken as indicating the extent of a student's factual knowledge regarding a particular non-routine problem. As in our 1994 study, a student was considered to have *substantial factual knowledge* for solving a non-routine problem if that student scored at least 66% on the corresponding routine questions. A student was considered to have *full factual knowledge* for solving a non-routine problem if that student's answers to the corresponding routine questions were correct, except possibly for notation, for example, answering $(1, -1)$ instead of -1 to Question 6. All others were considered as having *insubstantial factual knowledge*.

Table 4 gives the number of completely or substantially correct solutions for non-routine problems by solver's factual knowledge (as demonstrated on the corresponding routine questions). For example, on Problem 1, 15 (of 28) students had full factual knowledge; of these, five gave completely correct and two gave substantially correct solutions. An additional ten (of 28) students had substantial factual knowledge for Problem 1, but none of them gave completely or substantially correct solutions. That is, the performance of these ten students on the routine questions seemed to indicate they had sufficient factual knowledge to solve, or at least make substantial progress on, Problem 1; yet they either did not access it or were unable to use their knowledge effectively to make progress. The remaining three students demonstrated insubstantial factual knowledge. Thus, a total of seven students gave completely or substantially correct solutions on Problem 1.

Taking another perspective, in the 59 routine test solution attempts in which students demonstrated full factual knowledge, they were able to solve the corresponding non-routine problem 14 times (24%) or make substantial progress towards its solution six times (10%). Thus, on slightly more than a third of their attempts (34%), these students accessed their knowledge effectively. Students with substantial, but not full factual knowledge, did so on less than a quarter of their attempts. These results are summarized in Table 5. Furthermore, six students showed *no* factual knowledge of the components necessary for a non-routine problem and they each had a score of zero on the corresponding non-routine problem.

In order to compare overall student performances on the routine and non-routine tests, we introduce the notion of a *score pair*, denoted $\{a, b\}$, where a is the student's score on the routine test and b is the student's score on the non-routine test. In every case, $a > b$. Figure 1 shows students' routine test scores in ascending

TABLE 4. Number of correct solutions for non-routine problems by solver's factual knowledge.

	Non-routine problem				
	1	2	3	4	5
Full factual knowledge	*15*	*3*	*5*	*14*	*22*
Problem completely correct	5	1	1	0	7
Problem substantially correct	2	0	1	0	3
Substantial factual knowledge	*10*	*19*	*12*	*1*	*2*
Problem completely correct	0	2	4	0	0
Problem substantially correct	0	3	1	0	0
Insubstantial factual knowledge	*3*	*6*	*11*	*13*	*4*
Problem completely correct	0	0	0	0	0
Problem substantially correct	0	0	1	0	1
Total completely or substantially correct	7	6	8	0	11

TABLE 5. Percentage of correct solutions to non-routine problems from students with the requisite factual knowledge.

	Full factual knowledge (59)	Substantial factual knowledge only(44)
Completely correct non-routine solution	24% (14/59)	14% (6/44)
Substantially correct non-routine solution	10% (6/59)	9% (4/44)

order (from left to right); superimposed below each student's routine test score is her/his non-routine test score.

The three students with the highest non-routine scores had score pairs of {90,69}, {85,59}, and {83,59}; the first of these subsequently obtained a B.S. in civil engineering, summa cum laude with a cumulative grade point average of 3.948. The three mathematics majors in the study, all of whom had completed at least one additional mathematics course at the time of the study, had score pairs {95, 34}, {89, 18} and {87, 21}. That is, they scored in the top quarter on the routine test, and taken together, scored slightly higher than the mean non-routine score of 21.3. The last of these three subsequently obtained a Ph.D. in mathematics from a major state university.

Student performance on the routine and non-routine tests was not improved by having studied additional mathematics. The respective mean scores for the eleven students who had done so were 73.1 (vs. 75.3 for all of the students) and 15.3 (vs. 21.3 for all of the students).

Figure 1 shows a positive correlation (coefficient $r = 0.68$) between factual knowledge (resources) and the ability to solve novel problems. Nonetheless, having the resources for a particular problem is not enough to assure that one will be able to solve it. Two students had score pairs of {86, 4} and {80,3}, suggesting that having a reasonably good knowledge base of calculus and algebra skills (resources) is not sufficient to make substantial progress on non-routine calculus problems. One student, score pair {83,59}, lacked substantial factual knowledge on only those routine questions associated with Problems 2 and 4 (on which he scored zero) and

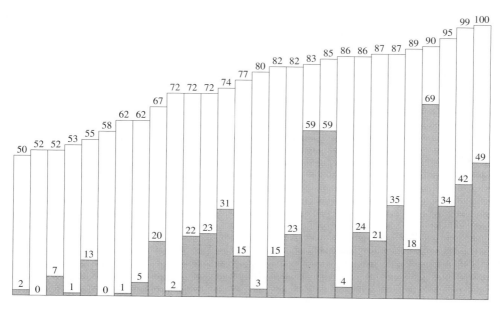

FIGURE 1. Upper score is for the routine test (factual knowledge). Lower score is for the non-routine test.

solved the three remaining non-routine problems (1, 3 and 5) completely correctly. This student and one other, the {90, 69} score pair, were the only students whose performance on the routine questions appeared to correspond closely with the non-routine problems they solved. An analysis of solution attempts indicates that some students were hampered by misconceptions. Indeed, the literature (Amit & Vinner, 1990; Eylon & Lynn, 1988) suggests that misconceptions are more likely to surface during attempts to solve non-routine problems, a phenomenon observed in this study and discussed in Section 4.

4. Favored Solution Methods

4.1. Non-routine Problem 1.

Find values of a and b so that the line $2x + 3y = a$ is tangent to the graph of $f(x) = bx^2$ at the point where $x = 3$.

Fifteen students rewrote the equation of the line in the form $y = \frac{a}{3} - \frac{2}{3}x$, set the right-hand side of this equation equal to that of the parabola and solved for either a or b. Eight of these ignored the tangency of the line to the curve. The remaining seven also set the derivative of f equal to the slope of the line to obtain a second equation so a and b could be fully determined. These were the seven with completely or substantially correct solutions. An additional seven students took a derivative, but abandoned it at some point.

The two most frequently occurring misconceptions on Problem 1 involved the meaning of tangent line. Six students conflated the ideas of the equation of the tangent line, the slope of the tangent line, and the derivative of the function. Two other students incorrectly claimed that the tangent line was perpendicular to the curve at the point of tangency and used the negative reciprocal of the derivative for the slope of the tangent line. Missing among these students was the error found among A/B calculus students of confusing a secant line, calculated using two points on the parabola, with the tangent line.

4.2. Non-routine Problem 2.

Does $x^{21} + x^{19} - x^{-1} + 2 = 0$ have any roots between -1 and 0? Why or why not?

On Problem 2, the three correct, and three substantially correct, solutions made no use of calculus. Instead, all used sophisticated arithmetic and algebraic reasoning to compare the relative sizes of the component terms in the given polynomial. The correct solutions made use of the observation that, for values of x in the open interval $(-1, \ 0)$, both $-x^{-1}$ and $x^{21} + x^{19} + 2$ would be positive and hence their sum must be positive. This type of first-principles argument was also used, although less successfully, by about the same proportion of the A/B first-calculus students.

Other solution attempts on Problem 2 suggest that these differential equations students were relatively comfortable with, and knowledgeable about, algebraic techniques. Yet their knowledge included some common misconceptions. Two students inappropriately applied Descartes' Rule of Signs, and eight erroneously concluded that no roots could exist when the Rational Root Test produced no solutions. Even this flawed use of algebra is an improvement upon the favored solution method of the C calculus students: substitute values for x until becoming convinced that no guess is ever going to work, hence no root exists (a method also used by four of the differential equations students). Only five of the differential equations students used any calculus on Problem 2, four taking derivatives but not using them. The fifth student did make effective use of the derivative but made an unrelated error.

4.3. Non-routine Problem 3.

Let $f(x) = \begin{cases} ax, & x \leq 1 \\ bx^2 + x + 1, & x > 1 \end{cases}$. Find a and b so that f is differentiable at 1.

Ten students set $ax = bx^2 + x + 1$, eight of them substituting $x = 1$ to get the relationship $a = b + 2$, and then stopped. Two of these ten students expressed doubts about the completeness of their solutions. One student had seven solution tries, all variations on the theme of matching for continuity. Eleven of the 26 who attempted a solution to Problem 3 took a derivative. Several differentiated ax and got x. Of the eleven who used calculus, eight made at least substantial progress towards a solution; five used the derivative to arrive at a completely correct solution and three used it to obtain a substantially correct solution.

4.4. Non-routine Problem 4.

Find at least one solution to the equation $4x^3 - x^4 = 30$ or explain why no such solution exists.

There were no completely correct or substantially correct solutions. Since these traditionally-taught calculus students were not allowed to use graphing, or other, calculators in this study, no easy graphical methods were available to them. Of the 27 students who attempted this problem, 44% (12 of 27) used the same method: narrow the domain of possible x values by eliminating those for which $4x^3 - x^4$ cannot be close to 30. Most of these students determined that no solutions could exist outside of the open interval $(1, \ 4)$ and some also eliminated the integer values of $1, 2$, and 3. This 44% who used sophisticated algebraic and arithmetic reasoning exceeds the 26% (5 of 19) of A/B calculus students who did so.

The next most popular solution method, used by six students (22%), was the Rational Roots Test. However, all six students incorrectly concluded that if a rational root could not be found from the factors of 30 then no roots at all existed. In our earlier study of A/B calculus students, this approach was also used by 20% of the students. Only two students used the method favored by C calculus students in the earliest study: factor $4x^3 - x^4$ and set each factor equal to 30. Four students took the first derivative, set it equal to zero and stopped. Several students used synthetic division to check whether particular values were roots.

4.5. Non-routine Problem 5.

Is there an a such that $\lim\limits_{x \to 3} \dfrac{2x^2 - 2ax + x - a - 1}{x^2 - 2x - 3}$ exists? Explain your answer.

On Problem 5, 39% of the solution attempts (11 of 28) were substantially or completely correct, similar to the 37% (7 of 19) of A/B calculus student attempts with substantial progress towards a solution.

Of the 28 solution attempts, 15 involved the use of L'Hôpital's Rule. Nine of these 15 attempts were at least partially successful. In the other six instances students failed to note that the numerator as well as the denominator must have limit zero before applying L'Hôpital's Rule. Five students substituted 3 for x, found the denominator of the expression to be zero and asserted that no limit could exist since the denominator was zero at the limiting value of the variable (two of these were mathematics majors). This was the favored method of the C calculus students (47% of them used it) and was used by 26% of the A/B calculus students. Here it was found in just 18% (5 of 28) of the solution attempts. Six students struggled, algebraically, with finding a way to factor the numerator so that the $(x - 3)$ in the factored denominator could be canceled and the limit taken; two of these resulted in completely correct solutions.

4.6. Summary.

4.6.1. *Use of calculus.* Inasmuch as Non-routine Problem 5 involved evaluating a limit, any solution attempt could be considered to involve calculus, we omit it from the following analysis. On Non-routine Problems 1, 2, 3, and 4, the differential equations students used calculus—often taking a derivative—on fewer than half (39%) of all solution attempts (42 of 108). Furthermore, in fewer than half of these (16 of the 42), students used calculus to produce potentially useful information, i.e., it could have been used to make progress towards a correct solution. Fifteen of the 16 potentially useful solution attempts led to substantially or completely correct solutions. That is, about three-fourths (15 of 21) of the completely or substantially correct solutions to Non-routine Problems 1, 2, 3, and 4 made effective use of calculus.

We observe that the differential equations students were no more inclined to resort to calculus than were the A/B and C calculus students of the previous two studies. They did not use calculus on 61% of these attempts (66 of 108); this is essentially the same percentage as in the previous studies (61% for the A/B and 59% for the C calculus students).

However, it appears that as students proceed through calculus to differential equations, the number of calculus misconceptions decreases. Slowly they become more proficient (or perhaps the less competent students drop out). For example, incorrectly asserting on Problem 5 that the limit of a quotient cannot exist when the denominator is 0 went from 47% for C calculus students, to 26% for the A/B calculus students, to just 18% for the differential equations students.

4.6.2. *More sophisticated, but still flawed, use of algebra.* Considering now all five non-routine problems, on 56% of the solution attempts calculus was not used; rather, a combination of guessing, trial-and-error, arithmetic techniques, and algebra was used. Eleven of the 32 substantially or completely correct solutions were almost purely algebraic, seven on Problem 2, one on Problem 4, and three (after the student observed the need to reduce the fraction in order to take the limit) on Problem 5. While these solutions demonstrated a level of algebraic competence not found in the first-year calculus students of our earlier studies, quite a few other attempts to use algebra were flawed. For example, fourteen solution attempts included an improper use of the Rational Root Test and two solution attempts used Descartes' Rule of Signs in an inappropriate setting.

4.6.3. *Use of graphs.* Most graphing was done by students on Problem 1— only four other graphs, three incorrect, appeared in all of the solution attempts for the other problems. Of the 27 students who attempted to solve Problem 1, 22 (81%) sketched at least one graph and six (22%) sketched three or more; this is substantially higher than the 12 of 19 (63%) who used graphs in the A/B study, where only 1 (5%) sketched more than two graphs. Fourteen of the 22 who used graphs in the present study first drew some version of the graph pictured in Figure 2. Of those 14 students, six also produced a correct graph like that in Figure 3. Three of these six rejected an incorrect graph (by striking through it); giving one substantially correct and two completely correct solutions to Problem 1, perhaps an example of monitoring their work (Schoenfeld, 1985, 1992). A fourth student drew only the correct graph, the one pictured in Figure 3, and gave a completely correct solution.

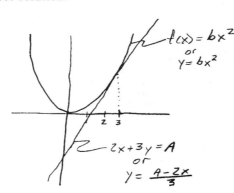

FIGURE 2. Example of frequently sketched incorrect graph for Problem 1.

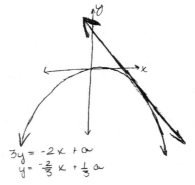

FIGURE 3. Example of a correctly sketched graph for Problem 1.

The fifth completely correct solution for Problem 1 came from a student who drew no graphs at all. Hence, in this study, four of the five completely correct solutions showed a graph while only one of the three completely correct solutions in the previous study of A/B students was accompanied by a graph. The other three differential equations students who drew both graphs used little calculus and did not indicate any rejection of incorrect graphs; in fact, all three seemed to be considering various cases for values of a and b by drawing a variety of graphs (between them, these three students produced ten graphs). The others who drew graphs were either misled by the assumption that the parabola was concave up (e.g. as in Figure 2) or gained no useful information from their graphs.

Whereas students who successfully solved Problem 1 in the present study were willing to draw and reject graphs, those in the A/B study who chose to use graphs generally had only one—either correct accompanied by a substantially or completely correct solution or incorrect and accompanied by an incorrect solution. In the A/B study, 16 graphs were produced by 12 students on Problem 1 whereas in the present study, 37 graphs were drawn by 22 students. Three of the 12 A/B students who used graphs (25%) rejected (crossed out) one of their graphs (including one who rejected a graph which was correct) while six of the 22 (27%) who used graphing on Problem 1 in this study rejected a graph. It would appear, then, that although the more experienced students were more willing to consider graphical ideas than the less experienced students (81% versus 63%), they were about equally likely to abandon the graphs they produced. This provides some evidence that extra experience in a traditional classroom environment does not necessarily increase effective self-regulation during problem solving (Boaler, 1999; Schoenfeld, 1992).

DeFranco's (1996) paper on expert problem solvers with Ph.D.'s in mathematics suggests that the skills possessed by experts which are often lacking in nonexperts might include a willingness to create, *abandon,* and *revisit* many ideas in the solution process. Thus it might be useful to know how students develop a willingness to risk committing graphical, and other ideas, to paper and to reject such ideas once they have been given life on paper.

5. Analysis

Reflecting on the three studies, one wonders when, if ever, traditionally taught students learn to use calculus flexibly enough to solve more than a few non-routine problems. Furthermore, how could it happen that students, who were more successful than average by a variety of traditional measures and who demonstrated full factual knowledge for a non-routine problem, failed to access and use their knowledge successfully on 76% of their attempts (as shown in Table 5)? Many of these students (the engineering majors) had almost completed their formal mathematical educations, except possibly for one or two upper-division mathematics courses, leaving them limited opportunity in future mathematics courses to improve their abilities to solve non-routine problems. Finally, does it matter whether students can solve such problems?

5.1. When do students finally learn to apply calculus flexibly? It would appear from this sequence of three studies that, at least for traditionally-taught calculus students in classes of 35–40, the ability to solve non-routine calculus problems develops only slowly. Performance for the best students went from one third who could solve at least one non-routine beginning calculus problem toward the end of their first year of college calculus to slightly less than half who could do so toward the end of the two-year calculus/differential equations sequence. In addition, the percentage of correct solutions increased slightly over the three studies (see Table 6).

For the sake of comparison, consider for a moment only those students in this study who had a grade of A or B in first-term calculus. There were 19 of them, nine

TABLE 6. Percent of correct solutions in all three studies.

Study	Completely correct	Substantially correct
DE	14% (20/140)	9% (12/140)
(A/B) Calculus	9% (9/95)	9% (9/95)
(C) Calculus	0% (0/85)	7% (6/85)

A and ten B. In the previous study of A/B calculus students there were ten A and nine B students. In the two groups of students there was very little difference in the number who did not answer any non-routine problem at least substantially correctly (8 in the previous study, 6 in this one). In Figure 4 we show the distribution of students according to how many non-routine problems each answered. For this figure, a substantially correct solution was counted as 0.5 and a completely correct solution as 1; e.g., the one student in the A/B study who gave two completely correct and two substantially correct solutions is represented by the single box to the left of the 3 in Figure 4. In Figure 5 we give similar histograms for just the completely correct solutions. Only two of the completely correct solutions in this (the DE) study came from students who had a C in first calculus, the other 18 were from A and B students. In both figures, the incremental shift upward is apparent, but not significant.[2]

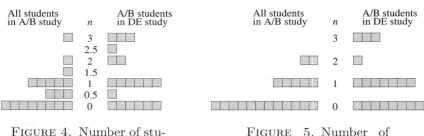

FIGURE 4. Number of students with at least n substantially correct solutions.

FIGURE 5. Number of students with n completely correct solutions.

Notably, by the time these students were coming to the end of their calculus/differential equations sequence their algebra skills seemed to be relatively sophisticated and readily accessible, albeit somewhat flawed. Such slow, incremental growth in mathematical capabilities may not be unusual. In a cross-sectional study of students' development of the function concept, Carlson (1998) investigated students who had just received A's in college algebra, second-semester calculus, or first-year graduate mathematics courses. She found that "even our best students do not completely understand concepts taught in a course, and when confronted with an unfamiliar problem, have difficulty accessing recently taught information. . . . Second-semester calculus students had a much more general view of functions

---

[2]The mean number of non-routine problems at least substantially correct per student for the A/B study was $\bar{x}_1 = 0.68$ (with standard deviation $s_1 = 0.82$) whereas for this study the mean was $\bar{x}_2 = 1.16$, (with $s_2 = 1.11$). The result of a Wilcoxon test on the hypothesis $H_0 : \bar{x}_2 = \bar{x}_1$ is $p \simeq 0.2$. For the comparison of completely correct solutions, a Wilcoxon test on $H_0 : \bar{x}_2 = \bar{x}_1$ gives $p \simeq 0.16$. These p-values are too large to say there is a statistically significant improvement.

[than college algebra students, but] . . . they were unable to use information taught in early calculus."

In addition, it is not unusual for students to fall back on earlier mathematical techniques with which they are perhaps more familiar and more comfortable. In a study of 900 Australian high school students in Years 9 and 10 ("in their third or fourth year of algebra learning"), Stacey and MacGregor (1997) found that when asked to solve three simple word problems, a large proportion wrote no equations, even though specifically asked to, and others tried to write equations but then switched to non-algebraic methods including trial-and-error and arithmetic reasoning to solve the problems.

In an analogous fashion, the differential equations students in this study relied more often (in 76 of 136 solution attempts) on a variety of arithmetic and algebraic techniques than on calculus. Taken together, our three studies suggest the folklore that one only really learns a course's material in the next course appears to be not quite accurate, rather several additional courses may be necessary. The differential equations students seemed most comfortable with algebraic methods—ideas first introduced to them several years before. Perhaps during those intervening years they had been exposed to numerous algebraic (sub)problems and built up a familiarity with algebraic techniques and habits of mind.

As in our previous two studies, over half of the students made no use whatever of calculus. This gives a negative answer to the question we posed in 1994: Would students at the end of a traditional calculus/differential equations sequence be more inclined to use calculus techniques in solving problems? Most of the students in this study had not learned to apply beginning calculus flexibly by the end of the calculus-differential equations sequence and many might well never do so.

The idea that students might or might not apply calculus, or any mathematics *flexibly* has been discussed by Dorier et al. (1998). They introduced three levels of applying mathematical knowledge to tasks: a *technical level* in which students are asked to apply calculus skills and use definitions, properties and theorems directly; a *mobilizable level* in which students adapt their knowledge to tasks which are not direct applications, require several steps, require some transformation or recognition that a property or theorem is to be applied; an *available level* in which students solve problems without being given an indication of methods, or must change representations. Dorier and colleagues tested one hundred French university students who had graduated (*Licence*) and were preparing for the CAPES competitive examination for teaching at secondary level (which just one in eight pass and which has curriculum the same as that for mathematics majors in the first two years of university). Dorier et al. found that whenever problems were anything but the technical level, the success rate was under 10%. These findings seem comparable to those in this study regarding students, who at the end of their calculus-differential equations sequence, produced just 14% completely correct solutions to our non-routine problems (Table 6).

5.2. How does it happen that students can have the knowledge, but not be able to access and effectively use it to solve non-routine problems? Part of the rationale for having our students take the routine test after the non-routine test was to determine whether they had the requisite algebra and calculus skills to solve the non-routine problems, but were unable to bring them to mind or to use them. This was a concern first raised by the study done with C calculus students (Selden et al., 1989). Adding the numbers in Table 4, one sees that for most non-routine problem attempts (103 out of a possible 140), the differential equations students had an adequate knowledge base (i.e., full or substantial factual knowledge), yet just 32 of their attempts were successful (i.e., produced completely or substantially correct solutions to non-routine problems). The inability of many otherwise successful students to access and effectively use their factual knowledge of calculus in non-routine problem solving is perhaps the most striking feature of our data.

Editorial comment on our second study (Selden et al., 1994) of A/B calculus students raised the question of whether the parameterizations in some of our non-routine problems might have caused them to be viewed as questions about families of functions, thereby rendering them inordinately difficult. Indeed, three of the five problems *could* be viewed in this way (Problems 1, 3, and 5). However, rather than being interpreted as indicating families of functions, the letters a and b appeared to have been interpreted by these students as fixed unknowns whose values they were expected to find. Much of their written work supports this idea. The a and b seemed to play much the same "arbitrary constant" role for these students as m and b do in the slope-intercept form of a line, $y = mx + b$. In fact, it was on these three problems (especially Problem 5) that the students did the best. Problems 1, 3, and 5 accounted for 85% (17 of 20) of the completely correct solutions.

We conclude that these problems were no more difficult for our students than Problems 2 and 4. If anything, Problems 2 and 4 proved to be the most difficult ones, and it was on these problems that students used algebraic and numerical trial-and-error methods. While it was possible to solve Problem 2 without calculus and three students did so completely correctly, Problem 4 was particularly intractable without calculus. Some students may not initially have accessed their calculus knowledge because this problem brought to mind algebraic techniques for solving an equation, namely $4x^3 - x^4 = 30$. Developing a calculus-based solution would have entailed considering $4x^3 - x^4$, alternatively $4x^3 - x^4 - 30$, as a function to be maximized, something they did not do.

In discussing Problem 4 we have mentioned bringing knowledge to mind, or accessing it, and the next section will depend on this terminology. The knowledge to which we refer is part of what has been called a person's knowledge base (Schoenfeld, 1992; Selden & Selden, 1995). This, in turn, is contained in what cognitive psychologists would call long-term memory (Baddeley, 1995). One is not aware of the contents of long-term memory, but knowledge there can be "activated," i.e., brought into short-term memory, and thereby brought into awareness. This appears to be close to what we mean by knowledge coming to mind. We do not mean just that the knowledge can be acted upon or used, but that it can be used in a special way—that it is conscious. There are a number of kinds of consciousness. Some arise from the external world through the senses after considerable lower-level processing and construction of meaning. Others include inner-speech and vision, or more subtle phenomena such as a feeling of understanding. Differing aspects of consciousness have been discussed by James (1910); for an example of more recent work see Mangan (1993).

5.2.1. *An additional kind of knowledge.* It may be that students who failed to solve our non-routine problems despite demonstrating full factual knowledge did not effectively employ some of Schoenfeld's hallmark categories of problem-solving, i.e., resources (including knowledge), control, heuristics, and beliefs (Schoenfeld, 1985). However, our problems are not *very* non-routine and their solutions are relatively straightforward. The unfruitfulness of false starts is not especially hidden, e.g., attempting to factor $4x^3 - x^4 - 30$ in Problem 4; also there is little need to invent multiple sub-problems. Solving such problems seems to call especially on resources, in particular on a kind of knowledge that "triggers" the use of factual knowledge appropriate to the specific problem situation. To put this another way, we suspect that what is required to solve our moderately non-routine problems is, mainly, some sense of the domain and what is important in it, above and beyond basic mastery of techniques, definitions, and their entailments. It is the way this sense of domain, regarded as a kind of knowledge, can bring to mind factual knowledge that we hope to explain.

The nature of this additional kind of knowledge cannot be fully established from an analysis of data such as ours, which does not emphasize the process aspect of problem solving. However, we will now frame a discussion of it in terms that might be useful in later, more process-oriented, research. A person who has reflected on a number of problems is likely to have seen (perhaps tacitly) similarities between them. He or she might be regarded as recognizing (not necessarily explicitly or consciously) several overlapping problem situations arising from problems with similar features. For example, experienced students would probably recognize a problem as one involving factoring, several linear equations, or integration by parts.

A problem situation seems to be much like a concept. While such a situation may lack a name, for a given individual it is likely to be associated with an image. This image is a mental structure possibly including strategies, examples, non-examples, theorems, judgments of difficulty, and the like, linked to the problem situation. Following Tall and Vinner's (1981) idea of concept image, we call this kind of mental structure a *problem situation image* and regard it as a part of one's knowledge base. When a problem situation is recognized, most of the features in its image do not immediately come to mind, i.e., into consciousness. Rather they seem to be partly activated (Baddeley, 1995), that is, they are very easily brought to mind as needed. Some such images may, and others may not, contain what we will call *tentative solution starts*: tentative general ideas for beginning the process of finding a solution. A tentative solution start might be seen as a strategy for solving the problem at hand. For example, a problem situation image involving the solution of an equation might include "try getting a zero on one side and then factoring the other." It might also include "try writing the equation as $f(x) = 0$ and looking for where the graph of $f(x)$ crosses the x-axis," or even "perhaps the max of f is negative so $f(x) = 0$ has no solution." We suggest that the problem-solving process includes the recognition of a problem as belonging to one or more problem situations and partly activates their associated images. This, in turn, easily brings to mind a tentative solution start or strategy (specific to the problem) contained in one of those problem situation images. This may thereupon mentally prime the recall of additional resources from one's knowledge base. Thus a tentative solution start may link recognition of a problem situation with recall of appropriate resources, i.e., what we have called accessing factual knowledge.

Although the kinds of calculus problem situations perceived by students do not seem to have been well examined, a number of studies of problem-solving performance support the idea that students recognize problem situations in various ways (Hinsley, Hayes, & Simon, 1977);Schoenfeld, 1985, Chapter 8). In research on physics problem solving, features focused on by solvers have been observed to correspond to the degree of expertise. Novices tend to favor surface characteristics (e.g., pulleys), whereas experts tend to focus on underlying principles of physics, like conservation of energy (Chi, Feltovich, & Glaser, 1981).

In this study a number of students tried factoring on Problem 4. When this algebraic approach did not work, had those students' images of such problems included "try looking at whether the graph crosses the x-axis," they might have recalled their knowledge of graphs and calculus to discover that the maximum was too small for the equation to be solvable. In viewing our data from this perspective, we are suggesting that problem situations, their images, and the associated tentative solution starts all vary from student to student and that the process of mentally linking recognition (of a problem situation) to recall (of requisite resources) through a problem situation image might occur several times in solving a single problem. We are not suggesting this is the only way accessing one's knowledge base might occur, just that it could play an important role in solving the kind of moderately non-routine problems discussed in this paper.

Recognizing a problem as associated with a particular problem situation image and bringing to mind a tentative solution start from that image results in what Mason and Spence have called "knowing-to act in the moment" (1999). Here the "act" is using the tentative solution start and *knowing to* should be contrasted with *knowing how* (procedural knowledge) and *knowing that* (conceptual knowledge). For example in Problem 4, reading the problem invokes a problem situation image whose richness depends upon the student's previous experiences. A possible tentative solution start that might come to mind from the student's image would be *knowing to* subtract 30 from both sides. However once the student establishes that attempting to solve the equation by algebraic methods is fruitless, the problem situation is changed. If the student's problem situation image is rich enough to contain a tentative solution start based on examining where the graph of $y = 4x^3 - x^4 - 30$ crosses the x-axis, knowing to do so would probably come to mind. We mean by this that the student could unhesitatingly respond if questioned about her/his intentions at that moment, and might report a conscious feeling of intending to examine the graph, or even report having articulated that intention in inner-speech. Alternatively, the student might simply begin to draw the graph. However, if the student's problem situation image lacked this or a similar tentative solution start, he or she might bring nothing to mind and would be reduced to searching her/his knowledge base for a serendipitous link to the problem. Many students may have neither time nor self-confidence enough for such a search. They might not solve the problem even if they know the relationship between roots of an equation and where the graph of the corresponding function crosses the x-axis because such information was not linked to the problem in a way that brings using it to mind.

Except while actually taking classroom tests, students in this study could have consulted worked examples from textbooks or lecture notes during their previous problem-solving attempts. Those who habitually consulted such worked examples before attempting their own solutions may have had little occasion or reason to reflect on multiple tentative solution starts. Such students might well have impoverished problem situation images with very few tentative solution starts, thereby reducing the usefulness of these images in priming the recall of factual knowledge. This could happen even if they realized a new idea was needed, that is, even if they exercised (metacognitive) control (Schoenfeld, 1985, 1992). In summary, we are suggesting that some of the students who were unable to solve our non-routine problems, although demonstrating full factual knowledge, may have lacked a particular kind of knowledge, namely tentative solution starts. This may have been due to a combination of the text, the way homework exercises were assigned and discussed, and our students' study habits. Much of our data concerning these students, who had adequate factual knowledge but did not access it to solve our non-routine problems, is consistent with the above analysis.

5.3. Does it matter whether students are able to solve non-routine problems? Perhaps surprisingly, the answer seems to be both yes and no. No, because the students in this study were among the most successful at the university by a variety of traditional indicators, both at the time of the study and subsequently, yet half of them could not solve a single non-routine problem. They had overall GPAs of just above 3.0 at the time of the study and almost double the graduation rate of the university as a whole. At least seven of them subsequently earned a master's degree and one a Ph.D. Furthermore, the idea that traditional academic success may not require very much non-routine problem-solving ability, including metacognitive control, is supported by DeFranco's problem-solving study comparing mathematicians of exceptional creativity (e.g., Fields medallists) with very successful published Ph.D. mathematicians. He found that while both were content experts, only the former were problem-solving experts (DeFranco, 1996). Thus, it seems possible to be academically successful in mathematics and related subjects without consistently being able to solve non-routine problems, especially the more difficult ones in which Schoenfeld's (1985, 1992) problem-solving characteristics (resources, heuristics, control, and belief systems) play a large role.

On the other hand, yes, it does matter. Most mathematicians in our experience seem to regard this kind of problem solving as a test of deep understanding and the ability to use knowledge flexibly. In addition, most applied problems that students will encounter later will probably be at least somewhat different from the exercises found in calculus and other mathematics textbooks. It seems likely that much original or creative work in mathematics would require novel problem solving at least at the modest level of the non-routine problems in this study. Thus, when other academic departments decry students' inability to apply their mathematics, the difficulty may lie partly with the ability to solve non-routine problems generally, rather than with particular kinds of applications.

In addition mathematicians often appear to view students' mathematical ability through the lens of problem solving, meant broadly to include a wide range from simple exercises to non-routine problems and the construction of proofs. As teachers they typically design problems whose solution requires deep understanding rather than gauging such understanding directly, for example, through evaluating an essay on the nature of continuity and its relationship to differentiability. Thus, in order to more precisely discuss mathematicians' views of which student abilities matter, it would be useful to have a way of analyzing kinds of problems. In what follows, we discuss only degrees of non-routineness and not the many other facets of problem solving.

5.3.1. *A non-routineness scale.* It may be helpful to consider problems (tasks together with solvers) arising in a mathematics course as arranged along a continuum according to routineness relative to the course. We regard the degree of routineness of a problem as dependent on what the solver knows so a task that is routine for one student may be non-routine for another. Nevertheless, the routineness of a problem, for most students, might be estimated by what has come up in the course. At one end of the continuum, we might have *very routine* problems which mimic sample problems found in the text or lectures, except for minor changes in wording, notation, coefficients, constants, or functions, that students view as incidental to the way the problems are solved. Such problems are often referred to as exercises and might not be regarded as problems at all in the problem solving literature (Schoenfeld, 1992). Moving toward the middle of the continuum, we might place *moderately routine* problems which, although not viewed as exactly like sample problems, can be solved by well-practiced methods, e.g., ordinary related rates or change of variable integration problems in a calculus course. Sandra Marshall (1995) has studied how schema can be developed to reliably guide the solution of such problems at and below the pre-calculus level. Moving further along the continuum, one might have *moderately non-routine* problems, which are not very similar to problems that students have seen before and require known facts or skills to be combined in a novel way, but are "straightforward" in not requiring, for example, the consideration of multiple sub-problems. The non-routine test in this study consisted of problems of this kind. Finally, at the opposite end of the continuum from routine problems, one might place *very non-routine* problems which may involve noticing unusual patterns, considering several sub-problems or constructions, and using Schoenfeld's (1985) characteristics of effective problem solving. For such problems a large supply of tentative solution starts, built up from experience, might not be adequate to bring to mind the knowledge needed for a solution, while for moderately novel problems such as ours it probably would.

Here is an example of what, for most undergraduates, would be a very non-routine problem.[3] *Given a point (a, b) with $0 < b < a$, determine the minimum perimeter of a triangle with one vertex at (a, b), one on the x-axis, and one on the line $y = x$. You may assume that a triangle of minimum perimeter exists.* This appears to be a calculus problem, but it only requires clever use of geometry. An elegant solution[4] involves extending the construction "outward" by reflecting across both the lines $y = x$ and the x-axis and noticing that the perimeter of the triangle equals the distance along the path from (b, a) to $(a, -b)$. Thus, minimizing the perimeter amounts to making that path straight. Probably only exceptionally experienced geometry problem solvers would have previously constructed images of problem situations containing a tentative solution start that would easily bring to mind such an unusual construction and "path straightening" technique.

Although in this paper we only discuss the degree of routineness of problems, there are other important aspects to problem solving. Some problems require an understanding of the underlying conceptual notions or the application of the core ideas behind the content. In addition, some problems call for the use of heuristics or problem-solving strategies. Furthermore, some problems may require a kind of

---

[3]Taken from the 59th Annual William Lowell Putnam Mathematical Competition (1998; see also Putnam Exam, 1991; Reznick, 1994).

[4]Posted by Iliya Bluskov to the sci.math newsgroup.

cleverness that goes beyond understanding content and beyond problem-solving heuristics. In particular, we observe that routineness differs from difficulty. The most routine of exercises can be quite difficult, e.g., requiring the fast execution of a long procedure, overloading working memory and leading to what students often call "dumb mistakes."

Most university mathematics teachers would probably like students who pass their courses to be able to work a wide selection of routine, or even moderately routine, problems. In addition, we believe that many such teachers would expect their better students to be able to work moderately non-routine problems such as ours. Many seem to think of this as functionally equivalent to having a good conceptual grasp and understanding of a course. In other words, the ability to work moderately non-routine problems based on the material in a course is often part of the implicit curriculum. Thus, yes, it matters if most students cannot work moderately non-routine problems because for the mathematical community the essence of the material in the course is not being successfully mastered, even by good students.

6. Implications for Teaching

The results reported here suggest that, at least in a traditionally taught calculus/differential equations sequence, many good students may not reach the level of understanding and moderately non-routine problem-solving ability that their teachers expect. In order that good students reach this level, non-routine problem solving may need to become an explicit part of the curriculum, that is, to be in some way explicitly taught. Furthermore, our explanation of the data—that students' problem situation images tend to lack a variety of tentative solution starts—suggests that the ability to solve moderately non-routine problems may depend partly on a rich knowledge of problem situations as well as on more general problem-solving characteristics (Schoenfeld, 1985, 1992).

We limit our comments on teaching mainly to how to improve student ability to solve *moderately* non-routine problems, e.g., problems similar to those on our non-routine test. For suggestions on developing student ability to solve *very* non-routine problems in which all of Schoenfeld's problem-solving characteristics are likely to play major roles, see Arcavi et al. (1998) and Schoenfeld (1985). We will focus on improving a part of what Schoenfeld calls resources, namely the richness of problem situation images. If our analysis in Section 5.2.1, or something close to it, is correct then encouraging students to build rich problem situation images should help bring to mind appropriate factual knowledge when needed. Of course, to bring to mind appropriate factual knowledge when needed, the student must *have* such knowledge. However, it was not factual knowledge that most of our differential equations students lacked, rather it was access to and knowing-to use it. That is what we address now.

For the purposes of this discussion we will assume students know a number of algebraic techniques for solving equations, quite a bit about graphs, the intermediate value theorem, how to find maxima and minima, which functions are continuous or discontinuous, that the real roots of $f(x) = 0$ are the points where the graph of $y = f(x)$ meets the x-axis, etc. The material on our routine test fits into this category. In addition, we assume they know things like solving $f(x) = g(x)$ amounts to solving $f(x) - g(x) = 0$, that the intermediate value theorem can be used to show there is a root to $y = f(x)$, and looking at maxima or minima can show $f(x)$ is entirely positive or negative and thus does not meet the x-axis—material covered in most traditionally taught calculus courses.

To build rich problem situation images that contain a number of tentative solution starts, the students would have to see a problem situation as an object (although probably tacitly and without name) which is associated with other objects. We see student construction of problem situations as objects as requiring a student to engage in multiple activities and experiences. We view this as being analogous to the construction of new concepts, such as the function concept, as described by Breidenbach et al. (1992) or Sfard (1991). Merely pointing out how a solution to a particular problem is started might not be any more effective than merely pointing out the features of a new concept, such as that of function.

One way of giving students multiple experiences that might lead to the construction of rich problem situation images would be to scatter throughout a course a considerable number of problems for students to solve without first seeing very similar worked examples. We mean to suggest a collection of problems that cover the course well and that most of the students really can eventually solve, albeit with some difficulty. The idea is that the students would struggle with these problems and reflect on their solutions more than they would with traditional exercises mimicking worked examples. However, we note that students often expect to be told precisely how to work problems. Thus, a change in the prevalent classroom culture that prefers tedium over struggle and reflection might be required. Such change might be difficult for some instructors, but there is some evidence it can be effected by reiterating in class that struggle and reflection are expected (Davis & Hauk, in preparation).

Alternatively, some problems might be solved in two phases. Problems could be introduced with the understanding that the first phase ends with students reporting, perhaps in writing, on their tentative solution starts, before going on to complete the solution in the second phase. As a practical matter, this kind of activity would probably be taken more seriously if it were in some way reflected in homework, classwork, and tests. For tests, adequate time for reflection would need to be allowed. Indeed, one might not ask for full solutions to such a problems, but only clear descriptions of how to begin. Another approach to providing the multiple experiences that could lead to rich problem situation images might be to ask students to justify the steps in solved problems, whether from the text, another student, or the instructor. Students might be asked to discuss alternative solution methods and the degree of promise of each. Such an approach might be incorporated into homework and tests and seems particularly appropriate for class presentations by students or group work. As an example, Santos-Trigo (1998) describes an aspect of Schoenfeld's problem-solving course where "homework assignments include problems in which students have the opportunity to discuss differences and qualities of each approach," i.e., to examine tentative solution starts and, "students search for multiple solutions, but they write their ideas clearly and support their methods with mathematical arguments."

A teacher might also use something akin to Socratic dialog with one or several students, or even with a whole class. From the point of view of building rich problem situation images, there are several benefits of such dialog: students can be brought to focus on various tentative solution starts, they can sometimes solve a greater variety of problems than otherwise, and they might eventually adopt the kinds of reflective questions the instructor asks in something like an "inner Socratic dialog." The questions in such an inner Socratic dialog might prompt bringing to mind appropriate factual knowledge. However, such reflective questions themselves *must* come to mind or at least be acted upon. The way inner Socratic dialog might be engendered in students, what form it might take, and to what degree it would be useful are interesting questions for future research. What we are talking about is distinct from the kind of habituation to questions such as "What are you doing?" discussed in Arcavi et al. (1998) and Schoenfeld (1985).

Having students work in groups might combine well with several of the above suggestions. There are a number of benefits to group work but two seem particularly pertinent. Students working in a group have additional sources of factual knowledge beyond those available when working alone. Thus, they may solve more and harder problems (Kieren, Calvert, Reid, & Simmit, 1995; Trognon, 1993). In addition, the discussion in a group may encourage individual students to reflect on their solutions or the groups' solutions. Indeed, discussion itself may be somewhat like reflection (Sfard, Nesher, Streefland, Cobb, & Mason, 1998). Thus, working in groups may help students enrich their problem situation images.

A college instructor's typical response might be, "But I don't have time for this!" We agree that implementing any of the above suggestions may call for considerable time, which is in short supply in college mathematics courses. The time might be obtained by assigning and discussing fewer problems: a reduced collection of problems which addresses a broader range of the non-routineness scale. Examination of a variety of problems in the ways we suggest, by teacher and students alike, might improve students' ability to solve moderately non-routine problems without reducing their command of routine exercises.

References

Amit, M., & Vinner, S. (1990). Some misconceptions in calculus: Anecdotes or the tip of an iceberg? In G. Booker, P. Cobb, & T. N. de Mendicuti (Eds.), *Proceedings of the 14th International Conference of the International Group for the Psychology of Mathematics Education* (Vol. 1, pp. 3–10). CINVESTAV, Mexico.

Arcavi, A., Kessel, C., Meira, L., & Smith, J. P. (1998). Teaching mathematical problem solving: An analysis of an emergent classroom community. In E. Dubinsky, A. H. Schoenfeld, & J. Kaput (Eds.), *Research in collegiate mathematics education III* (pp. 1–70). Providence, RI: American Mathematical Society.

Baddeley, A. (1995). Working memory. In M. S. Gazzaniga (Ed.), *The cognitive neurosciences* (pp. 755–764). Cambridge, MA: MIT Press.

Berkey, D. P. (1988). *Calculus* (2nd ed.). New York, NY: Saunders College Publishing.

Boaler, J. (1999). Participation, knowledge and beliefs: A community perspective on mathematics learning. *Educational Studies in Mathematics, 40*, 259–281.

Breidenbach, D., Dubinsky, E., Hawks, J., & Nichols, D. (1992). Development of the process conception of function. *Educational Studies in Mathematics, 23*, 247–285.

Carlson, M. P. (1998). A cross-sectional investigation of the development of the function concept. In E. Dubinsky, A. H. Schoenfeld, & J. Kaput (Eds.), *Research in collegiate mathematics education III* (pp. 114–162). Providence, RI: American Mathematical Society.

Chi, M. T. H., Feltovich, P., & Glaser, R. (1981). Categorization and representation of physics problems by experts and novices. *Cognitive Science, 5,* 121–152.

Davis, M. K., & Hauk, S. (in preparation). *Does the evidence of authority prevail over the authority of evidence?* (Contact: hauk@math.la.asu.edu)

DeFranco, T. C. (1996). A perspective on mathematical problem-solving expertise based on the performance of male Ph.D. mathematicians. In A. H. Schoenfeld, J. Kaput, & E. Dubinsky (Eds.), *Research in collegiate mathematics education II* (pp. 195–213). Providence, RI: American Mathematical Society.

Dorier, J. L., Pian, J., Robert, A., & Rogalski, M. (1998). A qualitative study of the mathematical knowledge of French prospective maths teachers: Three levels of practice. In *Pre-proceedings of the ICMI study conference on the teaching and learning of mathematics at university level* (pp. 118–122). Singapore, National Institute of Education.

Eylon, B. S., & Lynn, M. C. (1988). Learning and instruction: An examination of four research perspectives in science education. *Review of Educational Research, 58,* 251–301.

Hinsley, D. A., Hayes, J. R., & Simon, H. A. (1977). From words to equations: Meaning and representation in algebra word problems. In M. A. Just & P. Carpenter (Eds.), *Cognitive processes in comprehension* (pp. 89–106). Hillsdale, NJ: Lawrence Erlbaum.

James, W. (1910). *The principles of psychology.* New York, NY: Holt.

Kieren, T., Calvert, L. G., Reid, D. A., & Simmit, E. (1995). *Coemergence: Four enactive portraits of mathematical activity.* (Paper presented at AERA, ERIC #ED390706)

Mangan, B. (1993). Taking phenomenology seriously: The "fringe" and its implications for cognitive research. *Consciousness and Cognition, 2,* 89–108.

Marshall, S. P. (1995). *Schemas in problem solving.* New York, NY: Cambridge University Press.

Mason, J., & Spence, M. (1999). Beyond mere knowledge of mathematics: The importance of knowing-to act in the moment. *Educational Studies in Mathematics, 28,* 135–161.

Reznick, B. (1994). Some thoughts on writing for the Putnam. In A. H. Schoenfeld (Ed.), *Mathematical thinking and problem solving* (pp. 19–29). Hillsdale, NJ: Lawrence Erlbaum.

Santos-Trigo, M. (1998). On the implementation of mathematical problem solving instruction: Qualities of some learning activities. In E. Dubinsky, A. H. Schoenfeld, & J. Kaput (Eds.), *Research in collegiate mathematics education III* (pp. 71–80). Providence, RI: American Mathematical Society.

Schoenfeld, A. H. (1985). *Mathematical problem solving.* Orlando, FL: Academic Press.

Schoenfeld, A. H. (1992). Learning to think mathematically: Problem solving, metacognition, and sense making in mathematics. In D. A. Grouws (Ed.), *Handbook of research on mathematics teaching and learning* (pp. 334–370). New York, NY: Macmillan.

Selden, J., Mason, A., & Selden, A. (1989). Can average calculus students solve nonroutine problems? *Journal of Mathematical Behavior, 8*, 45–50.

Selden, J., & Selden, A. (1995). Unpacking the logic of mathematical statements. *Educational Studies in Mathematics, 29*, 123–151.

Selden, J., Selden, A., & Mason, A. (1994). Even good calculus students can't solve nonroutine problems. In *Research issues in undergraduate mathematics learning* (MAA Notes no. 33, pp. 19–26). Washington, DC: Mathematical Association of America.

Sfard, A. (1991). On the dual nature of mathematical conceptions: Reflections on processes and objects as different sides of the same coin. *Educational Studies in Mathematics, 22*, 1–36.

Sfard, A., Nesher, P., Streefland, L., Cobb, P., & Mason, J. (1998). Learning mathematics through conversation: Is it as good as they say? *For the Learning of Mathematics, 18*, 41–51.

Stacey, K., & MacGregor, M. (1997). Multi-referents and shifting meanings of unknowns in students' use of algebra. In E. Pehkonen (Ed.), *Proceedings of the 21st International Conference of the International Group for the Psychology of Mathematics Education* (Vol. 4, pp. 190–198). Gummerus, Finland.

Stewart, J. (1987). *Calculus.* Monterey, CA: Brooks/Cole.

Swokowski, E. W. (1983). *Calculus with analytic geometry* (Alternate ed.). Boston, MA: Prindle, Weber & Schmidt.

Tall, D., & Vinner, S. (1981). Concept image and concept definition with particular reference to limits and continuity. *Educational Studies in Mathematics, 12*, 151–169.

Trognon, A. (1993). How does the process of interaction work when two interlocutors try to resolve a logical problem? *Cognition and Instruction, 11*, 325–345.

The William L. Putnam Exam. (1991). *Mathematics Magazine, 64*, 143.

Zill, D. G. (1986). *Differential equations with boundary-value problems.* Boston, MA: PWS-KENT.

TENNESSEE TECHNOLOGICAL UNIVERSITY AND ARIZONA STATE UNIVERSITY

MATHEMATICS EDUCATION RESOURCES COMPANY

CHAPMAN UNIVERSITY AND ARIZONA STATE UNIVERSITY

TENNESSEE TECHNOLOGICAL UNIVERSITY

CBMS Issues in Mathematics Education
Volume **8**, 2000

Lasting Effects of the Integrated Use of Graphing Technologies in Precalculus Mathematics

William O. Martin

ABSTRACT. Technologies play a prominent role in current reforms of school and collegiate mathematics. Studies over the past decade have shown that technologies can be used in ways that promote enhanced learning of basic mathematical concepts such as *function*. However, there is little evidence about the robustness or durability of gains beyond an individual course. This study examined ways in which first-semester calculus students still showed the influence of a graphing-intensive college algebra course they had studied one or two semesters earlier. Eighteen students at a large research university were asked to solve a series of problems in individual, audiotaped interviews during the last month of a traditionally taught calculus course. Two matched groups of students participated: (a) GT (Graphing Technology) students who had studied college algebra in sections that integrated the use of graphing technologies and (b) NT (No Technology) students from traditionally taught sections of college algebra that did not use graphing technologies. The data showed that GT students did continue to use technologies but mainly in routine rather than advanced ways. Neither group of students showed strong preparation for calculus from the precalculus course. The main conclusions of this study were that (a) the use of graphing technologies does have a lasting impact on students, even when the use of these tools is discouraged or prohibited in subsequent courses, and (b) many students do not become sophisticated users nor do they appear to gain the expected lasting enhancements of conceptual knowledge during a one-semester course. The implication of these findings is that careful attention to pedagogical issues must accompany curricular integration of technologies if changes are to significantly improve students' conceptual learning in mathematics.

1. Potential Impact of Technologies on Learning and Retention

Many recent calls for the reform of mathematics education have suggested that technologies should be widely incorporated into mathematics courses. One reason that technologies are seen as important in mathematics education is as a reflection of the increasing importance of technologies outside school, in the "real world." Another reason is that technologies can be used to emphasize useful, conceptual mathematical knowledge, going beyond the routine procedural skills that have been so prevalent in mathematics courses. Steen (1990) clearly expressed the need for a new vision of mathematics education when he wrote "Forces created by computers, applications, demographics, and schools themselves are changing profoundly the

©2000 American Mathematical Society

way mathematics is practiced, the way it is taught, and the way it is learned" (p. iii).

Graphing calculators are increasingly being used in school and college mathematics classes. New ideas for uses of graphing technologies in mathematics classes are regularly featured at conferences and in the mathematics education literature; especially, in journals directed at school and college mathematics instructors (several examples include Curjel, 1993; Naraine, 1993; Demana & Waits, 1993; Vonder Embse, 1998). Curricular materials that draw on graphing technologies are common; many textbooks acknowledge the existence of graphing technologies; several textbooks for precalculus (e.g., Demana & Waits, 1990; Hungerford, 1997) and calculus (e.g., Dick & Patton, 1992; Hughes-Hallett et al., 1998) *require* student access to graphing tools. Is the movement toward increased use and integration of graphing technologies in mathematics, and particularly in precalculus, appropriate? Some research suggests that the answer is "yes" but the question needs much more study (Dunham & Dick, 1994). Unfortunately, much less has appeared in the literature about the real impact that graphing technologies have had on student learning in regular classrooms. A question of interest to many collegiate mathematics departments is "What happens when instructors in a mathematics department begin to use graphing technologies, supported by a text that requires such technologies and that has been organized to take regular advantage of them? Does this instruction influence students, even when instructors do not have a special commitment to the approach, just a willingness to 'give it a try' with minimal special support?" More specific questions that could be raised about student learning with the integrated use of graphing technologies in precalculus include:

- Do students who have used graphing technologies while studying mathematics draw on them in subsequent mathematics courses? If so, how? Is it thoughtful ways, such as to explore concepts and solve problems? Or is it in routine ways, to complete exercises or simply sketch graphs?
- Do differences between groups with graphing and traditional backgrounds, as have been found in research studies, persist beyond a precalculus course that uses a graphing approach?
- Do graphing approach college algebra students, because of decreased emphasis on traditional manipulative skills, appear disadvantaged during a subsequent traditionally taught calculus course?

Naturally, similar questions are of interest for other stages of the sequence of mathematics courses in schools and colleges. This particular transition, from algebra to calculus, may have special importance because it applies to so many programs and students.

The use of technologies in school and college mathematics instruction received considerable research attention, especially since the late 1980s with the development of general purpose personal computer mathematics software and the advent of graphing calculators. Technologies have been used in ways that promoted enhanced conceptual learning (Crocker, 1991; Palmiter, 1991; Park & Travers, 1996; Tufte, 1990), apparently with little or no detrimental effect on students' abilities to carry out necessary procedures (Heid, 1988; Schrock, 1989). In precalculus courses, graphing technologies fostered student learning of important graphical concepts (Browning, 1988; Harvey, Martin, Demana, Osborne, & Waits, 1994) while promoting positive changes in attitudes and classroom interactions (Dunham, 1991;

Farrell, 1996, 1990). Current views of learning and knowledge (see, for example, Hiebert & Carpenter, 1992) suggest that such changes should be long-lasting, but there is little research evidence to support the belief that these benefits are robust and enduring.

1.1. This Study. This research study explored the lasting impact of the integrated uses of graphing technologies in college algebra. In simplest terms, this was done by interviewing matched pairs of students who were completing a first-semester calculus course. The students had previously studied college algebra at the same institution. The pairs were chosen so that the main difference between students was whether or not the college algebra course had used an integrated graphing approach with graphing technologies. In this paper, the two groups will be referred to as GT (Graphing Technology) and NT (No Technology), based on the use of technology in their precalculus course.

During the individual interviews, students were asked to solve nine precalculus and calculus problems. This paper discusses the results of those interviews, looking for evidence that (a) GT students continued to use a graphing approach in a subsequent (traditional, non graphing) course and (b) there were differences apparent between the two groups. The main focus was on the college students at a major midwestern research university who had previously used one of the first precalculus textbooks written with graphing technologies and a functional theme at its core (Demana & Waits, 1990).

The decision to work with students who had studied precalculus with the C^2PC (Ohio State University Calculator and Computer Precalculus Curriculum project) materials was made for several reasons. A variety of studies, including a large-scale field test in 1988–89 (Harvey et al., 1994), had shown cognitive and affective benefits for high school and college students who had used preliminary versions of the materials with computer graphing software and handheld graphing calculators, so it was reasonable to expect that the use of this approach would have similar benefits for the participants in this study. Students who participated in this study had taken college algebra in many different sections and with a variety of instructors. The instructors, while committed to teaching at this level, were not especially committed to reform efforts or the use of technologies, so the study also investigated the extent to which graphing technologies and a textbook that integrated their use in the hands of regular instructors could influence learning and retention.

The department of mathematics at this university offered parallel college algebra courses for several years that were taught with and without graphing technologies. Course content was comparable in the sense that both versions were intended to serve as preparation for the regular, traditionally taught calculus courses at this institution. The graphing-technologies courses used the college algebra textbook by Demana and Waits (1990) while the traditional courses used Cohen's *College Algebra* (1989).

This mathematics department is very concerned about the role of precalculus courses as preparation for calculus. A faculty coordinator assisted by a graduate student develops common tests and grading curves for the many sections offered each semester. The coordinator provides a detailed syllabus, including teaching notes, to help ensure uniform coverage of material found in the text. The department makes strict use of placement test scores for initial placement in mathematics courses. The experimental sections did not share the common syllabus and tests because of the significant differences in texts and instructional use of technologies, but care was taken to ensure that the experimental sections (which where not specially identified in the timetable) provided comparable preparation for the institution's calculus sequences. For example, (a) Instructors of graphing-technologies sections used common examinations, met regularly to discuss course activities, and consulted about final grades; (b) a version of the MAA calculus-readiness placement test (non-calculator version) was used as pre- and post-test in the graphing approach sections; and (c) instructors of graphing-technologies sections were departmental faculty—including the college-algebra faculty coordinator—and experienced teaching assistants who had recent or concurrent teaching assignments in the department's regular college algebra courses.

1.2. Rationale. Because several earlier studies had shown gains in conceptual knowledge or higher levels of graphical reasoning for students who regularly used technologies, I expected to find several positive, lasting benefits of the graphical approach to precalculus. This expectation was based on a view of conceptual knowledge as richly linked networks of ideas (for example, Hiebert and Carpenter, 1992). By developing such connected knowledge, students would be expected to use ideas in new ways. The organization of ideas would also help them retain the knowledge longer. Because the text emphasized using graphing technologies to explore situations, to investigate connections between graphical and algebraic representations of functions, and to apply similar thinking across different families of elementary functions (such as polynomial, rational, exponential, logarithmic), students spent less time than in traditional precalculus classes learning a collection of special algebraic techniques relating to specific functions.

Underlying the expectation that lasting effects would be found is the idea that technologies provide tools that can increase the cognitive power of students by (a) reducing or eliminating the need for extensive development of routine manipulative skills before concepts can be studied, (b) providing multiple perspectives on important concepts that will help students to construct their mathematical knowledge, and (c) promoting a broader, more unified view of basic mathematical notions, particularly *function*, that can serve as a common thread running through courses (Philipp, Martin, & Richgels, 1993). "The mathematics of function acts as the unifying idea for the reformed curriculum," wrote Judah Schwartz (1993), echoing statements of other researchers when describing the direction being taken by members of the Algebra Working Group of the National Center for Research in Mathematical Sciences Education (NCRMSE). Graphing technologies give students easily accessible, different perspectives on this central theme that in the past has mostly been approached through algebraic representations. It is hoped that the use of various easily manipulated representations can promote the development of richly connected, meaningful knowledge in the domain.

The view that *function* can provide a useful organizing framework for student knowledge is not new. This view was advanced in earlier reform movements (examples are Buck, 1970; the Mathematical Association of America's National Committee on Mathematical Requirements, 1923/1970; Moore, 1903/1970; Smith, 1904/1970). Today, though, technologies increase the opportunities for successfully implementing such recommendations by providing easily manipulated graphical representations from algebraic equations (or from tabular data).

1.3. Related Research Perspectives. Although the body of research literature dealing specifically with studies of graphing technologies and their impact on learning of collegiate mathematics is not large, four areas of existing mathematics education research helped to provide a context for the study: (a) research on students learning to understand graphical representations of functions; (b) the influence of technologies on the balance between concepts and procedures in mathematics courses; (c) studies that investigated learning in courses using graphing utilities; and (d) the apparent impact of precalculus on subsequent mathematics course work. The fourth area, the focus of this study, contained almost no published research.

Research and theoretical discussion relating to students' understanding of functions and graphs is more widespread; one review including scores of studies was given by Leinhardt, Zaslavsky, and Stein (1990). An MAA Notes volume edited by Dubinsky and Harel (1992) contains a series of articles exploring this domain. Much of the literature shows that students have serious conceptual difficulties in the domain (for example, Carlson, 1998; Goldenberg, 1988); much less research shows how to remedy the situation (but see, as an example, Mokros and Tinker, 1987). More positively, several researchers have found thattechnologies can be used to enhance learning in precalculus and calculus classes (Beckmann, 1989; Dugdale, 1990; Heid, 1988; Schrock, 1989; Tufte, 1990). Studies suggested that students in the experimental courses performed just as well as traditional groups on routine computational tasks (Harvey, et al., 1994; Heid, 1988; Judson, 1988). Only one study looked for transfer of skills from graphing precalculus to a graphing unit in a subsequent physics course; in that study, Nichols (1992) detected no transfer of graphing skills from the prior use of graphing technologies in precalculus mathematics to the use of graphs in introductory physics. Penglase and Arnold (1996), Dunham and Dick (1994), and Dunham and Osborne (1991) compiled more detailed summaries of research in this area that I have not attempted to describe in this paper.

2. Design of the Study

2.1. Participants. For five years the Department of Mathematics at this University offered several sections of graphing approach precalculus (college algebra, trigonometry) along with many regular, traditionally taught precalculus sections. About 130 students, who studied college algebra in graphing approach sections over a two-year period, provided a pool of potential GT subjects for the study. Nine of these students, who were still enrolled in first-semester, traditionally taught (business or engineering) calculus courses after eight weeks of the semester, were paired with nine NT students from the same calculus discussion sections. The NT students—who were enrolled in the same sections of either regular (engineering) or of business calculus—had traditional college algebra backgrounds atthe same institution. The students were matched as closely as possible based on their grade for college algebra, cumulative grade point average, high school mathematics units, and gender (a complete listing of participants, with comparative data, is in Appendix A).

Participants in this study were from many different sections of precalculus, taught by a variety of instructors. Two points must be emphasized: (a) This is not a comparative study of carefully controlled experimental and traditional treatments; (b) No classroom observations of the precalculus instruction were available (having occurred up to a year earlier). Detailed questionnaires established that the graphing approach sections did include many activities that one would expect in such a class, including extensive use of graphing calculators in a variety of useful ways (see Appendix B for the questionnaires; Martin, 1994, contains details of the responses). Additionally, the researcher had extensive experience—including teaching both forms of the course but not any of this study's participants—with the precalculus program at this institution during these years. The graphing approach sections did take a different approach in terms of the content and the use of technologies. The instruction and class activities in those sections, however, was similar to that used in traditional sections. The experimental approach did not involve any dramatic changes in class organization, other than somewhat increased interaction between students and instructors because of the use of technologies during lessons. Instructors did not receive special instruction in the use of graphing technologies. They did regularly confer with other instructors about the course and the use of technologies.

A faculty member who taught sections of graphing precalculus over several years wrote the following about the use of technologies:

> When I teach a section of College Algebra I regard the graphing calculator as a mathematical tool. As such we use it a great deal to graph functions in different viewing rectangles, to observe-where the function achieves relative extrema, what the "general shape" of the graph is, where the graph crosses the x- and y-axes. Then we use the numeric abilities of the calculator to find values for the intercepts. Additionally the calculator is heavily used in helping students to understand what happens, for example, when $y = x^2$ becomes $y = a(x-b)^2 + c$. Students are encouraged to use their calculators whenever and however they like (even if I sometimes don't approve). But the use of the calculator as a tool to investigate and explore the problems is emphasized.

These remarks are representative both of my own experiences and of my impression of how others used technologies in parallel graphing-approach sections during the years up to this study.

2.2. Data. Data for the study came from individual, audiotape recorded, "thinking aloud" sessions; university records; and questionnaire responses from the 18 students and their calculus instructors:

(1) University records provided information about (a) the students' backgrounds in mathematics before calculus and (b) their final grades in the calculus courses they enrolled in during the study. These records were used to identify the matched group of NT students.

(2) Each student's calculus discussion section teaching assistant (TA) was asked to complete a short questionnaire about the student's apparent readiness for calculus based on their contact during the first three months of the semester. Copies of the questionnaires are given in Appendix B.

(3) Students completed two questionnaires. One was a parallel version of the instructor's questionnaire that asked about the student's own perceived

readiness for calculus. The second questionnaire dealt with their use of graphing technologies and, when appropriate, the use of technologies by their precalculus instructor.

(4) Participants provided written work on nine mathematics items from precalculus and introductory differential calculus along with an audiotape recording of the problem solving session. Students were asked to "think aloud" as they worked to provide insights to their thought processes.

2.3. Items. The problem solving session was the most important source of data for the study, so items were carefully chosen to provide a wide variety of opportunities for students to reveal their mathematical thinking. The use of graphing technologies as a regularly available tool in mathematics classes may have influenced several important mathematical capabilities of students. These capabilities included students' (a) use of graphical representations, (b) use of algebraic representations, (c) use and knowledge of functions, (d) use of mathematical models, and (e) problem-solving capabilities (meaning, for this study, their ability to deal with unfamiliar problems).

The items used in this study were selected with several criteria in mind:

(1) Items should cover important concepts from precalculus and calculus; that is, the items need to have content validity.
(2) The tasks should represent a wide range of mathematical capabilities.
(3) Some items should include tasks on which, in earlier studies, graphing students were found to be more successful than were their counterparts with a traditional background.
(4) Although graphing technologies could reasonably and effectively be used in solving problems, technologies should not provide an obvious, unfair advantage or a clearly easier solution.
(5) Several of the first items should be simple, routine ones so that students gain confidence and give the later, more complex tasks their best efforts.
(6) Some items should be solvable in more than one way. The choice of strategies should be real—that is, there is not one clearly superior approach.
(7) Some more complex problems should allow students to demonstrate more sophisticated mathematical capabilities. Several items should challenge all or most of the participants to reveal how they react when they reach an impasse.

A complete list of items, along with the rationale for the inclusion of each, is given in Appendix C. Several items were derived from problems used in previous studies. There were predetermined interview questions that were asked of each student once they appeared to have finished making progress on each item, which are also given in Appendix C. These interview questions were designed to encourage students to speak in greater detail about their reasoning and to consider alternate solution strategies for the item. They were selected in advance to avoid inadvertently and differentially influencing student responses and actions by asking leading questions or by making unplanned comments during the sessions.

The first four items were intended to start the session gently by testing precalculus material with which students should have been comfortable. The early items generally were unidimensional; that is, they required just one idea or action for their solution. The tasks increased in complexity and difficulty toward the end of the problem set where multidimensional tasks required the integration of several ideas.

In stating the rationale for choosing items I gave several criteria that guided the selection process. One objective was that the items draw on a range of mathematical capabilities. Without going into details, the selected items were carefully analyzed to ensure balance and spread of actions they might elicit from participants. As an illustration of the considerations that were involved, Item 3 could be solved by using a graphing utility to generate and manipulate graphical representations dynamically. However, this is neither necessary, nor is it inappropriate. The item provides a realistic choice of appropriate solution strategies. In contrast, Item 4 can only be solved by estimating values from the static graph presented with the item. This only requires static graph interpretation, not construction or manipulation that might favor the use of a graphing calculator with dynamic graphing capabilities.

The items used in this study were grouped into several (not necessarily disjoint) categories related to the main research questions. The categories that were used are listed below in Table 1.

TABLE 1. Item Categories

Categories		Items
Graph Interpretation	necessary	1, 2a–e, 4ab, 5d, 6ab, 7, 8a–c
	possible	3a–c, 5a–c, 9df
Algebraic Manipulation	necessary	5d, 9bce
	possible	3a–c, 5a–c, 9df
Graphing Calculator Possible		3a–c, 5a–d, 7, 9adf
Precalculus	routine	1, 2a–d, 3a–c, 4ab
	complex	7, 9a–c, 9e
Calculus	routine	2e, 6ab
	complex	5a–d, 8a–c, 9df

The most important data were the students' work on nine items drawn from precalculus and calculus. The free-response items ranged from routine precalculus (Item 3: solve a quadratic system of two equations in two unknowns) to conceptual calculus (Item 8: sketch a graph for a function based on a sign chart for the function and its first two derivatives). As illustrated above in the discussion of Items 3 and 4, graphing technologies could have contributed useful information for several items although they did not provide an obvious advantage. Other items involved graphical concepts, but did not offer any opportunity to use a graphing utility.

Several other items illustrate the range of mathematical ideas that were investigated. Item 5 is drawn from beginning differential calculus. It does not require the use of technologies but graphing tools could be useful in the search for answers to each of its four parts. It turned out that Item 5 produced the greatest variability of solution strategies and levels of success and, thus, some of the most useful data developed by the study. Item 1 involved graphical concepts without any possibility of using a graphing utility to obtain a solution. It was intended as a warmup exercise to help the students become accustomed to "thinking aloud." Surprisingly, more than half of the students were unable to solve it correctly.

2.4. Data Analysis. Two characteristics of the participants' work were of interest: (a) the *outcome*, or degree of success achieved on items; and (b) the *processes* by which students attempted to solve the items. Information about both came from students' written work and from protocol transcriptions of the audio-taped interviews. I developed several coding schemes to record this information; the codes were used to help identify patterns in the written and spoken data. The coding schemes are given in Appendix D. The outcome codes rated the degree of success achieved on the item on a 0–5 scale while the process codes were used to generate frequencies of various solution strategies and behaviors. During the coding process I also wrote extensive notes about the students' work. A random selection of papers were double coded, and the codes were carefully analyzed. This analysis showed that the coding was reliable. Details are available in Martin (1994).

Several statistical techniques, including factor and cluster analysis, were used to explore the data. In cases where there were noticeable differences between the two groups I used the nonparametric Wilcoxon matched-pairs signed-ranks test. A significance level of $p < 0.05$ was adopted for statistical tests in the study.

3. Results and Conclusions

3.1. Results. The data showed that there were statistically significant differences ($p < 0.05$) between the groups in the ways that students approached items and in their uses of graphing technologies. GT students not only made more use of graphing technologies and strategies as they worked on the items; they also reported greater use of technologies in their traditionally taught calculus course. Two NT students reported slight prior exposure to graphing technologies and used graphing technologies in calculus, but the GT students reported usi ng technologies in a much wider range of ways. The extent to which students used technologies was significant because students reported either indifference to, or even outright prohibition of, the use of technologies in their traditionally taught calculus courses. For example, when asked how the graphing calculator was used in calculus, one GT student wrote "Never. Not allowed to use it." Several GT students remarked that they believed they should avoid dependence on technologies if they were to be successful in regular mathematics courses. In this regard one wrote "... [U]sing the grapher was efficient, but it's very easy to depend on it too much. For that reason, I try not to use it as anything more than a check ...I can't use the calculator on tests so I try not to become too dependent on it." Detailed information about the questionnaire responses are available in Martin (1994).

There were no significant differences between the groups in their readiness for calculus (their own and their instructors' impressions), final calculus grades, or degree of success on most of the research items. Signs of differences reported by earlier studies could be seen in the pattern of results, but all but one of the differences were not statistically significant. For example, the GT students were more successful with all tasks that required graphical interpretation. They also had greater success solving an inequality, an item that could be solved graphically or algebraically.

The one significant difference in success rates ($p < 0.05$) was on Item 1: Six GT background and two NT students correctly solved the problem. A similar pattern, but with smaller differences between groups, had been observed on a similar multiple-choice problem administered to all college algebra students at the University of Wisconsin–Madison. There, success rates for 865 students in regular college algebra sections and 80 students in graphing-technologies sections were 59% and 74%, respectively.

The largest differences found by the study were in comparisons of success rates on different items instead of comparisons between the two groups of students. It was striking that, aside from the mentioned differences, the groups were so similarly successful with routine, precalculus tasks and had little or no success with more conceptualtasks drawn from calculus. Summaries of success rates on subtests are provided below in Table 2. These statistics reveal some information about the students' performance on the items and illustrates some of the patterns of success on different tasks.

TABLE 2. Test and Subtest Scores for Degree of Success

Subtest	Mean Score [a] Graph	Trad	Wilcoxon ($n = 9$ pairs) $G > T$ (mean rank)	$G < T$	p
All Items	3.18	3.11	5 (5.00)	4 (5.00)	0.7671
Precalculus	3.79	3.59	6 (5.00)	3 (5.00)	0.3743
Calculus	2.52	2.70	4 (4.25)	5 (5.60)	0.5147
Graphical Interpretation Necessary	3.43	3.21	6 (4.92)	3 (5.17)	0.4069
Algebraic Manipulation Necessary	2.83	3.08	4 (3.50)	4 (5.50)	0.5754
Choice of Alg. or Graph. Strategy	3.07	2.97	4 (6.25)	5 (4.00)	0.7671
Graphing Calculator Possible	2.97	2.80	5 (4.80)	3 (4.00)	0.4008

[a]The scores reported in this table are *mean score per item* for each test or subtest. This facilitates comparisons between tests and interpretation of absolute levels of scores. The codes awarded for the students' degree of success on items had the following interpretations:

5 = correct   4 = essentially correct   3 = good progress
2 = little progress   1 = nothing relevant   0 = blank

Table 3 provides more specific information about success rates, use of technologies, and consideration or description of alternate strategies by group. Technology use was rated on a scale of 0–4, with high scores reflecting more sophisticated usage (refer to Appendix D for details of coding schemes). Consideration of alternate scores is a count of the number of valid alternate strategies that students suggested, with and without prompting, for an item during the interview. I carefully kept track of which work was completed before and after questioning by the interviewer

None of the GT students could be characterized as *advanced* users of graphing technologies. Instead, they used the graphing calculators in *routine* ways such as to find the intersection of graphs of two functions. For example, GT students used their calculators effectively (mean use rating 3.0) to have greater success with item 3c (mean final success 4.33). This type of item was representative of problems solved with graphing technologies in their precalculus course. They provided somewhat better graphs for Item 5d (mean success 2.56 compared to 1.89 for NT students), but this often reflected copying of an image from the graphing calculator.

The GT students had little success using technologies to gain insights to items that they could not solve. For example, even though some students wanted to factor the numerator and denominator of the function in Item 5, they did not think to use

WILLIAM O. MARTIN

TABLE 3. Mean Outcome Score by Item and Group

	Initial		Final [a]		Tech. Use		Alt. Strat.	
Item	Gra	Tra	Gra	Tra	Gra	Tra	Gra	Tra
1	3.89	1.78	3.89	2.33	0.22	0.00	0.78	0.22
2a	3.78	3.78	3.78	3.78	0.00	0.00	1.00	1.22
2b	3.22	3.00	3.56	3.22	0.00	0.00	0.56	0.56
2c	5.00	5.00	5.00	5.00	0.00	0.00	1.00	1.11
2d	4.44	4.22	4.44	4.22	0.00	0.00	0.89	1.00
2e	3.56	4.00	3.56	4.00	0.00	0.00	0.67	1.00
3a	4.22	3.78	4.22	4.22	1.33	0.00	1.67	1.22
3b	4.89	4.00	4.89	4.00	0.78	0.00	2.00	1.33
3c	4.11	2.89	4.33	3.22	3.00	0.11	1.00	0.89
4a	3.67	3.56	4.11	3.56	0.00	0.00	1.11	0.78
4b	4.44	3.44	4.44	3.44	0.00	0.00	0.89	0.78
5a	2.67	3.89	2.67	3.89	0.67	0.00	0.56	0.89
5b	2.89	3.67	3.22	3.67	1.56	0.00	0.78	0.89
5c	3.11	2.33	3.22	2.56	1.67	0.00	0.11	0.11
5d	2.56	1.89	2.56	1.89	2.33	0.22	0.56	0.67
6a	3.00	3.67	3.22	3.67	0.00	0.00	0.78	0.67
6b	4.11	3.67	4.11	3.67	0.00	0.00	0.33	0.33
7	3.11	2.56	3.11	2.56	1.89	0.00	0.22	0.22
8a	2.00	2.33	2.00	2.33	0.00	0.00	0.11	0.33
8b	1.78	2.89	1.78	2.89	0.00	0.00	0.00	0.00
8c	1.89	1.67	1.89	1.67	0.00	0.00	0.22	0.22
9a	2.44	2.33	2.44	2.56	0.00	0.00	0.78	0.78
9b	3.67	3.67	3.67	4.00	0.00	0.00	0.44	0.56
9c	2.44	3.11	2.44	3.11	0.00	0.00	0.33	0.33
9d	1.33	1.22	1.44	1.22	0.56	0.11	0.56	0.67
9e	2.67	3.33	2.67	3.33	0.00	0.00	0.11	0.11
9f	0.56	1.00	0.56	1.00	0.00	0.00	0.67	0.44
2E [b]			3.56	3.56				
5E			2.56	2.56				
9E			2.11	2.11				

[a] Initial scores were for work completed before any questioning by the interviewer. Final scores were for all work on the item, including changes or additions made after the interviewer asked the predetermined questions and after students were permitted to review their work at the end of the session.

[b] On items 2, 5 and 9 students were given a final progress code reflecting their degree of success on the entire item.

the graphing calculator (mean use 0.67, 1.56, 1.67 for parts 5a–5c, respectively) to assist when they were uncertain about the factoring identities. The use of graphical representations to aid with factoring had received considerable attention in the graphing approach precalculus course, but apparently this did not transfer to an item expressed with the limit notation from calculus. Even though GT students had somewhat better graphs for part 5d, many did not see connections between those images and the first three parts of the item. Several even had responses in parts 5a–5c that contradicted the graphs produced in part 5d.

Neither did GT students use technologies (low mean use rating of 0.56 for part 9d) to help with the optimization Item 9. None of the 18 students completely solved this problem. Again, use of graphical representations for optimization was an important component of the graphing approach precalculus course, but students apparently did not transfer that knowledge to this calculus context.

Process codes were assigned for work on Items 3, 5, 7, 8, and 9. Items 1, 2, 4, and 6 were uni-dimensional tasks with one-step solutions that were not suited to this type of analysis. Table 4 below lists the process codes for which there were significant differences between the groups for the aggregated frequencies across all five items. A complete summary of the process codes used in this study is given in Appendix D.

TABLE 4. Significantly Different Process Codes by Group

Process Code	Sum of codes for Items 3, 5, 7, 8, and 9		
	GT	NT	p
Strategy-SGa [a]	27	6	0.0180
Check-G	15	1	0.0425
Check-N	1	12	0.0431
Relate-Rag	21	8	0.0251
Representation-GH	0	5	0.0431
Representation-GC	42	0	0.0077
Technology-3	23	0	0.0117
Technology-4	13	0	0.0117

[a]Process codes are described in Appendix D. For example, Strategy-SGa means that an *a*ppropriate *S*tandard *G*raphing *Strategy* was used or described by the student while working on the item.

The differences in the distributions of codes by item part (e.g., 3a or 5c) were generally not statistically significant. The exception to this was in frequencies of codes assigned for the use of technology and calculator-produced graphical representations; there were several significant differences between the groups for these scores on Items 3 and 5. The process codes for which there were significant or near significant differences between the groups are reported below in Table 5. These data show that GT students were more inclined to use graphical information to solve and check solutions than the NT students, who were more likely to use algebraic representations.

Qualitative analyses of the data revealed interesting contrasts between the two groups of students that warrant furtherinvestigation. Several apparent differences between the groups that, although not statistically significant (partly, at least, because of the small sample size), followed a pattern that could reflect expected patterns of thinking due to differences between graphing and traditional approaches. For example, and not surprisingly, the data above showed large differences between the groups in their inclinations to use graphical representations and strategies.

TABLE 5. Selected Number of Process Codes by Group

Item	Process Code	Sum of scores		p
		GT Total	NT Total	
3b	Check-N	0	4	0.0679
3c	Representation-GC	7	0	0.0180
3c	Representation-GH	0	4	0.0679
3c	Technology-4	5	0	0.0431
5a	Representation-GC	5	0	0.0431
5b	Check-G	4	0	0.0679
5b	Error-I	5	0	0.0431
5b	Representation-GC	5	0	0.0431
5c	Relate-Rag	4	0	0.0679
5c	Representation-A	4	8	0.0679
5c	Representation-GC	6	0	0.0277
5c	Strategy-SGa	4	0	0.0679
5d	Representation-GC	8	0	0.0117
5d	Representation-T	0	4	0.0679
5d	Strategy-SGa	8	0	0.0117
5d	Strategy-SNa	0	4	0.0679
5d	Technology-3	8	0	0.0117
7	Representation-GC	6	0	0.0277
7	Technology-3	5	0	0.0431
9f	Strategy-none	8	4	0.0679

Perhaps the most notable of these differences was in students' inclination to use *inappropriate standard strategies*, since this did not directly reflect use of technologies and graphs (something that one would expect students to use more often after a graphing precalculus course). A simple example of this is the use of l'Hôpital's rule on item 5c; several students used the rule after appropriately using it on 5a and 5b. Students apparently used the rule because of the appearance of the item but without checking that it was appropriate in this situation. A more revealing illustration is given by the work of one NT student, identified as T8, on the precalculus Item 3 shown below in Figure 1.

Here is the transcription of this student's comments while working on the item:

Item 3

T8: Um, I just believe you fill in negative one for g of x. So it equals seven I think. [Reads part (b)]

T8: Just set the two functions equal to each other. And factor. And then x would equal one and four. [Reads part (c)]

T8: Hm. Um, I'm not real sure about this one.

Interviewer: Remember to try to tell me what you're thinking as you work.

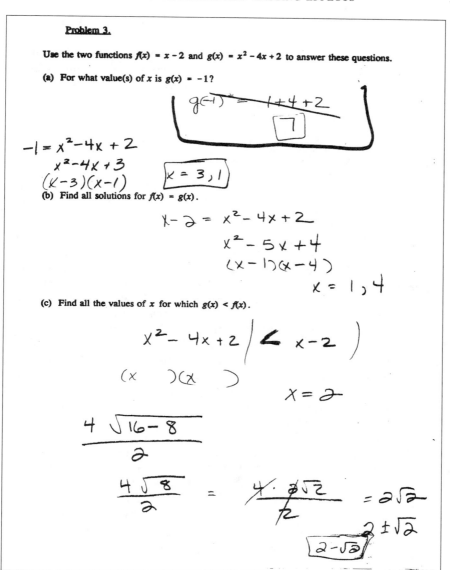

Problem 3.

Use the two functions $f(x) = x - 2$ and $g(x) = x^2 - 4x + 2$ to answer these questions.

(a) For what value(s) of x is $g(x) = -1$?

$$g(-1) = 1 + 4 + 2$$

$$\boxed{7}$$

$$-1 = x^2 - 4x + 2$$
$$x^2 - 4x + 3$$
$$(x - 3)(x - 1) \qquad \boxed{x = 3, 1}$$

(b) Find all solutions for $f(x) = g(x)$.

$$x - 2 = x^2 - 4x + 2$$
$$x^2 - 5x + 4$$
$$(x - 1)(x - 4)$$
$$x = 1, 4$$

(c) Find all the values of x for which $g(x) < f(x)$.

$$x^2 - 4x + 2 \, / < \, x - 2 \,)$$

$$(x \quad)(x \quad)$$

$$x = 2$$

$$\frac{4 \sqrt{16 - 8}}{2}$$

$$\frac{4 \sqrt{8}}{2} = \frac{4 \cdot 2\sqrt{2}}{2} = 2\sqrt{2} \qquad 2 \pm \sqrt{2}$$

$$\boxed{2 - \sqrt{2}}$$

FIGURE 1. Work of traditional group student T8 on Item 3

T8:   Okay. Let's see. I mean, usually when you use less than or greater
        than, you know what, when you divide by negative and the sign
        changes. But I don't know, that's not going to do any good in this
        problem. I don't think. Or you just solve for x. Um ... Either that
        or just solve each one individually. So x would be equal to two and
        this one ... you can't factor, you'd have to use ... negative, that,
        oh I can't think what it's called. Maybe, I can't, I don't know
        what the name of the formula was. Negative b plus or minus b
        squared minus four ac over two a.

Interviewer: Okay, I know what you mean.

T8:   Okay. Square [silence, writing sounds] So ...I think that's plus
             or minus, two plus or minus square root two, I think. So ...the
             values for x in which g of x is greater than, or less than f of x
             would be two minus the square root of two, I think. Would be the
             only one. Not that. I don't know.

Interviewer: Okay. Do you believe you've correctly solved each part of this
             problem? Again, go through each part.

T8:   Um, the first one I think so. Second one, for sure. And third one
             ...possibly. [laughs]

Interviewer: You're not so sure on the third one. Okay.

T8:   Right.

The student T8 seemed to solve problems by choosing a remembered technique
that was based on the surface appearance of the item. In solving parts 3a and 3b T8
was successful (after correcting an initial error in the first part) because the chosen
technique was appropriate. The work on part 3c, however, is unrelated to the content of the item; instead, it applies standard manipulative techniques, which were
sometimes carried out incorrectly, used to solve linear and quadratic equations. The
remarks made while working on the item indicate that T8 also considered—because
the inequality symbol appeared in the item—and rejected other techniques. The
use of standard manipulative procedures in an invalid context, such as illustrated
by T8's work on Item 3, provides important evidence of the student's beliefs about
the nature of mathematical problem solving. If this only happens once we might
attach no more significance to it than carelessness (especially in a case such as 5c).
However, in this study several students, both from the GT and the NT groups,
repeatedly used inappropriate standard strategies, suggesting an undesirable view
of mathematical problem solving.

NT students, including T8, used inappropriate standard procedures half again
as often (36 times) as did GT students (25 instances). Four of the nine NT students
each used inappropriate strategies at least four times, accounting for 25 of the 36
cases; only two GT students used inappropriate strategies at least four times for
a total of 9 instances. One might expect that weaker students would be most
inclined to apply procedures inappropriately but this was not the case. Several
of the students who used many inappropriate strategies did have lower grades for
college algebra and calculus (C's), but this group also included students who did
quite well in both courses: One of these two GT students earned an AB[1] in calculus
and had a BC for college algebra; two of the aforementioned NT students had AB's
and B's for the two courses. (T8 had a BC for calculus and a B for college algebra.[2])

This pattern fits with ideas about differences between traditional and graphing
approaches in precalculus. Were similar differences found to hold in further studies,
it could reflect that many students in traditional precalculus courses learn to recognize problem types and apply specific techniques, while students from graphing
approach courses focus more on understanding problem situations from a variety of
perspectives. When students with these backgrounds face an unfamiliar problem,

[1]In addition to the usual A–F grading scale, the institution records intermediate grades of
AB and BC between the regular grades of A, B, and C, respectively. The grade-point values of
AB and BC are 3.5 and 2.5, respectively, on the usual 4-point scale where A is 4 points.

[2]One interesting sidelight to the findings from this study was the lack of any apparent relationship between performance on these problems and grades in precalculus or calculus.

a traditional background student may be more likely to try to impose what seems like an appropriate solution strategy from his or her repertoire. Graphing students, on the other hand, might be more inclined to explore the problem, perhaps using technologies and various representations. If true, this would indicate that the use of technologies can, in fact, contribute to the development of more powerful and flexible problem-solving capabilities as hypothesized. The results of this study hint that this could be the case, but are not strong enough to offer confirmation.

3.2. Summary and Conclusions. The purpose of this study was to investigate the lasting effects that the use of graphing technologies in precalculus. Based on a wide variety of data gathered from university records, questionnaires, and a problem solving interview, it is apparent that while the graphing approach did influence the calculus students in ways distinct from a comparable group of students with a more traditional precalculus background, the impact of the precalculus course was less than one would hope. GT students in this study did not display the expected enhanced conceptual knowledge of important calculus concepts. The graphing precalculus course appeared to have had a lasting influence, but not to the extent nor in the conceptual domain that one might have hoped for. GT students did not appear disadvantaged in the traditional calculus course and they continued to draw on a graphical perspective in their work even though this was not encouraged by their instructors. In a search for possible explanations of these findings I considered three groups of issues that could account for the weaker than expected results: (a) issues surrounding the design of the study, (b) issues relating to the mathematics courses that were studied, and (c) issues relating to the broader educational context of the study and courses.

Several noticeable differences between the groups were in the expected direction but were not found to be statistically significant. One problem was that the group of students enrolled in calculus provided a sample that was considerably smaller than had been anticipated during the planning stage. This reduced the power of the statistical tests used to detect significant differences between the groups. For example, GT students were noticeably less inclined to unquestioningly apply inappropriate strategies. Similarly, GT students had more success with graph interpretation tasks, although not significantly so. Perhaps true differences, in the direction one would predict, were not detected because of the lack of statistical power.

I do not believe that the lack of statistical power accounts for the small number of significant differences detected between the two groups. Instead, the extensive information obtained about each student by this study suggests that both groups were, in fact, highly comparable calculus students, despite their differing backgrounds. The impression I gained from the data and my contact with the students during the interview sessions was that neither group had the sort of strong, conceptual mathematical knowledge, applicable in new or different situations, that one would hope for at this level.

Why, then, had the graphical precalculus not developed the conceptual power that I had expected? I believe that (a) the nature of the precalculus and calculus courses and (b) their position in the broader educational context accounted for the pattern of results found by the study. Dugdale's work (1990) emphasized that the way in which technologies are used is at least as important as whether or not they are used. Although the Demana and Waits textbook (1990) emphasizes the use of technologies to develop conceptual knowledge of functions, their textbook was used in a very traditional way by the students involved in this study. That is, instructors would explain new material and the students would work on similar problems. While technologies provided more opportunities for discussion and interaction in class, the approach did not give students important experiences with independent work in unfamiliar problem situations. In contrast, students who study mathematics in more radically restructured courses, such as the calculus courses studied by Park and Travers (1996) and Crocker (1991), may show greater lastingdifferences in their approach to mathematics because technology is used to restructure instruction, not just course content. The importance, and a means, of relating instructional treatment to careful analysis of student thinking has been described in more recent collegiate mathematics education research programs (Asiala et al., 1996). The studies conducted under that research framework have also used technologies as a tool to enhance student learning. However, the treatments were based on much more careful analyses of the mathematical ideas being studied and how individuals build their understandings. The results of this present study emphasize (a) the inherent complexity of appropriately modifying instructional treatments to enhance student learning of mathematics, and (b) the importance of considering all components of instruction, not just the content and tools.

One also must question how much a single course can be expected to change students with at least 12 years' of experience in predominantly, if not entirely, traditional courses. We should expect that students will mostly organize new experiences within the existing structure of their mathematical knowledge. Taking a structural view of conceptual knowledge, it would take a remarkable intervention to have students dramatically reorganize their mathematical thinking over several months. I raised the question earlier of whether instructors were committed to the graphing precalculus approach, which could influence teaching and learning—one should also question the impact of student motivation. To what extent do the motivations that students bring to mathematics courses at this level moderate the impact of new approaches? Many students who take precalculus mathematics in college seem more concerned with grades and credits than with developing mathematical knowledge. Students also have beliefs about the nature of mathematics that they probably do not question. This, too, would act to limit the lasting impact of any course, especially if the course goals did not include efforts to modify student or instructor beliefs.

The main conclusions of this study were that (a) graphing technologies do have a lasting impact on students, even when the use of these tools is discouraged or prohibited, and (b) many students do not become sophisticated users nor do they gain lasting enhancements of conceptual knowledge during a one semester course delivered in a fairly traditional manner. The students' inclination to continue using technologies, even without encouragement, is a positive sign that technologies can be used to favorably influence their learning and attitudes toward mathematics. Still, our real hope is that we can find ways to help students develop much better mathematical knowledge than demonstrated by the participants in this study. Technologies offer many potential benefits that could help to improve conceptual learning and mathematical power that last into subsequent courses. But for this to happen, technologies must be used in more varied and thoughtful ways so that the benefits, such as found by earlier studies, persist.

The results of this study, although disappointing for their lack of conclusive support for, or against, the use of technologies in precalculus mathematics, may still have served a useful purpose. This study provides no evidence that the graphing approach is detrimental even for students who will move to traditionally taught courses; that they will become unthinking "button pushers" lacking important algebraic skills. Instead, we see here that curricular materials and technologies, even in the hands of regular instructors (as distinct from technology advocates or enthusiasts), can have a lasting impact on students' approaches to mathematics. Most important, though, is the clear need to further study how we might change and improve instruction of precalculus mathematics so that students develop levels of mathematical power that were evident in almost none of the participants in this study, either GT or NT. While technologies may be a tool thatcan aid such instructional changes, this study shows that technologies and curricula are only part of the issue. It seems that real changes in learning will require more careful attention to the interplay of curricula, instruction, and learning.

References

Asiala, M., Brown, A., DeVries, D., Dubinsky, E., Mathews, D., & Thomas, K. (1996). A framework for research and curriculum development in undergraduate mathematics education In J. Kaput, A. H. Schoenfeld, & E. Dubinsky (Eds.), *Research in collegiate mathematics education II* (pp. 234–283). Providence, RI: American Mathematical Society.

Beckmann, C. E. (1989). Effect of computer graphics use on student understanding of calculus concepts. Doctoral dissertation, Western Michigan University. *Dissertation Abstracts International, 49*, 1974B.

Browning, C. A. C. (1989). Characterizing levels of understanding of functions and their graphs. Doctoral dissertation, Ohio State University. *Dissertation Abstracts International, 49*, 2957A.

Buck, R. C. (1970). Functions In E. G. Begle & H. G. Rickey (Eds.), *Mathematics education (69th yearbook of the National Society for the Study of Education)* Chicago: University of Chicago.

Carlson, M. (1998). A cross-sectional investigation of the development of the function concept. In A. H. Schoenfeld, J. Kaput, & E. Dubinsky (Eds.), *Research in Collegiate Mathematics Education III* (pp. 114–162). Providence, RI: American Mathematical Society.

Cohen, D. (1989). *College algebra* (2nd ed.) St. Paul, MN: West Publishing Company.

Crocker, D. A. (1991). *A qualitative study of interactions, concept development and problem solving in a calculus class immersed in the computer algebra system Mathematica.* Unpublished doctoral dissertation, Ohio State University, Columbus.

Curjel, C. R. (1993). Classroom computer capsule: A computer lab for multivariate calculus *The College Mathematics Journal, 24*(2), 175–177.

Demana, F., & Waits, B. (1993) The particle-motion problem. *Mathematics Teacher, 86*(4), 288–292.

Demana, F., & Waits, B. (1990). *College algebra and trigonometry: A graphing approach.* Reading, MA: Addison-Wesley.

Dick, T. P. & Patton, C. M. (1992) *Calculus* (Vol. 1). Boston: PWS-Kent Publishing.

Dubinsky, E., & Harel, G. (Eds.) (1992). *The concept of function: Aspects of epistemology and pedagogy* (MAA Notes no. 25). Washington, DC: Mathematical Association of America.

Dugdale, S. (1990). Beyond the evident content goals, part III. An undercurrent-enhanced approach to trigonometric identities. *Journal of Mathematical Behavior, 9*, 233–287.

Dunham, P. H., & Dick, T. P. (1994) Connecting research to teaching: research on graphing calculators. *Mathematics Teacher, 87*(6), 440–45.

Dunham, P. H. & Osborne, A. (1991). Learning how to see: students' graphing difficulties *Focus on Learning Problems in Mathematics 13*(4), 35–49.

Dunham, P. H. (1991). Mathematical confidence and performance in technology-enhanced precalculus: Gender-related differences. Doctoral Dissertation, Ohio State University. *Dissertation Abstracts International, 51*, 3353A.

Farrell, A. (1996). Roles and behaviors in technology-integrated precalculus classrooms. *Journal of Mathematical Behavior 15*(1), 35–53.

Farrell, A. M. (1990). Teaching and learning behaviors in technology-oriented precalculus classrooms.). Doctoral Dissertation, Ohio State University. *Dissertation Abstracts International, 51*, 100A.

Goldenberg, E. P. (1988). Mathematics, metaphors, and human factors: Mathematical, technical and pedagogical challenges in the educational use of graphical representation of functions. *Journal of Mathematical Behavior, 7*(2), 135–173.

Harvey, J., Martin, W., Demana, F., Osborne, A., & Waits, B. (1994). *Effectiveness of graphing technologies in a precalculus course: The 1988-89 field test of the C^2PC materials.* Unpublished technical report. University of Wisconsin–Madison: Author.

Hiebert, J., & Carpenter, T. P. (1992). Learning and teaching with understanding. In D. C. Grouws (Ed.), *Handbook of research on mathematics teaching and learning* (pp. 65–97). New York: Macmillan.

Heid, M. K. (1988). Resequencing skills and concepts in applied calculus using the computer as a tool. *Journal for Research in Mathematics Education, 19*, 3–25.

Hughes-Hallett, D., Gleason, A., McCallum, W., Flath, D., Lock, P., Gordon, S., Lomen, D., Lovelock, D., Mumford, D., Osgood, B., Pasquale, A., Quinney, D., Tecosky-Feldman, J., Thrash, J., Thrash, K., & Tucker, T. (1998). *Calculus: Single and multivariable* (2nd ed.). New York: John Wiley & Sons.

Hungerford, T. (1997). *Contemporary precalculus: A graphing approach* (2nd ed.). Philadelphia: Harcourt Brace.

Judson, P. T. (1988). Effects of modified sequencing of skills and applications in introductory calculus. Doctoral dissertation, The University of Texas at Austin. *Dissertation Abstracts International, 49*, 1397A.

Leinhardt, G., Zaslavsky, O., & Stein, M. K. (1990). Functions, graphs, and graphing. *Review of Educational Research, 60*(1), 1–64.

Martin, W. O. (1994). Lasting effects of the integrated use of graphing technologies in precalculus mathematics. Doctoral Dissertation, University of Wisconsin–Madison. *Dissertation Abstracts International, 54*, 3694A.

The Mathematical Association of America's National Committee on Mathematical Requirements. (1970). The reorganization of mathematics in secondary education. In J. K. Bidwell & R. G. Clason (Eds.), *Readings in the history of mathematics education* (pp. 382–459). Washington, DC: National Council of Teachers of Mathematics. (Original work published in 1923)

Mokros, J. & Tinker, R. (1987). The impact of microcomputer-based science labs on children's ability to interpret graphs. *Journal of Research in Science Teaching, 24*, 369–383.

Moore, E. H. (1970). On the foundations of mathematics. In J. K. Bidwell & R. G. Clason (Eds.), *Readings in the history of mathematics education* (pp. 246–255). Washington, DC: National Council of Teachers of Mathematics. (Reprinted from *Science*, 1903, *17*)

Naraine, B. (1993). An alternative approach to solving radical equations. *Mathematics Teacher, 86*(3), 204–205.

Nichols, Jeri Ann (1992). *The use of graphing technology to promote transfer of learning: The interpretation of graphs in physics.* Unpublished doctoral dissertation, Ohio State University, Columbus.

Palmiter, J. R. (1991). Effects of computer algebra systems on concept and skill acquisition in calculus. *Journal for Research in Mathematics Education, 22*, 151–156.

Park, K., & Travers, K. (1996) A comparative study of a computer-based and a standard college first-year calculus course. In J. Kaput, A. H. Schoenfeld, & E. Dubinsky (Eds.), *Research in collegiate mathematics education II* (pp. 155–176). Providence, RI: American Mathematical Society.

Penglase, M. & Arnold, S. (1996). The graphics calculator in mathematics education: A critical review of recent research. *Mathematics Education Research Journal 8*(1), 58–90.

Philipp, R., Martin, W., & Richgels, G. (1993). Curricular implications of graphical representations of functions. In T. Romberg, E. Fennema, & T. Carpenter (Eds.), *Integrating research on the graphical representation of function.* Hillsdale, NJ: Lawrence Erlbaum Associates.

Ruthven, K. (1990). The influence of graphic calculator use on translation from graphic to symbolic forms. *Educational Studies in Mathematics, 21*, 431–450.

Schrock, C. S. (1989) Calculus and computing: an exploratory study to examine the effectiveness of using a computer algebra system to develop increased conceptual understanding in a first-semester calculus course. Doctoral dissertation, Kansas State University. *Dissertation Abstracts International, 50* 1926A.

Schwartz, J. (1993). The teachers and algebra project. *NCRMSE Research Review: The Teaching and Learning of Mathematics, 2*(1), 5–8.

Smith, D. E. (1970). The teaching of elementary mathematics, chapter VII: Algebra, what and why taught. In J. K. Bidwell & R. G. Clason (Eds.), *Readings in the history of mathematics education* (pp. 211–219). Washington, DC: National Council of Teachers of Mathematics. Original work published in 1904

Steen, L. A. (Ed.). (1990). *On the shoulders of giants: New approaches to numeracy.* Washington, DC: National Academy Press.

Tufte, F. W. (1990). The influence of computer programming and computer graphics on the formation of the derivative and integral concepts (derivative concepts). Doctoral dissertation, University of Wisconsin–Madison. *Dissertation Abstracts International, 51*, 1149A.

Vonder Embse, C. (1998). Multiple representations and connections using technology. In *Mathematics Teacher 91*(1), 62–67.

Appendix A. Participant Characteristics

Selection of participants. For about five years, the department of mathematics at this university offered several sections of graphing approach precalculus (college algebra, trigonometry) along with many regular, traditionally taught precalculus sections. About 130 students who studied college algebra in graphing approach sections during two years provided a pool of potential GT subjects for the study. Using campus transcripts, I found 13 students enrolled in first-semester calculus who had taken a graphing version of precalculus on this campus. Using transcript data for all students enrolled in the same calculus recitation sections, I identified a closely matched group of 13 NT students. Naturally, perfect correspondences were not possible. In selecting potential subjects I gave the highest priorities to the grade earned in college algebra, gender, and cumulative GPA. By choosing students from the same calculus sections I could assume that the GT and NT students had received, or at least had the opportunity to receive, comparable instruction in calculus during the semester.

All students identified for the study were still enrolled in the calculus course after eight weeks. Initial contact was made then by asking their TA to deliver a note that explained that I was conducting a study of students who took Math 112 (college algebra) and then enrolled in calculus. The note said nothing about graphing calculators or the specific aims of the study. My note to them said that they would be asked to solve some math problems and to complete a questionnaire. I emphasized in the note that because of my sampling method it was very important that they all participate—I said I would be unable to replace any of them. I also said that I would pay them for their time. I spent several weeks contacting the students by phone and ultimately all but two of the original 26 students agreed to participate. The interviews were conducted during a one-month period.

During the interviews I discovered that several students had dropped calculus. In each case I conducted the interviews anyway and retained the students in the study since the calculus questions I used would have been covered during the first six weeks of the course. One GT and one NT student retained in the study dropped calculus—they were from different pairs. I also discovered during the interview that three students I had identified as graphing background actually had a traditional background in Math 112. Two of the students were in a section that was changed from graphing to traditional because of low enrollments at the start of the semester. The third student started out in a graphing section but changed to a traditional section after several weeks. This change did not show up on the university records I used to identify the subjects. Because of this, I interviewed 24 students but only used the data for nine pairs in this study. Characteristics of the 18 students finally included in this study are given in Table 6. In mid-November I asked each TA to complete the questionnaires about his or her students. After the end of the semester I obtained the final calculus grade for each participant.

TABLE 6. Student Characteristics

ID (Gender)	Pair	Group	GPA	112 Grade	Calc Grade	HS Math
DM (M)	B	GT	3.24	A	DR	ALG1 GEO 1 ADV 2[a]
LM (M)	B	NT	3.58	A	A	ALG 1 GEO 1 ADV 2
DR (M)	D	GT	3.65	B	BC	ALG 1 GEO 1 ADV 1.7
MK (F)	D	NT	3.52	AB	B	ALG 1 GEO 1 ADV 2
KM (F)	F	GT	2.88	B	C	ALG 1 GEO 1 ADV 3
CM (M)	F	NT	2.78	BC	DR	ALG 1 GEO 1 ADV 2
MS (M)	G	GT	3.18	B	B	ALG 1 GEO 1 ADV 2
JP (M)	G	NT	3.34	B	AB	ALG 1 GEO 1 ADV 2
MW (M)	H	GT	1.30	C	F	ALG 1 GEO 1 ADV 2
BH (M)	H	NT	2.48	BC	B	ALG 1 GEO 1 ADV 2
CC1 (M)	I	GT	2.37	BC	AB	ALG 1 GEO 1 ADV 2
JG1 (M)	I	NT	2.69	AB	A	ALG 1 GEO 1 ADV 2
HS (F)	J	GT	2.84	C	C	ALG 1 GEO 1 ADV 1.5
RM (M)	J	NT	2.33	C	BC	ALG 1 GEO 1 ADV 2
AS (F)	K	GT	3.40	B	AB	ALG 1 GEO 1 ADV 3
DC (F)	K	NT	3.03	B	BC	ALG 1 GEO 1 ADV 2
FI (F)	L	GT	2.84	C	C	ALG 1 GEO 1 ADV 1
MH (M)	L	NT	2.18	C	C	ALG 1 GEO 1 ADV 2

[a] Years of high school mathematics: algebra (ALG), geometry (GEO), advanced (ADV)

Appendix B. Questionnaires

Calculus Instructor Questionnaire. _____ is participating in a longitudinal study of precalculus here at the University. Based on your contact with this student since the start of Math 221 [regular calculus I], how would you rate [his/her] readiness for calculus? (Circle the appropriate response for each statement)

(1) Overall readiness and preparation for calculus:
    Very high   Above average   Average   Below average   Very low   I can't tell

(2) Manipulative skills—ability to carry out algebraic manipulations accurately:
    Very high   Above average   Average   Below average   Very low   I can't tell

(3) Conceptual readiness—degree to which the student has the mathematical maturity or concepts necessary to learn calculus:
    Very high   Above average   Average   Below average   Very low   I can't tell

(4) Graphical ability—to what extent is the student able to construct, read and interpret graphs of functions:
    Very high   Above average   Average   Below average   Very low   I can't tell

(5) What grade is the student currently earning in calculus?
    A   AB   B   BC   C   D   F

(6) Please make any other comments about the student's apparent readiness for calculus:

Student Questionnaire. Based on your work during this semester in Math 221 how would you rate your readiness for calculus? (Circle the appropriate response for each statement)

(1) Your overall readiness and preparation for calculus is:
    Very high   Above average   Average   Below average   Very low

(2) Manipulative skills—your ability to carry out algebraic manipulations accurately:
    Very high   Above average   Average   Below average   Very low

(3) Conceptual readiness—degree to which you have the mathematical maturity or concepts necessary to learn calculus:
    Very high   Above average   Average   Below average   Very low

(4) Graphical ability—to what extent are you able to construct, read and interpret graphs of functions:

Very high   Above average   Average   Below average   Very low

(5) What grade do you believe you are currently earning in calculus?

A   AB   B   BC   C   D   F

(6) Please comment on the preparation you feel Math 112 gave you to study calculus:

Student Questionnaire—Calculator Experience. The following questions are intended to make you think about how you used calculators in mathematics classes. Some questions ask you to respond separately for when you took Math 112 and now that you are studying calculus. If you have difficulty responding to a question, please write a comment that explains why it was hard to answer.

(1) Do you own a graphing calculator? ___ Yes ___ No

(2) Did you have access to a graphing calculator...
In Math 112: ___ Yes ___ No
In Calculus: ___ Yes ___ No

(3) Do you have access to computer software that graphs functions?
___ Yes ___ No

Only answer the remaining questions if you said Yes to at least one question above. In the remaining questions graphing technology refers to either a graphing calculator or computer graphing software.

(4) How often did/do you bring graphing technology to class?
In Math 112:   Very often   Often   Sometimes   Rarely   Never
In Calculus: Very often   Often   Sometimes   Rarely   Never

(5) How often did/do you use graphing technology in class?
In Math 112:   Very often   Often   Sometimes   Rarely   Never
In Calculus: Very often   Often   Sometimes   Rarely   Never

(6) How often did/do you use graphing technology outside of class?
In Math 112:   Very often   Often   Sometimes   Rarely   Never
In Calculus: Very often   Often   Sometimes   Rarely   Never

(7) Check any of the following ways you use(d) graphing technology (respond separately for Math 112 and calculus):

	112	Calculus
Help to sketch a graph	--	--
Find intercepts	--	--
Find where curves intersect	--	--
Find values of a function	--	--
Solve inequalities	--	--
Look for ideas about a problem	--	--
Other (describe)		

(8) How useful do you find graphing technology in Calculus? How do you use it?

Complete the following statements about your Math 112 instructor's use of technology:

(1) The instructor spent time in class explaining how to use the calculator
Very often   Often   Sometimes   Rarely   Never

(2) The instructor demonstrated how to use the calculator as a tool for solving problems
Very often   Often   Sometimes   Rarely   Never

(3) The instructor encouraged us to solve problems using standard algebraic methods
Very often   Often   Sometimes   Rarely   Never

(4) The instructor encouraged us to use the calculator to help in solving problems
Very often   Often   Sometimes   Rarely   Never

(5) The instructor assigned problems that required the use of graphing technology
Very often   Often   Sometimes   Rarely   Never

(6) The instructor asked us to check something using graphing technology during class time
   Very often   Often   Sometimes   Rarely   Never

(7) The tests had questions that required the use of graphing technology
   Very often   Often   Sometimes   Rarely   Never

(8) The tests had questions that had to be solved using standard algebraic methods
   Very often   Often   Sometimes   Rarely   Never

(9) The instructor asked students to see if they could figure out how to use the technology to solve a problem in class
   Very often   Often   Sometimes   Rarely   Never

(10) The instructor indicated that graphing technologies are very useful tools
   Very often   Often   Sometimes   Rarely   Never

(11) In class, I talked with other students about using graphing technology
   Very often   Often   Sometimes   Rarely   Never

(12) The instructor indicated that algebraic skills were more important than were graphing technology skills
   Very often   Often   Sometimes   Rarely   Never

Appendix C. Items

Item 1. *Here is the graph of a function $y = f(x)$. Find a value of c so that $f(x) = c$ has three real solutions.*

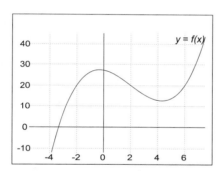

FIGURE 2. Item 1

This relatively simple problem was intended to put students at ease and give them practice "thinking aloud" as they worked. A similar multiple-choice item appeared on a college algebra test at this university. It was used to compare the performance of GT and NT approaches. GT students were more successful then but I expected high success from both groups on this task (Success rates for 865 students in regular college algebra sections and 80 students in graphing sections were 59% and 74% respectively). A graphing utility was not useful for this graph interpretation item that tested the student's ability to relate the solutions of an equation to a graphical representation. It could only be solved graphically.

The questions asked about Item 1 were:

(1) Do you believe you know how to solve this problem?
(2) Tell me what you know about the problem.

Item 2. *A furniture factory produces rocking chairs. This graph shows P, the profit made in a week, if they make and sell x rocking chairs. The factory can produce at most 2500 chairs each week.*

(a) According to the graph, what is the fewest chairs they must produce and sell each week to avoid losing money?

(b) What weekly production levels (that is, number of chairs produced and sold) are profitable for this manufacturer? Explain your answer.

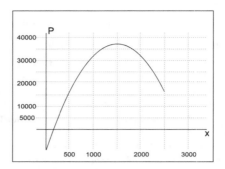

FIGURE 3. Item 2

(c) The plant has been producing 2000 rocking chairs per week. Should they continue produc-ing 2000 chairs per week to earn the largest profit? Give specific reasons for your recommendation.

(d) Over which production level(s), if any, is the profit decreasing?

(e) At which of the following production levels is the profit increasing at the greatest rate? Explain how you can tell.

(i) 0   (ii)200   (iii) 500   (iv) 1000   (v)1500   (vi)2500   (vii) insufficient data

This graph interpretation item examined students' abilities to extract information from the graphical representation of a function. Most parts of the item cover material from precalculus, though part (e) involves the basic calculus concept of rate of change. The item was included to test how well students could interpret simple, important characteristics of graphical representations. Graphing technologies were not useful since no algebraic function was given.

The questions asked about Item 2 were:

(1) How confident are you about your answer to each part of this question?

(2) Suppose I also told you the algebraic function that produced this graph. Would that have led you to solve any part of the problem in a different way? If so, how?

Item 3. *Use the two functions $f(x) = x-2$ and $g(x) = x^2 -4x+2$ to answer these questions.*
(a) For what value(s) of x is $g(x) = -1$?
(b) Find all solutions for $f(x) = g(x)$.
(c) Find all the values of x for which $g(x) < f(x)$.

This standard precalculus item can easily be solved algebraically or graphically; neither approach is clearly superior. There is nothing in the item statement to suggest a graphical solution method; so, I only expected students in the GT group to recognize this alternative strategy. This question is closely tied to Item 4, an item that can only be solved graphically. Item 3 canbe solved several ways; so, it shows something about a student's willingness to consider alternate strategies.

The questions asked about Item 3 were:

(1) Do you believe you have correctly solved this problem?

(2) Could you describe a different way to solve this problem?

(3) Have you ever seen this type of problem before? What do you remember about the method used to solve it then?

(4) (If not already described or used.) Would a graph be useful for solving this problem? How?

Item 4. *This is the graph of two functions, $y = f(x)$ and $y = g(x)$. Use the graph to answer these questions.*
(a) For which values of x is $g(x) = 9$?
(b) Solve the inequality $f(x) < g(x)$.

This item is a graph interpretation version of the previous item—the functions graphed are different, but the questions are similar otherwise. It was included to determine whether students

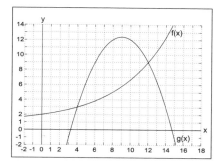

FIGURE 4. Item 4

could make the necessary connections between algebraic and geometric representations even if they did not use or describe such an approach in Item 3. I expected GT students to be more successful on this item. Both groups should have high success rates since it covered basic ideas studied in many mathematics courses before calculus.

The questions asked about Item 4 were:

(1) Are you confident in your solution to this problem?
(2) Suppose I gave you the algebraic rule for each function on the graph. Would you use them to solve any part of the problem? How?

Item 5. *For each of the following, decide whether or not the limit exists. If not, give a reason. If so, find the value of the limit.*

(a) $\lim_{x \to 2} \frac{x^3-8}{x^2-4}$ (b) $\lim_{x \to +\infty} \frac{x^3-8}{x^2-4}$

(c) $\lim_{x \to -2} \frac{x^3-8}{x^2-4}$

(d) *Carefully sketch a complete graph for the function* $f(x) = \frac{x^3-8}{x^2-4}$

This item was designed to show how well students integrated information from algebraic and geometric representations without explicitly suggesting that they should do so. Students who made connections among parts of this item demonstrated the ability to integrate the algebraic and geometric representations of a function in a calculus setting. The first three parts could be solved algebraically, numerically, graphically, or by integrating information from several sources. No approach is particularly easier or better than another since the item does not ask for proof of the answers. Students needed information from earlier parts to sketch the graph in part (d). A graphing utility would not provide enough information to obtain a completely correct graph because of the removable discontinuity at (2,3) and the function's asymptotes. This item provided considerable insight into the mathematical sophistication of the student; the item's complexity allowed them to demonstrate a good understanding of relationships between various functional representations. Although the graphing precalculus course touched on related ideas—for example, end behavior for complete graphs of functions—this item as stated was a new situation that allowed students to, but did not obviously indicate that they should, use ideas from the graphing approach precalculus course. Any use of graphing utilities, other than copying a displayed graph when sketching in part (d), indicated a willingness to apply skills from the graphing precalculus course in a new setting.

The questions asked about Item 5 were:

(1) How confident are you in your solutions for each part of this problem?
(2) Could you describe a different way to solve this problem?
(3) Are any parts of this problem related? If so, how?
(4) Have you ever seen this type of problem before? What do you remember about the method used to solve it then?
(5) (If not already described) Would a graph be useful for solving this problem? If so, explain how.

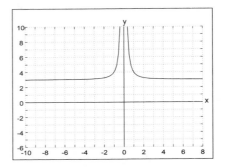

FIGURE 5. Item 6

Item 6. *Here is the graph of a function, $y = f(x)$. Use the graph to evaluate these limits.*
(a) $\lim_{x \to 1} f(x)$ (b) $\lim_{x \to -\infty} f(x)$

This short item was intended to clarify whether students understood the graphical interpretation of limits that could have been useful in Item 5. Responses also helped to show whether students had misconceptions about graphical representations since the interpretations that were required differed from those in Items 1, 2, and 4.

The question asked about Item 6 was:

   (1) How confident are you in your solution to this problem?

Item 7. *Find an expression for y in terms of x that could produce this graph.*

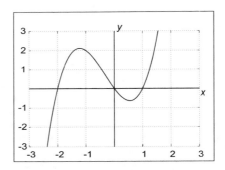

FIGURE 6. Item 7

Answer: y =
Reasoning:

This is similar to one of six *symbolisation* (Ruthven's term, 1990) problems that Ruthven used with high school students. He found that graphing students who regularly used graphing calculators in mathematics classes were significantly more successful on this group of tasks. Ruthven found for a similar graph that graphing students were better than students without access to graphing technologies at identifying the cubic family it belongs to but not better at finding the equation for the graph. He found larger differences favoring the graphing approach students on the task when graphs of linear, quadratic, and trigonometric functions were used. Understanding the relation of algebraic and geometric representations, specifically x-intercepts and the factors of polynomial expressions, was crucial for success on the item. This item might have been solved with persistence by using trial-and-adjustment with a graphing utility. This item was not typical of those found in most precalculus courses; so, it gave some insight to the student's resourcefulness in an unfamiliar situation. The most direct solution path was to recognize the cubic form and then use the correspondence of zeros of the function to factors of its polynomial expression. The point $(-1, 2)$ with integer coefficients not on an axis made it easy for students to check whether they had obtained the correct function.

The questions asked about Item 7 were:

(1) Do you believe you have correctly solved this problem?
(2) Could you describe a different way to solve this problem?
(3) Have you ever seen this type of problem before? What do you remember about the method used to solve it then?

Item 8. *Here is a sign chart for a function, $y = f(x)$, and its derivatives (recall that $f' = \frac{dy}{dx}$):*

	$x < -1$	$x = -1$	$-1 < x < 1$	$x = 1$	$x > 1$
f	+	undefined	+	0	−
f'	+	undefined	−	0	−
f''	+	undefined	+	0	−

(a) Sketch the graph for a function satisfying the conditions in this table.
(b) For which values of x is the function decreasing?
(c) Could this function have any linear asymptotes? You should consider the possibility of vertical, horizontal, and oblique linear asymptotes. For each kind that is possible, state where it could occur.

This item not only investigated the student's understanding and interpretation of derivatives in a geometric context, but also their ability to reason graphically without an algebraic representation. Success depended on correctly making connections between analytical and graphical information about a function and recall of the interpretations of derivatives. This type of item could be a normal part of beginning differential calculus, though most students had probably not seen something exactly like this; more typically, they are given an algebraic function and are asked to construct a sign chart as an aid to producing a sketch graph of the function. I expected GT students to be quite successful with this item because of their experiences in thinking about graphical representations. It seemed that their graphical perspective could transfer easily to this geometric interpretation of derivatives.

The questions asked about Item 8 were:

(1) Do you believe you have correctly solved this problem? Why or why not?
(2) Suppose I told you the algebraic form of a function that corresponded to this table. Would that influence your approach to any part of the problem? If so, how?

Item 9. *A circular sheet with radius 5 cm has a sector removed, leaving a sector of θ radians. As shown by the arrow in the figure, the gap can be closed so that a cone with slant height 5 cm is formed.*

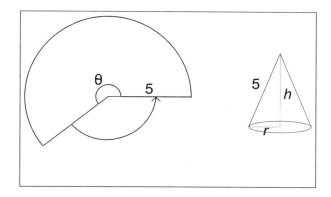

FIGURE 7. Item 9

(a) Carefully describe how the shape and volume of the cone will change as the angle θ decreases from 2π to 0 (that is, the sector remaining gets smaller).

(b) Express the height of the cone in terms of r, the radius of its base.

(c) The volume of a circular cone is given by the formula $V = \frac{1}{3}\pi r^2 h$. Write a function that gives the volume of the cone in terms of the radius of its base, r.

(d) Find the radius of the base of a cone with slant height 5 cm that has the largest volume.

(e) Notice that the circumference of the base of the cone is the same length as the arc of the circle, $\frac{\theta}{2\pi} \times 10\pi = 5\theta$. Express r, the radius of the cone's base, in terms of θ. (The circumference and radius of a circle are related by the formula $C = 2\pi r$).

(f) Find the angle θ of a sector of a circle with radius 5 that forms a cone with the greatest volume.

This typical calculus item was included because it asked students to represent a problem algebraically and because algebraic or graphical methods both could be used. I provided students with manipulatives—one circle with a sector removed and another with a slit cut along a radius—in order to help them understand the physical situation on which the item was based. The decision to state the item so that students were guided toward a solution was deliberate. One reason for this was the time constraints on the interview. Also, having students work through steps gave me the opportunity to see how they handled algebraic representations. This is the only item that required such manipulations for its solution. I was interested in the student's ability to devise functions based on a physical situation and his or her choice of methods for finding a maximum value—approximation using calculated numbers, graphical approximation, or exact solution using a derivative. This item was most important to this study for what it showed about each student's ability to relate a situation and mathematical model, and his or her willingness to use, or at least awareness of, alternative solution paths. I presented a graph of the correct function without stating its algebraic form after they finished the item to see if they could use it to check their work.

The questions asked about Item 9 were:

(1) Do you believe you have correctly solved this problem?

(2) Could you describe a different way to solve this problem?

(3) Have you ever seen this type of problem before? What do you remember about the method used to solve it then?

(4) (If not already described.) Would a graph be useful for solving this problem? How?

(5) Here is the graph of the function showing the volume of the cone as a function of the size of the sector of the circle used to form the cone. How can you use this to check your work?

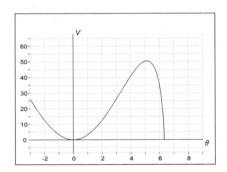

FIGURE 8. Item 9 Questioning

Appendix D. Coding Schemes

D.1. Outcome Codes. The outcome coding scheme was designed to reveal the degree of success achieved by participants on the items, their recognition of alternate solution strategies, and their use of graphing technologies. Each of the 27 parts, from the nine items, was scored using a scale of 0–5 to indicate the degree of success that the student achieved before any questioning took place in the interview. A second score was assigned to reflect the success achieved by the end of the interview session on the item. Additionally, on Items 2, 5, and 9 a code was assigned for the degree of success finally achieved on all parts of the item considered together. The meaning of the codes are given below in Table 7. It was important to distinguish between work done before and after questioning because the questioning probably influenced what the student did on the item. For example, in several cases after I asked the question "Would a graph be useful for solving this problem?" students tried something else.

TABLE 7. Outcome Codes

Score	Interpretation
5	Completely correct—the solution is fully acceptable
4	Basically correct—the student carried out an acceptable solution, but made a minor error preventing a score of 5
3	Flawed response—the solution has at least one significant error, but the student did make an appropriate start that included at least one substantial appropriate step toward a solution
2	Little progress—at least one appropriate action was taken toward a solution, but no significant progress was made
1	No progress—nothing relevant to a solution is written
0	Blank

The use of graphing technology was coded using the scale given below in Table 8. This scale indicates the level of sophistication of calculator usage. During the outcome coding stage only the highest score that applied to an item part was recorded. Thus, if a student used his or her calculator to compute some numerical values and to produce a graph that was used to gain insight but not to solve the item, the code of 3 was assigned.

A fourth numerical score was assigned for each part of every item. The score represented the number of valid alternative strategies used or described by the student for that item. A nonnegative integer score represented the number of valid alternative solution strategies identified by the student by the end of the problem session. In determining this score students had to have said or written enough to convince the coder that they understood the strategy; they did not need to show that they could carry it out successfully. For example, on Item 9(d) a student would not be given credit for an alternative strategy just for saying "I know you would just take a derivative to find the max." They would, however, be given credit for saying "I need to solve $\frac{dV}{dr} = 0$ or $V' = 0$" even if they did not show that they could take the derivative or solve the resulting equation.

The four numeric codes were assigned based on what students wrote and said during the interview. As I coded the outcome variables I wrote accompanying comments about what the student had written or said. These written remarks were used to help find comments or behaviors that seemed to give useful insights related to the research questions.

D.2. Process Codes. The process coding scheme developed for my study recorded quantitative and qualitative information about seven aspects of the students written work and recorded statements: (a) strategies used, (b) mathematical representations used, (c) relationships made

TABLE 8. Technology Use

Score	Interpretation
4	Graphing utility used to obtain a correct solution
3	Graphing utility appropriately used to provide insight to the problem, but not to actually solve it
2	Graphing utility used, but not in a way that contributed any insight to the problem or its solution
1	Calculator used to compute numeric values
0	Technology was not used

between representations or between parts of an item, (d) use of technologies, (e) checking of work, (f) nature of errors, and (g) reactions to impasses. Information about these characteristics of the students' work provided detailed, quantified information directly related to the research questions. The analysis and description of these data facilitated the development of characterizations of the strategies used by students during their enrollment in a traditional first-semester calculus course.

Six of the seven aspects that were coded contained several variables that may or may not occur in a given student's work on each item. The seventh aspect, reactions to impasses, was not coded; instead, it was described using a brief written statement. Information about the other six aspects was recorded by means of codes. Each distinct code could be thought of as a variable. Coding involved recording each observed occurrence of the variable in the student's work. These occurrence frequencies were tallied for each item part.

The following codes were assigned for all parts of Items 3, 5, 7, 8, and 9. Multiple codes could be assigned for work on one item. A code of O, meaning other, was recorded in any situation where the planned codes did not seem to apply. Copies of the process code summary sheet and the coding sheet that was used are included below.

   (1) Verification: When the student checked an answer the method used was coded as numerical check (N), algebraic check (A), graphical check (G), or reasonableness of answer check (R).

   (2) Technology: Each use of technology was coded using the same 0–4 point scale as was used for outcome scoring (Table 8 above), but not just the highest level as recorded during the first stage. This time, each part of an item received as many codes as there were distinct uses of a calculator.

   (3) Errors: Errors were coded as faulty numerical computation (N), algebraic manipulation (A), interpretation of item (P), strategy (S) or interpretation of a representation (I). This code was only given for actions taken, not for errors of omission (that is, a blank item or incomplete correct work was not assigned an error code). A blank error field on the coding sheet indicated that either there was partial work that was correct or that nothing had been attempted. A code of (C) was used to indicate that the work was correctly carried out to obtain a solution.

   (4) Representations: Each reference to a graphical representation was coded as based on a calculator produced image (GC), hand drawn image (GH), or imagined image (GI)– that is, one that was not physically produced but was mentioned. The use of tabular (T), geometric figure (F), and algebraic (A) representations was also coded.

   (5) Connections: Each instance of the student successfully and appropriately relating different parts of the item was coded P; this code letter was followed by two lower case letters for the parts related. This code was recorded for the part of the item where the student used information from another part. For example, Pbc recorded in Item 3(c) indicated that parts (b) and (c) were correctly related in the solution for part (c). This would happen if the student used the points of intersection found in part (b) when solving the inequality in part (c). Each time information from different representational forms was related correctly a code of R was recorded. The R code was followed by lower case letters indicating the representations that were related: algebraic (a), geometric figure (f), graphical (g), tabular (t), or description (d). Contradictory answers given by the student in different parts or the same part of an item were coded using an X followed by three lower case letters: The first two indicated the item parts involved and the third showed whether the contradiction was ignored (i) or recognized (r). Each contradiction between parts was coded twice: once for each part containing contradictory answers. For example, in Item 5 the code Xadi was often recorded in parts (a) and (d) because the student's graph in part (d) did not correspond to the limit found in part (a) and the student showed no awareness of the contradiction.

   (6) Strategy: Each attempt to use a standard (S) or nonstandard (N) procedure or algorithm was coded; the type of strategy was also recorded as graphical (G), algebraic (A), or numerical (N); a third code indicated whether the strategy was appropriate (a) or inappropriate (i) for the item. If no strategy was apparent "none" was recorded. A standard procedure for this coding scheme is one that could have been studied or learned in prior course work, an admittedly vague definition. I wrote a brief description of each strategy that was attempted or used. This written description gave something to check in relation to the question of just what is a standard strategy; thus, the coding

	AGN R	1-4	C N A P S I	G(CHI) T A F	P Xir (part) R a g t d	(D) S - N G A C N a i none	Description
	Chk	Tech	Error	Rep'n	Relate	Strat	Description
3a							
3b							
3c							
5a							
5b							
5c							
5d							
7							
8a							
8b							
8c							
9a							
9b							
9c							
9d							
9e							
9f							

Process Coding
(Impasse description on back)

Student Code ____
Number ____

FIGURE 9. Process Coding Form (Side 1)

produced a list of standard and nonstandard strategies as defined for this study. It was not possible to anticipate the large number of "standard" procedures that could be used by students; so, the coding relied on the judgment and mathematical experience of the coder. The nature of the items made it likely that it would not be difficult to recognize standard strategies. These codes were not an indication of whether the procedure was successfully executed. If a student described but did not attempt to carry out the procedure the code was preceded by a (D).

(7) Reaction to Impasse: An impasse was defined as any time that the student stopped or changed what they were doing and indicated that they knew they had not finished what was needed for a solution. I did not use a time criterion for this. An asterisk was recorded to indicate impasses, and a statement was written describing what the student did when they reached an impasse. Examples of these statements included:

Process Variables Coding Scheme

The code O-other should be recorded in any case where no existing codes are appropriate. This seems most possible in the codes for *Chk, Error, Rep'n,* and *Strat.* The use of an O code indicates a gap in the coding scheme, so it is preferable to avoid its use if possible. Each use of the O code should be accompanied by a written description of the reason for its use to help with devising a modification of the existing coding scheme.

Chk (Checking of work)

    A: Algebraic method
    G: Graphical method
    N: Numerical method
    R: Analyzes answer for ``reasonableness"

Tech (Use of technology)

    4:   Graphing utility used to obtain a correct solution
    3:   Graphing utility appropriately used to provide insight to the problem, but not to actually solve it
    2:   Graphing utility used, but not in a way that contributed any insight to the problem or its solution
    1:   Calculator used to compute numeric values
blank: Technology was not used

Error (Classification of errors)

    C: Correct solution—no errors made
    N: Numerical error—computation
    A: Error in algebraic manipulation
    P: Incorrectly interpret problem or task
    S: Strategic error—use an invalid strategy, including faulty recall of procedure
    I: Interpretive error—interpret a representation incorrectly

Rep'n (Representations used during solution attempt)

G(CHI):   GC = Graphical on calculator, GH = Hand drawn, GI = Imagined or described
    T: Tabular (including a series of computations not necessarily in a table)
    A: Algebraic
    F: Geometric figure (other than a graph)

Relate (*Correctly* relate parts of a problem or different representations—multi-letter code)

    P: Correctly relate parts—follow P by lower case letters of parts related
    X: Gives contradictory answers—follow by lower case letter for parts of problem involved followed by a lower case i or r to indicate whether the contraction was i-ignored or r-recognized
    R: Correctly relate representations—follow R by lower case letter to indicate the representations that were related: a-algebraic, g-graphical, t-tabular, d-description

Strat (Strategy attempted or described—not necessarily correctly—3 or 4 letter code)

    D: Precede the code by D if the strategy was described but *not* attempted
S or N: Code each strategy as S-standard (procedure taught in math class) or N-nonstandard
G, A, C or N: Follow the S or N with the code for the type of strategy: G-graphical, A-algebraic, C-Calculus, N-numerical
   a or i: The last letter is a lower case a or i to indicate whether the strategy was a-appropriate or i-inappropriate
   none: No apparent strategy for the problem (including blank page)—use in place of letter codes

Write a succinct description of each strategy (e.g. ``used factoring to solve a quadratic equation" which would be coded SA or ``correctly used graphing calculator to solve inequality" which would be coded SG).

Impasse (record * in strategy column and describe on back of page, labelling problem)

FIGURE 10. Process Coding Form (Side 2)

(a) quit, (b) tried a different approach with the same representation, (c) calculated several function values, (d) tried another part of the item or (e) tried to use a graph.

These codes and descriptions were intended to help identify patterns of problem solving used by students. For each item I used the codes and written descriptions to produce a summary description of the different strategies used. While coding, I also looked for: (a) patterns of thinking that appeared to have been influenced by the access to graphing technologies and (b) a variety of levels of use and integration of graphical information in problem solving for the students from graphing sections of precalculus. I kept notes on things that struck me as significant, interesting, or important with the coded data for each part and in a general summary statement written about each student. This summary statement was based on their work on all the items considered as a whole and sought to provide a characterization of their mathematical capabilities as demonstrated during the interview. The process coding sheet is exhibited in Figure 9 and Figure 10.

DEPARTMENT OF MATHEMATICS, NORTH DAKOTA STATE UNIVERSITY, 300 MINARD, PO BOX 5075, FARGO ND 58105-5075

E-mail address: wimartin@plains.nodak.edu

CBMS Issues in Mathematics Education
Volume **8**, 2000

Visual Confusion in Permutation Representations

John Hannah

1. Introduction

Consider the dihedral group D_4 of all symmetries of a square. Zazkis and Dubinsky (1996) discuss two ways of representing D_4: one as motions of the square with composition of transformations as the operation, and the other as a set of permutations with composition of permutations as the operation. Figure 1 is typical of the diagrams which they use to construct these representations. The eight symmetries consist of four rotations along with four reflections, one about each of the lines indicated in the diagram. The numbers 1 to 4 were used to construct the permutation representation in the obvious way. (Or is it obvious? You may find it helpful to construct these representations for yourself before going any further.)

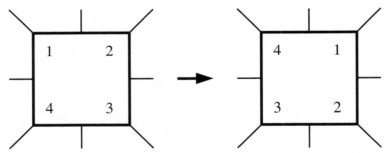

FIGURE 1. A 90-degree clockwise rotation of a square, adapted from Zazkis and Dubinsky (1996, p. 63).

These two representations of D_4 should presumably be isomorphic, but almost all (nine out of ten) of Zazkis and Dubinsky's students made the same "error" and effectively obtained representations which were anti-isomorphic copies of one another. It turns out that the students must make two key choices of interpretation (object or position, and global or local; see the definitions below) when tackling the above task. The students who erred all chose the same "incorrect" position/global combination (although one student escaped detection by evaluating his permutation product from left to right). Zazkis and Dubinsky go on to say

> Such behavior suggests that the position interpretation is not just
> an occasional preference—it may be a very salient interpretation
> by novices in group theory. If so, then why? Is it an artifact

 ©2000 American Mathematical Society

of instruction or is it based in deeper perceptual and conceptual factors? (1996, p. 74)

This paper investigates the effect on students of altering some of the visual cues in diagrams like Figure 1. We shall see that some of the students' errors seem to be caused by confusion about the roles of the labels in such diagrams. We shall also see that remarkably similar difficulties occur in a much more algebraic setting, the calculation of permutation products.

The paper is laid out as follows. Section 2 recalls the key ideas from Zazkis and Dubinsky's paper. Sections 3 to 8 describe two cycles of teaching activity and the data which were collected during these cycles. Section 9 discusses possible influences on the students' choices of interpretation, and Section 10 looks at examples of students' difficulties with permutation calculations.

2. Background

In this section we shall review the sources of possible confusion which have a bearing on the following work. In each case we need to make a choice between two possible points of view or conventions. None of these choices is wrong but there are (possibly unexpected) connections between the choices and inconsistent decisions can lead to contradictions.

Object or position? The first choice involves the way we assign a permutation to a geometric transformation. Zazkis and Dubinsky call this the object/position dichotomy. Following Zazkis and Dubinsky (1996, p. 66), one way of assigning a permutation to a transformation (the *object* interpretation) is to look at the vertices of the square as objects being moved. In performing a 90-degree clockwise rotation, vertex 1 is moved to vertex 2, vertex 2 is moved to vertex 3, 3 is moved to 4, 4 is moved to 1. Therefore we may represent this rotation as

$$\begin{pmatrix} 1 & 2 & 3 & 4 \\ 2 & 3 & 4 & 1 \end{pmatrix}$$

(see Figure 1). Another way (the *position* interpretation) is to look at the environment of the square and think, not about vertices moving, but about positions and which vertices they contain. In this interpretation, after a 90-degree clockwise rotation, position 1 contains vertex 4, position 2 contains vertex 1, position 3 contains 2, and 4 contains 1. This is represented by a permutation

$$\begin{pmatrix} 1 & 2 & 3 & 4 \\ 4 & 1 & 2 & 3 \end{pmatrix}$$

and this is, of course, the inverse of the permutation we found using the object interpretation.

Global or local? The second choice to be made involves different ways of describing geometric transformations (Zazkis & Dubinsky, 1996, pp. 68–70). In the *global* interpretation all transformations are described relative to a reference frame fixed in space. This is the interpretation adopted in Figure 2, where the reference frame is provided by the axes which define the reflections D_L, V, D_R and H. So, for example, regardless of the orientation of the square (indicated by the positions of the labels 1 to 4), V indicates a reflection in the vertical axis.

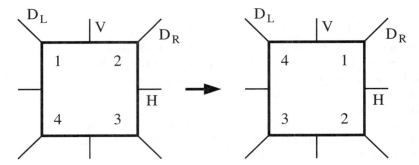

FIGURE 2. A 90-degree clockwise rotation: the "global" interpretation, adapted from Zazkis and Dubinsky (1996, p. 63).

This may be contrasted with the *local* interpretation where transformations are described relative to a reference frame which moves with the object. (The term "local" is borrowed from Turtle Geometry.) Zazkis and Dubinsky illustrate this interpretation as in Figure 3. Here the reference frame is again provided by the axes defining D_L, V, D_R and H, but this time the name V is given to the reflection about the axis labelled by V in the diagram, wherever this happens to lie.

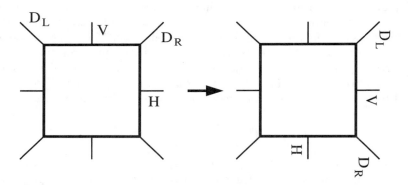

FIGURE 3. A 90-degree clockwise rotation: the "local" interpretation, adapted from Zazkis and Dubinsky (1996, p. 69).

It turns out that these interpretations are again mutually inverse. Choosing the object/global or position/local combination gives isomorphic copies of the group but, since the map $x \to x^{-1}$ is an anti-isomorphism for any group, choosing the position/global or object/local combination gives anti-isomorphic copies of the group (Zazkis & Dubinsky, 1996, pp. 70–72).

Functions on the left or right? Another source of possible confusion (alluded to by Zazkis and Dubinsky at the bottom of pages 62 and 72) lies in the conventional function notation. Most students prefer to write their functions on the left of the arguments (as in an equation like $f(x) = x^2$). From the students' point of view, this convention has the unfortunate consequence that a product $\alpha\beta$ of permutations (or transformations) is read as "do β first, then α," and this runs counter to the usual rule for reading text from left to right.

There is no way out of this dilemma, of course, and the student will have to adjust to an unfamiliar situation either for evaluating permutation products as above, or else for the function notation. It is perhaps worth mentioning that some algebra texts do write some of their functions on the right of their arguments, but without always bothering to explain their apparent inconsistencies. See Birkhoff and Mac Lane (1965, pp. 51, 61, 113) and Herstein (1975, pp. 11, 55, 173) for examples.

Algebraically, these two possibilities again result in anti-isomorphic copies of the group involved. This becomes relevant when the application of two successive anti-isomorphisms leads to an isomorphism which masks a student error (see Sections 5 and 10).

3. Method

This research was carried out in the form of two cycles of action research: planning the relevant teaching episode, implementing it, and then evaluating the results and incorporating the lessons learned into the planning step of the next cycle. See Kemmis (1994) or Angelo and Cross (1993, pp. 33-36) for discussions of this approach to classroom research.

Although this is possibly an unusual methodology to use for such an investigation, it has two distinct advantages here. Firstly, it acknowledges the fact that, as the teacher, I myself was learning, looking for more comprehensible ways of presenting the ideas of abstract algebra. Although my research training is in ring theory, this was the first time I had taught a beginning course in group theory, and so there were many issues which I had never had to analyze from the beginner's point of view. Secondly, this methodology gives me the chance, as the researcher, to distance myself from any particular interpretations which I, as teacher, adopted during the two cycles which are being reported on here. In particular, I do not need to defend any such interpretation as being the "right way" of looking at things, as that is simply the way that things were done in that particular class.

On the other hand there is one disadvantage to this methodology. Since the focus of this study is the students' difficulties with permutation representations of groups, some of the data collected (dealing with the teacher's developing understanding of the problems involved) will be of at best peripheral interest. In what follows I have included only enough of this data to indicate the kind of influences which were brought to bear on the students being studied.

The study was conducted in two successive years of a beginning course on abstract algebra at a research university in New Zealand. All the students had done the equivalent of a two-semester first year course on calculus and linear algebra, and most of them intended to major in mathematics or physics. The classes were quite small (14 in the first year, and 16 in the second) with about two-thirds of each class having obtained an A– or better in their first year course.

Evidence was collected during ordinary teaching activities: incidents which occurred during lectures, tutorials or when students came for individual help, were all recorded in note form immediately afterwards, while assignments were photocopied before they were handed back to students. Information about my behavior as the teacher comes mainly from letters or e-mails which I wrote at the time, and from a diary I kept during the study.

4. Planning the first cycle

The experiment consists of two cycles of planning, implementing, and evaluating a certain teaching segment. In this section I shall describe the relevant parts of my course preparation for the first cycle.

I was not keen to spend much class time discussing Zazkis and Dubinsky's different interpretations (at the time it didn't seem to me to be "core" abstract algebra). I hoped to lead my students to consistent interpretations by equipping them with the tools they would need (how to represent symmetries as matrix transformations or as permutations), by eliminating any misleading "artifacts of instruction" which I could find in Zazkis and Dubinsky's approach, and by presenting a consistent approach to the two ways of representing symmetries.

Separating the roles of object and position. When examining Zazkis and Dubinsky's approach, I focused my attention on their diagrams (see Figures 2 and 3, or Zazkis and Dubinsky (1996, pp. 63, 64, 69). It's not clear from Zazkis and Dubinsky (1996, pp. 65–66) what diagrams their students eventually used, but in any case their Figures 2 and 3 were the starting points for my own preparation. There were two related aspects of these diagrams which I thought might have confused their students: firstly, the labels 1, 2, 3, 4 are attached to the object (and so might be inclined to move when the object moves) and secondly, the separate roles of object and position are not well distinguished (the one set of labels serving to indicate both features). I decided to use separate labels for the background and the object, and in this first year of the experiment I used numbers for both features, as in Figure 4.

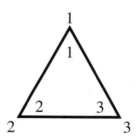

FIGURE 4. Separating the roles of object and position (first cycle).

I presented a matrix transformation view with all transformations taking place relative to fixed axes (or background). So a transformation involves moving the object and using the background to identify the corresponding permutation. For example, rotating the triangle 120 degrees anticlockwise gave Figure 5, so the corresponding permutation for this transformation was

$$\begin{pmatrix} 1\ 2\ 3 \\ 2\ 3\ 1 \end{pmatrix}.$$

A consistent approach. Not having any experience of Turtle Geometry myself, and not expecting my second year students to have any memories of it even

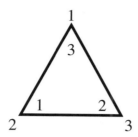

FIGURE 5. A 120-degree anticlockwise rotation: labeling position and object with numbers.

if they had met it at high school, I sought another description of Zazkis and Dubinsky's local interpretation. I described the global interpretation as being characterized by fixed reference axes and the local interpretation as being characterized by moving reference axes. (It wasn't until the second cycle that a student pointed out the drawbacks of this description. Just as Figures 4 and 5 actually have two reference frames, so Figure 3 has a hidden secondary frame consisting, for example, of the up-down and left-right directions on the printed page. Once this secondary frame is admitted, the phrase "fixed axes" becomes ambiguous.)

I expected that my students, like those of Zazkis and Dubinsky, would prefer the global interpretation for their transformations. So for consistency I needed to make sure that my presentation of permutation representations encouraged the object interpretation. As part of this consistency, I described the object interpretation as a fixed background interpretation too, along the lines indicated above. This common description of both interpretations (global and object) was part of my way of explaining why they give isomorphic representations.

5. Implementation (first cycle)

Classroom activities. All the necessary material was covered in the first six lecture hours of the course. To give some idea of the influences on the students, I'll give a brief description of this first part of the course.

In the first session we discussed what we mean by symmetry, and the students were encouraged to think of examples from the world around them, or from their mathematical experience. In the second session I reminded them about 2×2 matrices, interpreting them as geometric transformations of the plane, and pointing out my own preference for fixed axes and moving objects (the global interpretation). However I also signalled the possibility that they might have cause to adopt the other (moving axes) interpretation in the forthcoming assignment, and told them that I was perfectly happy with this alternative.

The next session was devoted to the dihedral group as the set of symmetries of a regular polygon together with the operation of composition (however, I had not yet given the formal definition of a "group" at this stage). We looked firstly at the case of an equilateral triangle, giving the reflections the provisional names V_{up}, V_{left}, V_{right} after the manner of Figures 2 and 3 (see Figure 6). My hope was that these names would retain their global meaning even after one transformation had been applied. The corresponding matrix names for these transformations were developed in an accompanying tutorial exercise.

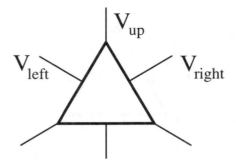

FIGURE 6. Naming the reflection symmetries of an equilateral triangle.

The students used some physical triangles to calculate the multiplication table for D_3, the elements being identified by comparing the final orientation of the triangle with a standard orientation displayed on the board. Thus, for example, Figure 7 shows the manipulations which yielded the relation $V_{\text{up}} R_{240} = V_{\text{left}}$ (all our rotations were measured anticlockwise).

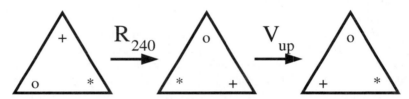

FIGURE 7. Calculating the product $V_{\text{up}} R_{240}$.

In the next session we discussed how the Platonic solids can be constructed from equilateral triangles, squares and regular pentagons, and why there aren't any more regular solids besides the Platonic ones.

I then introduced permutations, showed the correspondence between elements of S_3 and D_3 (as described in the previous section; see Figure 5), and used the resulting multiplication tables to motivate (still informally) the idea of isomorphic groups. To reinforce the correspondence, I showed them the embedding of D_4 in S_4, and elicited from them why this could not be an isomorphism.

To encourage the object interpretation of a permutation's action, I included a class activity (Haines, 1997) where permutations were interpreted as instructions for students to move around numbered chairs in the classroom. So in Figure 1, for example, the numbers 1, 2, 3, 4 would be labels attached to chairs at the corners of a room, and the students would follow the instructions which are encoded as the permutation

$$\left(\begin{array}{cccc} 1 & 2 & 3 & 4 \\ 2 & 3 & 4 & 1 \end{array} \right).$$

So the student sitting in chair 1 goes to chair 2, and so on.

The main task. There were several opportunities during the course to gauge the students' understanding of symmetries and permutations, but my own version of Zazkis and Dubinsky's experiment was embedded in the following excerpt from my students' first assignment.

> In this assignment you will construct an example of a group based on the symmetry of a three-dimensional object. . . .
>
> Make a model of a regular polyhedron (a tetrahedron, cube, octahedron, dodecahedron or icosahedron). You will need to compare the positions of your model before and after a transformation has been applied to it. So you should make sure that the faces or edges or vertices look different from one another (label them if you have to), and record the original orientation of the shape. . . .
>
> The next two exercises look at two ways of representing the symmetries.
>
> (4) . . . Choose two [rotational] symmetries which are not rotations about the same axis, and describe their product as a rotation (what is its axis of rotation and the angle of the rotation?)
>
> (5) Label the original positions of the vertices of your model 1, 2, 3, and so on. If your model has fewer faces than vertices, then you may prefer to use its faces instead.
>
> Describe each symmetry as a permutation of 1, 2, 3, and so on. . . .
>
> Using the same two symmetries as in the last part of exercise 4, find the product of the corresponding permutations. How is the resulting permutation related to your answer to the last part of exercise 4?

I told the students what my own preferences were (the fixed background interpretations, and evaluating transformations from right to left), and that if they made different choices then their answers might be different from mine and yet still be correct.

Explanatory comments. Of course, by this stage there was little point in getting the students to replicate Zazkis and Dubinsky's experiment using D_4. I had just used D_3 and D_4 to introduce the students to the appropriate representational tools, and to exhibit the kind of agreement (isomorphism) which the different representations should yield. So I needed a different example of symmetry for my version of the experiment. In the circumstances it seemed natural to turn to a three-dimensional model. This also suited my pedagogical goals for the course quite well, as I wanted to give the course a distinctly geometric flavour, including both two- and three-dimensional objects.[1]

It should also be noted that, because the geometry here is three-dimensional, the introductory lecture about 2×2 matrices as transformations didn't actually give the students any mathematical tools for this assignment. In fact the lecture merely served to implant the idea of the global interpretation. In this sense, at least, my

---

[1] It may be worth remarking here that, from my point of view, the three-dimensional model had several other advantages: actually making the model is fun and students begin what is usually a hard course with an enjoyable experience; the students' three-dimensional intuition is traditionally weak and this experience might help them improve their intuition; and finally the model became a useful resource later in the course giving, for example, interesting geometric examples of cosets and the proof of Lagrange's Theorem.

students would be in the same boat as Zazkis and Dubinsky's students, in that they would have no convenient mathematical symbolism (such as 3×3 matrices) with which to describe the various symmetries. (I don't discuss 3×3 matrices until later in the course. By then the students have met eigenvalues and eigenvectors in a parallel course, and I can use these ideas to discuss the classification of orthogonal matrices in terms of rotations and reflections.)

Problems with error detection. The final part of the task hints at the isomorphism of the two representations of the group being discussed. Most of Zazkis and Dubinsky's students came to grief when they chose inconsistent interpretations for the two representations. Zazkis and Dubinsky (1996, p. 66) explain how this error will sometimes escape detection when the group has many involutions (elements of order 2). However this error can actually escape detection every time if the student simultaneously chooses to evaluate permutation products from left to right. To see this, suppose that, in the first part of the task, the student calculates the effect of the transformation x followed by the transformation y. If they swap interpretations in the second part, then they will actually use the permutations corresponding to x^{-1} and y^{-1}. Hence if they also evaluate the permutation product in the wrong order, they will actually find $x^{-1}y^{-1} = (yx)^{-1}$ which they will interpret as the desired yx. Since this combination of choices never draws attention to itself by producing wrong answers (Zazkis & Dubinsky, 1996, bottom of p. 72), both the student and the teacher may be totally oblivious to the fact that they are doing things differently (see, for example, the discussion in Section 10).

There is an extra complication in this three-dimensional version of the task. As the rotations are described in terms of axes and angles of rotation, there is a further choice facing students because the sense of the angle will be clockwise or anticlockwise depending on which end of the axis they are looking from. Again these two choices produce inverses of one another, and in each cycle a couple of students thwarted reliable diagnosis by not indicating their actual choice of viewpoint.

6. Results (first cycle)

Of the 14 students in this first cycle, roughly half used the position instead of the object interpretation (I couldn't follow one student's answer), and all but one of these students came to the same inconsistency as Zazkis and Dubinsky's students. The remaining student, Clare,[2] adopted the local interpretation of her transformations and so got the desired consistency.

Given my attempts to guide the students in the "right" direction, we see that Zazkis and Dubinsky's conclusions are remarkably robust, and that the inconsistent combination of global and position interpretations seems to spring up naturally even in relatively unfriendly soil. On the other hand the prevalence of this error had dropped by almost a half. Why were students still making the wrong choices?

There was some evidence that using numbers for both sets of labels (as in Figure 4) caused confusion. Some of this evidence surfaced in later sessions involving permutation calculations, where a type of object/position confusion seemed to occur independently of any physical model. For example, different students could take a diagram like Figure 8 to mean either of the permutations

---

[2]Students have been given pseudonyms that preserve gender.

$$\left(\begin{array}{ccc} 1 & 2 & 3 \\ 2 & 3 & 1 \end{array} \right) \quad \text{or} \quad \left(\begin{array}{ccc} 1 & 2 & 3 \\ 3 & 1 & 2 \end{array} \right).$$

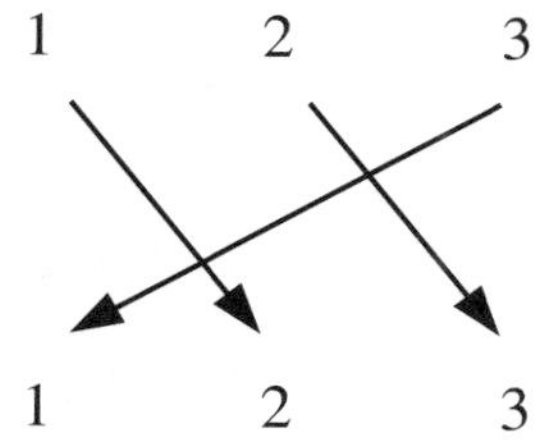

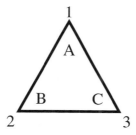

FIGURE 8. A potentially ambiguous picture of a permutation.

As one student explained to me, the second option comes from looking at the diagram as a set of instructions for where, on the lower deck of the double decker form, you should write each of the numbers. So for example, with this interpretation, Figure 8 tells you to write the 1 under the 2, and so on.

One student, Mark, suggested that I could remove some of the potential for confusion if I changed diagrams like Figure 4 by using different labels, as in Figure 9. Here A, B, C are merely distinguishing marks to show us how the triangle is oriented, while 1, 2, 3 are the fixed reference positions on the background.

FIGURE 9. Separating the roles of object and position even further.

Clare's example seemed to offer a hint as to why students might be attracted to the local interpretation. She had constructed an icosahedron but found it difficult to keep track of the transformations. In the end she drew partial nets of the model (essentially, what she could see of the solid from one viewing position with a little bit of peeping round the corner), recording on these nets the positions of the labels which she had fixed to the vertices of her model. (See Figures 15 and 16 for a similar idea used for a dodecahedron.) Her only reference points were these labels, so her names for the transformations were things like "rotate 72 degrees about vertex 3." After applying a transformation to such a complex model, it is easier to identify a second transformation by using these vertex labels rather than by referring back to the original position. So it is natural to use the local interpretation, keeping the moving labels as your reference frame.

7. Planning the second cycle

Distinguishing between the roles of object and position. As we have seen, there was some evidence that using numbers for both sets of labels (as in Figures 4 and 5) caused confusion. With hindsight this is not very surprising. The object and background play different roles in each interpretation and, by using the same labels for both features, I missed a valuable opportunity to signal which reference frame I intended to use for constructing the permutations. So in the second year I followed Mark's suggestion and used diagrams like Figure 9 to try to make the position and object interpretations more clearly different.

With this new convention, we can reword Zazkis and Dubinsky's definitions of the object and position interpretations. In the next two diagrams, the letters A, B, C, D are distinguishing features which we follow during the transformation, while the numbers 1, 2, 3, 4 are the labels used to record the transformation (a discrete version of the Cartesian axes, if you like).

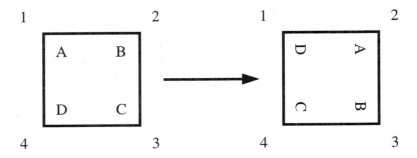

FIGURE 10. A 90-degree clockwise rotation: the object interpretation.

Translating Zazkis and Dubinsky (1996, p. 66) for this new diagram (Figure 10), we have: the *object* interpretation assigns a permutation to a transformation by looking at the vertices of the square as objects being moved. In performing a 90 degree-clockwise rotation, vertex A goes from position 1 to position 2, vertex B goes from position 2 to position 3, C goes from 3 to 4, and D goes from 4 to 1. Therefore we may represent this rotation by the permutation

$$\begin{pmatrix} 1 & 2 & 3 & 4 \\ 2 & 3 & 4 & 1 \end{pmatrix}.$$

On the other hand, the *position* interpretation looks at the environment of the square and thinks, not about vertices moving, but about positions and which vertices they contain. So performing a 90-degree clockwise rotation again, as in Figure 11, we see that the vertex in position A has changed from 1 to 4, the vertex in position B has changed from 2 to 1, the vertex at C has changed from 3 to 2, the vertex at D from 4 to 3. This time the rotation is represented by the permutation

$$\begin{pmatrix} 1 & 2 & 3 & 4 \\ 4 & 1 & 2 & 3 \end{pmatrix}.$$

As in the first cycle, I was trying to encourage the object interpretation, so the class saw only Figure 10, not Figure 11.

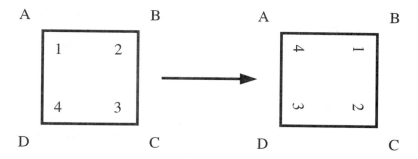

FIGURE 11. A 90-degree clockwise rotation: the position interpretation.

Fixing the background. Another refinement, in this second cycle, involved the names for the reflections in the two-dimensional examples which we used in the introductory lectures.

The names which Zazkis and Dubinsky gave to their reflections seemed intended to conjure up global or "fixed axes" ideas (for example, H represents a reflection in a horizontal axis, and D_L is a reflection in the diagonal going from the top left-hand vertex), but once the objects had been moved students seemed happy to let these meanings vary (Zazkis & Dubinsky, 1996, p. 67). Perhaps these names didn't have the same cue value for the students as I had hoped. Some of my own students had been uncertain, the first time I used H, whether "horizontal" referred to the position of the mirror or to the direction of "motion" of the reflection. D_L may be an even weaker cue, as it is not immediately obvious why the diagonal referred to is a "left" or a "right" diagonal. Again some of my own students had failed see what was "left" or "right" or "up" about the reflections in Figure 6. To get round these problems, I used names for the reflections which were tied to a global interpretation, namely, M_θ for the reflection in a mirror placed on the line $y = x \tan \theta$. I thought that students would be unlikely to move the x-axis when they performed a rotation or reflection.

8. Results (second cycle)

Apart from the above changes, my presentation was more or less the same as in the previous year. This time however, the possibility of different reference frames surfaced in my very first lecture, before any "mathematics" was mentioned. When I asked the class what the word "symmetric" means, I got two answers:

- the object looks the same after we do something to it;
- the object looks the same if we look at it from a different point of view.

I explained that I would be viewing transformations in terms of fixed axes, but that I would not mind if, for some reason, they adopted a different convention.

After Clare's experience in the first cycle, I suggested that students making a complicated model, like her icosahedron, could consider making a labeled cradle for it to sit in. This could act as a fixed reference frame for the calculations in the assignment mentioned earlier.

This time, out of the 15 students who attempted the main task, 5 students definitely chose the position interpretation for their permutations, and 4 of these also chose (like Clare the previous year) the local interpretation for their transformations. Although there was some uncertainty about some of the other students' choices (because of their failure to indicate the sense of their rotations), there was only one student who made inconsistent choices, and he failed to complete the calculation and so never realized his error.

Of course, one shouldn't attach too much significance to the improvement in the proportion of students with inconsistent choices (80% for Zazkis and Dubinsky's students, 50% for my first class and 7% for my second class). All the classes are small, and there are many uncontrolled variables: for example, the task was different, the presentation was different and there were different students and teachers involved. However, there are fewer variables if we compare the results between my first and second cycles. Here the same teacher set the same task and, furthermore, the students appeared to be of more or less the same ability. Perhaps in this case it is plausible that the improvement can be attributed to the change from Figures 4 and 5 to Figure 10 or, in Dreyfus's words, to "the adaptation and correction of features of these visual representations on the basis of student reaction to them" (Dreyfus, 1994, p. 119).

Interestingly, there was still some confusion with numerical labels, but now it appeared in the students' evaluation of permutation products. There were several more opportunities during this second cycle to interview students about these difficulties, and we'll look at this in more detail in Section 10.

9. What influences the choice of interpretation?

As I indicated earlier, I favoured the object/global combination of interpretations when I was presenting this material to the students, so it is inevitably going to be difficult to isolate other influences when students also made that particular choice. We can however make some educated guesses when the student deviated from the expected choice. One consequence of using three-dimensional models was that my students had to find and name their symmetries before they could even start the crucial calculation of the two products. So they had a relatively free choice not just for the object/position decision but also for the global/local decision.

Choosing the object or position interpretation. Given my redrawn versions of Zazkis and Dubinsky's diagrams in Figures 10 and 11, we might expect students to choose the object interpretation if they had numerical labels fixed to the background, and to choose the position interpretation whenever they had numerical labels attached to, and moving around with, the object. Nine students left detailed enough evidence to check this hypothesis, and all except one behaved as expected.

David is typical of those who used the object interpretation. When listing all the symmetries of his tetrahedron, he used diagrams like Figure 12, where the labels 1, 2, 3, 4 are fixed to the background, and four other symbols are attached to the model to show which rotation has occurred.

For the rotation shown in Figure 12, David found the permutation

$$\alpha = \left(\begin{array}{cccc} 1 & 2 & 3 & 4 \\ 1 & 3 & 4 & 2 \end{array} \right)$$

in accordance with the object interpretation. Interestingly, when he came to the main task mentioned earlier, he described his two rotations, and calculated their

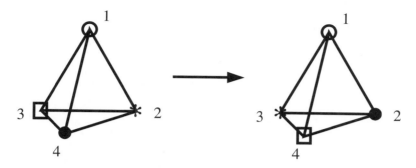

FIGURE 12. Numerical labels fixed to the background, giving the object interpretation.

product, using diagrams in which the numbers followed the vertices around (thus effectively using the local interpretation). Because he simultaneously changed the way he assigned permutations to the position interpretation, he now had consistent interpretations and got the desired equality. He didn't explain why he changed his interpretations.

Declan was typical of those adopting the position interpretation. Although he had described the symmetries of his cube relative to fixed Cartesian axes (the global interpretation), he then put numbers on the faces of the cube and let these labels move as the symmetries were applied. Figure 13 shows his diagram for one rotation to which he assigned the permutation

$$\alpha = \left(\begin{array}{cccccc} 1 & 2 & 3 & 4 & 5 & 6 \\ 1 & 5 & 3 & 6 & 4 & 2 \end{array} \right).$$

This agrees with the position interpretation (boldface numbers indicated a visible face, italics a hidden face). Declan failed to complete his calculation, and so he never had to resolve his inconsistent choice of interpretations.

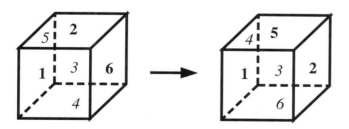

FIGURE 13. Numerical labels moving with the object, giving the position interpretation.

Only one student bucked the expected trend. Graeme managed to use numbers which were attached to, and moved with, his tetrahedron without falling into the trap which caught so many of Zazkis and Dubinsky's students: adopting the position interpretation. Figure 14 shows his diagram for a rotation to which he assigned the permutation

$$\alpha = \begin{pmatrix} 1 & 2 & 3 & 4 \\ 1 & 3 & 4 & 2 \end{pmatrix}.$$

This is consistent with the object interpretation.

FIGURE 14. Numerical labels moving with the object, giving the object interpretation.

So it appears that the object/position choice is usually, but not always, determined by whether the numerical labels are attached to the background or to the model.[3]

Choosing the global or local interpretation. The global interpretation is implicit in the way both Zazkis and Dubinsky and their students describe the transformations: "the square is flipped across its vertical axis" and "then I rotated it" (1996, p. 67). It is perhaps because Zazkis and Dubinsky's students made this choice unconsciously that the effect of the "wrong" choice between the position or object interpretation was so disconcerting.

Nevertheless, students sometimes seem to be aware of the possibility of another choice here. As I mentioned earlier, the possibility of different reference frames surfaced in my very first lecture, before any "mathematics" was mentioned. At this verbal level, it may be that the size of the object determines which view is adopted. With a "small" object (such as a square or a cube, and so on) it is possibly easier to move the object than to move one's own viewpoint. On the other hand, in that first lecture, when we discussed the symmetry of a large table, it was perhaps easier to imagine walking around the object rather than picking it up and rotating it.

Whether this verbal form is converted into its mathematical equivalent is another matter altogether. "Seeing" a fixed background, which can be used to describe the symmetries, may play an important role in enabling the student to choose the global interpretation. Clearly much of my presentation was driven by this consideration, but the students could sometimes find a fixed background by themselves. For example, some of my students noticed that their cube or octahedron could be placed

---

[3]Diagrams like Figure 10, representing the object interpretation, have appeared occasionally in textbooks. Gallian (1994, pp. 87, 95) uses the idea only for three-dimensional objects but, in an earlier generation, Hall (1972, pp. 235, 240) uses the idea for both two- and three-dimensional objects. Hall (1972, p. 234) actually constructs the desired isomorphism in the case of an equilateral triangle, giving the following advice to the reader "Label the positions of the vertices by 1, 2, 3. (Not the vertices themselves, which move with the triangle, but the positions in space.)" but without bothering to explain why this advice may be necessary. On the other hand, diagrams for the position interpretation do not seem to have appeared in the literature at all, presumably because they are not consistent with the preferred global interpretation of transformations.

so that some of its axes of symmetry were parallel to the standard three-dimensional Cartesian axes. This opened up the possibility of describing the symmetries using a fixed axes interpretation, as in "90-degree rotation about the x-axis" (usually, but not always, accompanied by a diagram with an arrow to indicate the sense of the rotation). Not all students "saw" this possibility. Of the eight students who chose to make a cube or octahedron in the second cycle, five followed this line of thought.

It is not so easy to place a tetrahedron, dodecahedron or icosahedron conveniently in relation to the Cartesian axes, and none of my students tried this idea. This meant that students using these models (along with the ones using a cube or octahedron who had not seen the Cartesian axes) had to rely on labelled features as in Zazkis and Dubinsky's study. As David's example (Figure 12) shows, it is relatively easy to draw pictures like Figure 10 for a tetrahedron, and the cube and octahedron can be handled in the same way. So most of these students still managed to adopt the global interpretation. However Becky put numbers on the vertices of her tetrahedron (very much as in Figure 14) and then adopted the local interpretation.

Diagrams like Figure 12 are quite difficult to draw for the dodecahedron or icosahedron. Four students used such models, and three of them left enough evidence of their thinking for us to draw some conclusions. (Students could use their models to do the calculations but I didn't usually see them do this.) Jack adopted a similar strategy to Clare's in the previous year, and kept track of his transformations by using labels fixed to the faces of his model. He showed the original position of his model with a double view, front and back as in Figure 15.

FIGURE 15. Jack's record of his dodecahedron in its original position.

For his rotations he recorded simply the new view of the left hand diagram in Figure 15. For example, he said, "Choose axis going from 12 to 11, rotating 72 degrees anticlockwise (looking down on 12)" and then drew Figure 16.

As his name for the transformation was tied to the labels 11 and 12, and as these labels moved around with the model, he was using the local interpretation described in Figure 3. As we would expect from the earlier discussion, he also adopted the position interpretation for his permutations.

On the other hand Moira invested some extra effort in constructing a numbered cradle for her icosahedron which could be used as a fixed background for describing rotations or permutations. She used the same numerical labels for her cradle and for the vertices of her model (as in Figures 4 and 5). Despite the consequent potential for confusion, she adopted the global interpretation (and the consistent object interpretation for her permutations).

FIGURE 16. Jack's record of his dodecahedron after a rotation.

It seems reasonable to conclude that students choose the global interpretation when they can "see" a fixed background, whether in the form of fixed Cartesian axes or as numerical labels fixed to the background. On the other hand, they usually choose the local interpretation if they have described their transformations in terms of labels fixed to the object.[4]

10. Is the double decker notation ambiguous?

The group structure for permutations is often introduced by explaining that the double decker notation

$$\alpha = \left(\begin{array}{ccc} 1 & 2 & 3 \\ 3 & 1 & 2 \end{array} \right)$$

means that α is the function given by $\alpha(1) = 3$, $\alpha(2) = 1$, $\alpha(3) = 2$; and by then giving an example of a product calculation. See, for example, Armstrong (1988, pp. 26–27) or Gallian (1994, pp. 85–86). But in practice this definition appears to be soon forgotten and the notation takes on a life of its own. When this happens it seems that the double decker notation is itself subject to conflicting object and position interpretations. Once again the student seems to be led astray by visual cues, not in the model or pictures this time, but rather in the numerical arrays which some students construct to evaluate permutation products.

Using arrays to calculate permutation products. To begin, here are two students apparently using the same visual information but in fact doing quite different things.

To find the product (1 2 3)(2 4 3), Joe wrote down the array

$$\begin{array}{lcccc} & 1 & 2 & 3 & 4 \\ (243) & 1 & 4 & 2 & 3 \\ (123) & 2 & 4 & 3 & 1 \end{array}$$

and read off the final answer as the cycle (1 2 4). Here a permutation operates on the previous row by writing down under each number its image under the permutation, so to find the product you simply write down the top and bottom lines of the array as the decks of a double decker.

---

[4] There is a little circumstantial evidence to suggest that earlier generations have learned the same lesson. For example, Birkhoff and Mac Lane (1965, p. 110) offer the following warning: "Caution: The plane of Figure 1, which contains the axes of reflection, is not imagined as rotated with the square."

The same idea of simply writing down the top and bottom lines of an array can lead to problems however. Jack had been getting the right answers for his calculations, but became worried when he realized that he had been multiplying permutations in the "wrong" order (working left to right, instead of right to left). Presumably there was a compensating error somewhere, but he hadn't been able to find it. His calculation of the product (4 5)(1 4) came from the following array.

$$
\begin{array}{ccccc}
1 & 2 & 3 & 4 & 5 \\
1 & 2 & 3 & 5 & 4 \\
5 & 2 & 3 & 1 & 4
\end{array}
$$

Here Jack has gone from the first row to the second by applying the transposition (4 5), in the sense that he swapped the numbers in the fourth and fifth positions. Similarly, to go from the second row to the third row, he applied the transposition (1 4), swapping the numbers currently in the first and fourth positions. He then read off the answer as the permutation

$$
\begin{pmatrix}
1 & 2 & 3 & 4 & 5 \\
5 & 2 & 3 & 1 & 4
\end{pmatrix}
$$

and this is the correct double decker expression for the original product. (Incidentally, Jack's error went undetected because he actually calculated the inverse of (1 4)(4 5) and this happens to be the desired product (4 5)(1 4) because each transposition is its own inverse.)

What has happened here? Certainly Jack has applied the permutations (4 5) and (1 4) in the wrong order. On the other hand, he has essentially adopted the object interpretation while applying each of these permutations. We can see this if we place objects (say, the letters A, B, C, D, E) in positions 1, 2, 3, 4, 5 and perform his operations on these objects, as in the following array.

$$
\begin{array}{ccccc}
1 & 2 & 3 & 4 & 5 \\
A & B & C & D & E \\
A & B & C & E & D \\
E & B & C & A & D
\end{array}
$$

This is a reasonable interpretation of his thinking because it corresponds precisely to how permutations were interpreted in the previously mentioned classroom exercise (which he took part in), where the numbers were labels fixed to the chairs which the students moved among (see Figure 10). Of course Jack did not use letters but numbers, and so he found himself in the same situation mentioned in connection with Figures 4, 5, and 8, where the dual role of the numerical labels also led to confusion for some students in the first year of this project.

At the final step, where Jack read off the answer, he abandoned the meaning of these calculations (which said, for example, that what had been in the first position originally was now in the fourth) and settled instead for reading off the top and bottom rows of his array (what seemed to be the initial and final states). This inverted the permutation and effectively adopted a position interpretation of the double decker notation.

Of course, given the completely formal way in which Jack extracted the answer, it may be doubted whether he consciously adopted either interpretation. However, we can see the same strategy being used by Moira, successfully this time, in the following calculation. To find the product (4 5 6 8)(1 2 4 5), Moira drew the following diagram (Figure 17) and then explained that "1 goes to 2" with an accompanying wave of her hand from the circled 1 back to the circled 2.

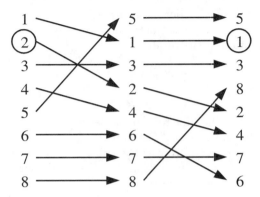

FIGURE 17. Moira's calculation diagram for the permutation product (4 5 6 8)(1 2 4 5).

So she evaluated the product as

$$\alpha = \left(\begin{array}{cccccccc} 1 & 2 & 3 & 4 & 5 & 6 & 7 & 8 \\ 2 & 5 & 3 & 6 & 1 & 8 & 7 & 4 \end{array} \right).$$

These product calculations would probably be classified by Zazkis and Dubinsky as "analytic" rather than "visual" (1996, p. 61). Certainly there is no geometry involved here. However, such arrays are really just schematic descendents of the visual representations which we saw in Figures 4 and 5 (the grandparents, if you like), and in Figure 8 (the parent). Furthermore, there is a strong pictorial element in the way such arrays are used, with answers being deduced purely from the spatial arrangement of the numbers involved. So it is natural to see Jack's error as a continuation of the same "visual confusion" which we met earlier in more geometric settings. It would seem that the distinction between visual and analytic approaches is not so clear cut in this situation.

It is possible, in this particular case, that Jack's and Moira's approach to these calculations was conditioned by their experience of the earlier task involving Platonic solids. As we saw in Section 9, Jack used labels fixed to his model and assigned permutations according to the position interpretation. Moira, on the other hand, had used numerical labels attached to both her model and its cradle, and had assigned permutations according to the object interpretation. Her calculations for this could easily have looked like Figure 17, and Jack's different interpretation of the same diagram could just reflect his different choice of the position interpretation in that earlier task.

Position or object? Although the above discussion dealt with a product calculation, some of the difficulties may actually stem from the double decker notation itself. Consider the permutation

$$\alpha = \left(\begin{array}{cccc} 1 & 2 & 3 & 4 \\ 2 & 3 & 4 & 1 \end{array} \right).$$

It seems natural to read the right hand side as saying that "1 goes to 2, 2 goes to 3, and so on" and, as Joe's calculation shows, this verbalization can be used successfully to evaluate products.

However, problems arise almost as soon as the permutation is allowed to act on an ordered array. If α acts on the array ABCD, then the object interpretation says that α sends the object in the ith position to the $\alpha(i)$th position. So the above permutation applied to ABCD produces DABC (see Figure 10). Similarly, if α is applied to the array 1234 we get 4123, and a natural way to illustrate the effect of α is to write down the array of initial and final states.

$$
\begin{array}{cccc}
1 & 2 & 3 & 4 \\
4 & 1 & 2 & 3
\end{array}
$$

Suddenly we have some sympathy for Jack!

Of course you could argue that Jack simply made a mistake, and that there is no question of him adopting a position interpretation of the double decker notation. However, this behavior is consistent with the alternative interpretation of Figure 8 which we saw earlier. Here are two more students who seem to have adopted the same interpretation.

Sean talked his way correctly through the calculation of the permutation product $(4\ 5\ 6\ 8)(1\ 2\ 4\ 5)$, saying for example that "4 goes to 5 which goes to 6," but then he wanted to record this result as

$$
\alpha = \left(
\begin{array}{cccccccc}
1 & 2 & 3 & 4 & 5 & 6 & 7 & 8 \\
  &   &   &   &   & 4 &   &
\end{array}
\right).
$$

Sean's thinking here seems to be that he wants to know where in the bottom row to put the 4. After tracing the path of 4 as above, he sees that it should be put in the sixth position. This idea is stated explicitly in the next example.

Explaining the following product calculation,

$$
\left(
\begin{array}{ccc}
1 & 2 & 3 \\
2 & 1 & 3
\end{array}
\right)
\left(
\begin{array}{ccc}
1 & 2 & 3 \\
3 & 2 & 1
\end{array}
\right)
=
\left(
\begin{array}{ccc}
1 & 2 & 3 \\
2 & 3 & 1
\end{array}
\right)
$$

Brian said (with accompanying gestures)

> Where do I put the 3 [points to the bottom row]? the right hand factor says 3 goes to the 1 position. Where [on the bottom row] does the 1 go in the first factor? [Under the 2 and so] put the 3 there.

Instead of calculating a product $\alpha\beta$, Brian's method actually finds $\beta^{-1}\alpha$. This ought to give quite different answers, of course. However Brian was originally calculating permutation products from left to right, and he became suspicious only when he realized that I had been working from right to left (as in his commentary above). Presumably all the examples he had worked on up to that point had involved cases where β was an involution.

It seems that Brian had used worked examples of permutation products as a sort of Rosetta Stone to be deciphered to discover the intended method of calculation. Some might be tempted to accuse students like Brian of not reading their text carefully enough. It would be wrong, however, to dismiss Brian too quickly: he was quite an able student and eventually got an A+ for this course. So perhaps there is an ambiguity, a possibility of two different interpretations, which could explain why students like Brian would find it hard to decipher the method of calculation.[5]

11. Conclusions

Zazkis and Dubinsky's study concerned two different ways of representing the symmetry group of an object, a visual or geometric way, using motions of the object, and an analytic or symbolic way, using permutations of some characteristic features of the object. Almost all their students came up with conflicting answers when comparing these two representations. Zazkis and Dubinsky explained this behavior by pointing out the different reference frames which students tended to use in the two situations: a fixed reference frame when using motions of the object, and one which moved with the object when using permutations.

In this study we have seen how the author sought to guide his students towards a consistent use of fixed reference frames for both situations. Although various influences were brought to bear on the students, the author's main efforts went into producing new diagrams to clarify the respective roles of object and background in each situation. Successive clarifications over two years were followed by successive decreases in the proportion of students making inconsistent choices, which suggests that visual cues in diagrams like Figure 1 may be a contributing factor for the errors observed in Zazkis and Dubinsky (1996).

Zazkis and Dubinsky wondered if there were "deeper perceptual and conceptual factors" which led students to choose the position interpretation (p. 74). The global interpretation of transformations seems to be an almost automatic choice for their example, and so an inconsistent choice in Zazkis and Dubinsky (1996) always involved the position interpretation. The situation in three dimensions is not so clear cut, and it is reasonable to ask what leads students to choose any of the possible interpretations.

We saw in Section 9 that the object/position choice is usually, but not always, determined by whether the numerical labels are attached to the background or to the model. On the other hand, the global/local choice seems to be determined by whether the student can "see" a fixed background. For some more complicated shapes (such as the dodecahedron or icosahedron) the local interpretation remains simpler and more attractive to students despite instructional pressure to choose the global interpretation.

Although most students made consistent choices in the second year of this study, remarkably similar difficulties persisted in a much more algebraic (or perhaps, in Zazkis and Dubinsky's terminology, analytic) setting, the calculation of permutation products. These difficulties can be traced to another kind of visual confusion, not in pictures this time, but in the numerical arrays which some students construct to evaluate permutation products.

Finally we have seen that, although the double decker notation is officially completely unambiguous, in practice it too is subject to conflicting interpretations. So the confusion observed by Zazkis and Dubinsky in fact spans the full visual-analytic spectrum.

---

[5]There is a hint in some textbooks that these problems have been noticed before. For example, Gallian (1994, p. 86) decorates his first permutation product calculation with arrows tracing the "path" followed by one particular integer, a device which might have been aimed at students like Jack, Sean and Brian.

References

Angelo, T. A., & Cross, K. P. (1993). *Classroom assessment techniques* (2nd ed.). San Francisco: Jossey-Bass.

Armstrong, M. A. (1988). *Groups and symmetry.* New York: Springer Verlag.

Birkhoff, G., & Mac Lane, S. (1965). *A survey of modern algebra.* New York: MacMillan.

Dreyfus, T. (1994). Imagery and reasoning in mathematics and mathematics education. In *Selected lectures from the 7th International Congress on Mathematical Education* (pp. 107–122). Quebec: Les Presses de l'Université Laval.

Gallian, J. A. (1994). *Contemporary abstract algebra* (3rd ed.). Lexington: Heath.

Haines, M. J. (1997). A classroom activity on permutations. *Notices of the American Mathematical Society, 18*, 216. (Abstract 918–H1–869)

Hall, F. M. (1972). *An introduction to abstract algebra* (Vol. 1, 2nd ed.). Cambridge: Cambridge University Press.

Herstein, I. N. (1975). *Topics in algebra* (2nd ed.). New York: Wiley.

Kemmis, S. (1994). Action research. In *International encyclopedia of education* (Vol. 1, pp. 42–48). Oxford: Pergamon Press.

Zazkis, R., & Dubinsky, E. (1996). Dihedral groups: A tale of two interpretations. In J. Kaput, A. H. Schoenfeld, & E. Dubinsky (Eds.), *Research in collegiate mathematics education II* (pp. 61–82). Providence, RI: American Mathematical Society.

DEPARTMENT OF MATHEMATICS AND STATISTICS, UNIVERSITY OF CANTERBURY, CHRISTCHURCH, NEW ZEALAND

E-mail address: j.hannah@math.canterbury.ac.nz

CBMS Issues in Mathematics Education
Volume **8**, 2000

Factors, Divisors, and Multiples:
Exploring the Web of Students' Connections

Rina Zazkis

ABSTRACT. This study is a contribution to the ongoing research on preservice teachers' learning and understanding of introductory number theory. The focus of this article is on fundamental concepts of factor, divisor, and multiple; the meaning students construct of these three concepts; and students' links among the three notions as well as their connections to other concepts of elementary number theory, such as prime factors, prime decomposition and divisibility. Nineteen clinical interviews in which students in a course for preservice teachers were asked to explain and exemplify the concepts and to apply their understanding in several problem situations were analyzed focusing on the connections students made among the concepts. An examination of students' responses showed that the meaning they assign to the concepts is often different from the meaning assigned by mathematicians in the context of number theory, and that their links among the concepts are often weak or incomplete.

The mathematical knowledge of preservice elementary school teachers has been a focus of a variety of recent studies. However, investigation of preservice teachers' understanding of concepts related to introductory number theory has received little attention. Previous research indicated that students' understanding of the basic concepts in introductory number theory and their connections among these concepts have yet to be explored (Ferrari, 1997; Zazkis & Campbell, 1996a, 1996b). The objective of this study was to provide a fine grained analysis of preservice teachers' understanding of a limited set of fundamental concepts related to number theory— factor, divisor, and multiple.

1. Synopsis of Prior Research

Research on learning and understanding topics related to number theory has not been as extensive as research on other core topics of mathematics courses for preservice elementary school teachers, such as rational numbers or geometry. This research is concerned with preservice teachers' learning and understanding of concepts such as prime and composite numbers, divisors, factors, and multiples, greatest common divisor and lowest common multiple, divisibility and divisibility "rules," division with remainder and the Euclidean algorithm, and prime factorization and

This study was supported in part by grant # 410-96-0689 from the Social Sciences and Humanities Research Council of Canada.

©2000 American Mathematical Society

the Fundamental Theorem of Arithmetic, among others. Prior research (Campbell & Zazkis, 1994; Zazkis & Campbell, 1996a) has revealed that preservice teachers' understanding of many of these concepts is incomplete due to gaps in their understanding of "prerequisite" concepts, such as natural and rational numbers or distributivity. It also noted that preservice teachers had difficulty connecting divisibility viewed in terms of multiplication with divisibility viewed in terms of division. In addition, weak or missing connections among number theory concepts in preservice teachers' representations of their mathematical knowledge have been observed. For example, when a number was represented in its prime decomposition, its prime factors were not recognized by some students as its divisors; the divisibility of a by b was not recognized as entailing a remainder of zero in division of a by b.

This pioneering research pointed out the complexity of issues involved in learning number theory and the need to investigate students' understanding of specific concepts in further detail. Consequent studies focusing on students' understanding of specific themes, such as the Fundamental Theorem of Arithmetic (Zazkis & Campbell, 1996b) and even and odd numbers (Zazkis, 1998a) highlighted the influence of students' intuitive beliefs about numbers and their structure on their decision making in a problem solving situation. For example, many students believed that a "large" number must have a "small" prime factor, such as 3, 7 or 11. This belief co-existed with students' awareness of the existence of "very large" prime numbers.

Other studies investigated problem solving strategies related to the divisibility concept among first year computer science students (Ferrari, 1997). It has been observed that students frequently regress to computation in discussing divisibility and are often not able to base their justifications on the structure of a given number. Further, Brown, Thomas and Tolias (1999) deliberated on the complexities of students' understanding of multiples and least common multiple. Students' reluctance to operate on numbers presented in prime-factored form was reported, strategies encouraging students to attend to this representation and appreciate the information it offers were suggested (Brown, 1999).

In general, prior research indicates the need to study two general problems related to students' construction of knowledge in the domain of elementary number theory:

(1) The problem of re-learning:
   What is preservice teachers' knowledge of number-theory related concepts prior to entering a "Principles of Mathematics for Teachers" course? What are their intuitive beliefs about numbers and their structure? How do prior knowledge and intuition influence their learning of elementary number theory? What pedagogy can help replace intuitive misconceptions with appropriate representations of mathematics?

(2) The problem of connectedness of knowledge:
   How are components of knowledge connected? What links are missing or incomplete? What instructional approach can assist preservice teachers to make connections among concepts of number theory?

In this study I focus on the second problem, attending to connections among factors, divisors, and multiplies, and students' understanding of these connections.

2. A Web Perspective on Knowledge

Mathematical concepts are not studied in isolation but in relation to other mathematical concepts. This idea has been illustrated by researchers using the metaphor of a web. Hiebert and Lefevre described conceptual knowledge as "a connected *web* of knowledge, a *network* in which the linking relationships are as prominent as the discrete pieces of information. Relationships pervade the individual facts and propositions so that all pieces of information are linked to some *network*" (my italics, Hiebert & Lefevre, 1986, pp. 3–4). The ideas of web and network suggested not only connectedness but also a complexity of knowledge.

In the past decade, with the increasing popularity of and the increasing access to the World Wide Web, the ideas of network and linking have been revisited, introducing new meanings to familiar words. So was a web metaphor of knowledge. Borrowing a metaphor from the World Wide Web in discussing a web of knowledge, this metaphor has been extended. In addition to connectedness and complexity, the WWW provides an underlying structure, through which learning is navigated. Therefore, "the idea of webbing is meant to convey the presence of a structure that learners can draw upon and reconstruct for support—in ways that they choose as appropriate for their struggle to construct meaning for some mathematics" (Noss & Hoyles, 1996, p.108).

The acknowledgment of underlying structure shifts the discussion from students' "construction of knowledge" (Davis, Maher, & Noddings, 1990), which has been predominant in mathematics education research, to "construction of meaning" (Noss & Hoyles, 1996). Although Noss and Hoyles used and exemplified the idea of webbing in computerized environments, they suggested that it could be applied to non-computerized environments as well. This article attempts to analyze students' construction of meaning of several introductory concepts in number theory through a web perspective.

The underlying structure in this discussion is presumed by the mathematical content. In the domain of knowledge discussed in this paper, the mathematical links supporting the underlying structure of knowledge are simple and clear. The mathematical connection among a factor, a divisor, and a multiple is expressed in the equivalence of the following three statements, for any two natural numbers A and B:

> B *is a factor of* A
> B *is a divisor of* A
> A *is a multiple of* B

There are additional language forms to express the same relationship:

> B *divides* A
> A *is divisible by* B

Investigation of students' understanding of divisibility has been a focus of an earlier research reported in Zazkis and Campbell (1996a). It has been found that "insufficient connections between [the view of divisibility in terms of division and in terms of multiplication] were a source of cognitive discord for many of the participants" (p. 562). To follow up on this discord, this study focused on students' understanding of three fundamental concepts—factor, divisor, and multiple—and their webbing among these concepts. My initial intention in this study was to focus on students' understanding of mathematical links, that is, links embedded in the content. However, it became apparent in the analysis that it is impossible to ignore another set of links—connections to students' prior knowledge.

3. Methodology

3.1. Participants and setting. Participants in the study were 19 volunteers from a group of 72 preservice elementary school teachers, enrolled in a course called "Principles of Mathematics for Teachers." Despite the size of the class, course instruction focused on problem solving approach and small group work. New concepts were usually presented as a summary of students' experiences. Conversation among students, raising and verifying conjectures, and oral and written explanations of students' thinking were encouraged and supported. At the time of the interview, the students had just completed their work on the introductory number theory chapter. Topics included prime and composite numbers, prime factorization, the division algorithm, factors, multiples and divisors, divisibility and divisibility rules as well as least common multiple and greatest common divisor. Rigorous mathematical definitions for these concepts were provided, however the emphasis was put on meaning rather than on memorization of formal definitions. Students were expected to be able to explain and describe these concepts and to apply them in problem situations.

3.2. Clinical interview. Interview results from two question sets explicitly pertaining to divisors, factors and multiples are analyzed here. Occasionally, additional evidence from students' written work is provided to support the data from the interviews.

> **Question Set 1**
> Look at the following terms:
> whole number, natural number, sum, product, divisor, factor, prime number, composite number, multiple (a card with this list was presented to students)
> Do they sound familiar? What does each one of the words mean?
> Can you exemplify?
> **Question Set 2**
> What are the prime factors of ? Can you list all of them?
> What are the factors of $117 = 3^2 \times 13$? Can you list all of them?
> What are the divisors of $117 = 3^2 \times 13$? Can you list all of them?
> Of what numbers is 117 a multiple?
> Is 117 a multiple of 26?
> Can you give an example for a multiple of 117?
> Any observations? Can you think of a factor which is not a divisor? Can you think of a divisor which is not a factor? Can you explain?
> Can you think of a number which is both a multiple and a divisor of 117?
> Any others?

Specific questions in the question sets presented guidelines, rather than established wording or order. Not all the questions were presented to all the interviewees, and different levels of clarification were requested in different interviews. Also, the interviewer made an attempt to tie questions to students' prior experience in the interview. Therefore the actual questions varied. Following Ginsburg (1981, 1997) it was understood that contingent manner of questioning is an integral part of the method of clinical interviewing.

Question Set 1 intended to outline the meanings that students had constructed for fundamental number theory concepts. It was chosen to present the terms altogether and not one by one to let the interviewee choose the order in which to address the concepts and also to provide an opportunity to demonstrate connections among the concepts by discussing them together.

Question Set 2 intended to investigate students' understanding of factors, divisors and multiples and their connections among those concepts. The Question Set 2 was designed following the findings of a previous study, which suggested that equivalence between factors and divisors may not be well grasped (Zazkis & Campbell, 1996a).

Students' ability to address questions in Set 2 depended on the meanings they had constructed for the concepts. It became evident early in the interviewing process that in many cases the interviewer and the interviewee did not share a common meaning, especially for the concepts of a divisor and a multiple. In these cases the interviewer reminded the participant of their classroom activities and of the mathematical meaning intended in the number theory context. Such minimal intervention appeared necessary to proceed with the interview.

3.3. Guidelines for analysis. Question Set 2 was designed so that it was effortless to answer by attending to connections among different mathematical concepts, but could also be approached without attending to these connections. Faster and more elegant approaches result from attending to relations among concepts, so the strategy chosen by a student served as an indication of whether or not the student had constructed a link among the concepts.

For example, in considering the prime decomposition $117 = 3^2 \times 13$, the prime factors of 117 are transparent. All the factors of 117 can be listed by considering combinations of the prime factors, by drawing factor trees, or by trial and error. The first approach was taken as a demonstration of students' understanding of a link between factors and prime factors. Further, in listing divisors of 117 one can rely on the previously made list of factors or attempt different computations in order to find what division results in a whole number. The first approach was taken as an indication of a link between a factor and a divisor in the meaning that students constructed. Similarly, considering whether 117 is a multiple of 26, one can rely on the previously generated list of factors of 117 or ignore this list and address the question by observing evenness or oddness of these numbers, or by regressing to calculations. Attention to the available previously derived information regarding factors and divisors of 117 served as an indication of the existence of a link between those concepts in a student's mind.

In general, from the perspective that individual's knowledge is connected, links can be seen as connections between different fragments or aspects of knowledge. As mentioned earlier, my initial intention was to describe to what degree students took advantage of mathematical links among the concepts in addressing the questions. Students' approaches demonstrated to what degree the meaning of links of mathematical structure have been constructed in students' minds. During the analysis another substantial set of connections became apparent, connections to students' prior understanding and their school experience.

4. Results and analysis

In the following I describe students' responses to Question Sets 1 and 2. I will discuss clusters of students' responses and students' mathematical meanings in their relation to the "underlying structure" of conventional mathematical meanings. I wish to make no claims on generality of the observed phenomena. However, some frequencies are reported to suggest that many of these responses may identify common trends rather than random appearances.

4.1. Assigning Meaning to a Factor. Factor appeared to be the least problematic of the three concepts discussed. The initial analysis of students' responses suggested that most students described and exemplified the concept correctly. The consequent in-depth analysis identified several common features in the students' interpretation of the concept and suggested that students' understanding may be incomplete.

All but one students presented in their examples a pair of factors rather than a single factor. The following excerpts from the interviews exemplify this tendency. Students' preference to discuss factors in plural seems significant because the concept was presented to the students in a singular form. In what follows I present a variety of students' explanations for the concept of a factor and analyze the possible sources of this preference.

Amy: And factor is (pause) it's hard to explain what a factor is. (pause) Not a fraction, but a factor, right?

Interviewer: Um hm ...

Amy: I'm not sure how to explain it, it's, (pause) I don't know how to explain that one.

Interviewer: Maybe you can give me an example.

Amy: OK, well if you take the number 6, the factors are 2 and 3. So the numbers that you can multiply together to get that number, are its factors.

Anne: A factor is that, when two numbers are multiplied, they are the factors of the answer.

Interviewer: What does it mean to you to say that one number is a factor of another number?

Robin: A factor? (pause) I can see it, but I just have to figure out how to say it. A factor, it's a, in a way it's a divisor. It's a number that can be either multiplied, if you take two factors you can multiply them together to give you a particular number that you're factoring, or dividing, you can divide a factor out of, of your number that you're looking to factor. Um, (pause) it's a, I can't exactly call it a part of that number, because it's not the number, but um, (pause) I always think of them as the factor trees (laugh).

Interviewer: What does it mean to you to say that a number is a factor?

Kate: If a number is a factor, that means it's (pause) a, something that could be a, when you get a product, it's the two numbers that lead to the product.

As mentioned above the reference to "factors," rather than "a factor" appeared in the majority of the interviews. There is additional evidence for this way of thinking from a midterm examination of a similar group of students, taking the course a year earlier. When students were asked to define and exemplify the word "factor," most examples (59 out of 68) discussed pairs of numbers. A closer look at students' examples helps to explain this phenomenon.

Consider the multiplication sentence $C \times B = A$. The number A, the result of the multiplication, is called the product. The numbers C and B can be referred to as multiplicand and multiplier, however English and American traditions disagree on which is which. In order to avoid this disagreement and also to emphasize the commutativity of multiplication, both numbers C and B can be referred to as factors. This describes their role in a multiplication number sentence.

In elementary number theory, there is a different, although related, definition for a factor: *Given two natural numbers A and B, the number B is called a factor of A if and only if there exists a natural number C such that $C \times B = A$.* In this sense a factor specifies a relation between numbers. (In fact, this is an incomplete relation of order, that is, reflexive, asymmetric and transitive.) This later definition was used in the course for the three weeks of exploring elementary number theory prior to the interviews. There is no agreement in mathematical texts with respect to the reference number set in this definition. Several texts define factors, divisors, and multiples with a reference to natural numbers, while others refer in their definitions to integers or whole numbers. Limiting the discussion in this study to natural numbers was a pedagogical choice: It excludes from the discussion the troublesome number zero and does not compromise the mathematical meaning.

In summary, there are two meanings for the word factor in the context of numbers and number operations—factor as a *role player* in a multiplication number sentence and factor as a *relation* between natural numbers. (Associations with the word "factor" related to the process of factoring in algebra are outside the scope of this discussion.) Acknowledging the two different meanings of the word factor, I suggest that students' reference in plural was more than a syntactic convenience. In the majority of students' definitions a factor was not seen as a relation between numbers, students were not discussing the property of one number being a factor of another number, even when a question had been explicitly posed this way. Students were talking about factors as the roles the numbers play in a multiplication sentence. Discussing the concept of a factor, this view does not cause confusion or false examples, since whenever a product of two natural numbers is considered, each one of the factors (as role players) in this product is a factor of the product (as a relation). I mention this here because the case of divisors, discussed in the next section, is rather different.

In addition to a procedural reference to multiplication, there are several other common features in students' descriptions of a factor. The following excerpts exemplify these features.

Marty: I was going to say a factor is when you've got it in its prime, or down to the lowest possible ...

Interviewer: How about the term factor, what does it mean to you to say that one number is a factor of another number?

Mary: I think of a tree's branches, I think about um, figuring out all the different numbers, when multiplied by each other go into that um, big number evenly.

Interviewer: Hmm, OK. Um, (pause) could you uh, could you give me an example of that?

Mary: If I had 24 as the number that I wanted to factor, I would go underneath that, beneath that, 2×12 and then underneath the 12 I would go 2×6 and then under that I would go 2×3 and those lowest numbers at the bottom would be my, my factors. So in order for me to remember that, I'd have to write it down and show them.

Interviewer: How about the term factor? What does that mean to you?

Wendy: (pause) Factor, um, is, well you have a factor tree where you, where you have certain numbers that multiply to, to make the number that you started out with ...

Interviewer: OK ...

Wendy: So it, it's kind, it's like multiplication.

Interviewer: Alright. Could you give me an example?

Wendy: The 2 and the 6, $2 \times 6 = 12$.

Interviewer: OK. How about if you had 2.5×6 ...

Wendy: That's also a factor. It doesn't matter if there's a decimal.

The reference to natural or whole numbers was missing in all the responses, even when a student had been confronted with a specific example, such as in the excerpt from Wendy's interview above. This can be seen as a further confirmation of students' seeing a factor as a role player in multiplication rather than a relation.

Another common feature, which appeared in six interviews, is a reference to prime factors or to "lowest possible" factors, as in the above excerpt from the interview with Marty. References to a factor as a prime factor will be discussed further in the next section.

Moreover, references to factor trees appeared in five interviews. This could be an indication of students' thinking of a familiar process as well as recalling the context in which the word was used, rather than seeing a relation between numbers. Mary's talk about a number she wanted "to factor" and Robin's reference to the number she was "factoring" present additional evidence of students thinking about procedures rather than relationships.

In summary, for the majority of students the concept of a factor appeared to be linked to a multiplication sentence and pairs of numbers multiplied. Some links suggested connections to students' experiences, like consideration of factor trees and prime factors. The essential links to the context of natural or whole numbers as well as to relations among numbers appeared weak or missing in most interviews.

4.2. Assigning Meaning to a Divisor. Divisor is a problematic concept in elementary arithmetic. On one hand the label divisor specifies the number's *role* in a division number sentence: When $A \div B = C$, we call A the dividend, B the divisor and C the quotient. On the other hand, the concept of a divisor specifies a *relation* between natural numbers. In a formal mathematical definition this relation is expressed as follows: For natural numbers B and A, B *is called a divisor of A if and only if there exists a natural number C such that $B \times C = A$.* In fact, this definition is identical to the definition of a factor. In mathematics textbooks for elementary school as well as in some texts for elementary school teachers, this relation between natural numbers is expressed in an equivalent way in terms of division: *B is called a divisor of A if and only if the division of A by B results in a natural number.* That is, if the division does not result in a natural number, the number that has the role of the divisor in the number sentence is not a divisor of the dividend. This polysemy, that is, a property of the word of having different but related meanings, creates lexical ambiguity. (For a detailed discussion on lexical ambiguity of divisor and quotient see Zazkis (1998b)). In the three weeks of learning the topic of elementary number theory by the interviewees the concept of a divisor was defined and used as a relation between natural numbers. However, as the following excerpts show, this was not the view taken by the majority of students when they were asked to explain the meaning of the concept. Below are several excerpts from the interviews in which students are attempting to describe the meaning of a divisor.

Anne: Um, divisor is what you are putting into your number, like 4 divided by 2, 2 is the divisor.

Dorothy: Divisor. That would be a number that you would be using to break down another number, that's how many times another number could be divided down. For example, 36 using a divisor 9, that 9 can be taken out of 36 four times before you have nothing left, before you can't take 9 out any more.

Vickie: Divisor, the divisor is the number which is (pause), that you're dividing into another number, 3 into 6, 3 would be the divisor of 6.
Interviewer: 3 into 6, 3 is a divisor of 6 ...
Vickie: Yeah.
Interviewer: And what about something like 5 into 6?
Vickie: 5 would be the divisor.
Interviewer: 5 is the divisor.
Vickie: Right.
Interviewer: Would you say that 5 is a divisor of 6?
Vickie: Yes.

Interviewer: What does it mean to you to say that one number is a divisor of another number?
Kate: That means that it divides another number. It um, wants you to take one number that is given and they want you to break it off into this many of it.
Interviewer: Could you give me an example of that?
Kate: (Laugh) Like 12 divided by 3 is 4.
Interviewer: Um hm ...
Kate: So, divisor is the 3 ...
Interviewer: OK.
Kate: Either that or it's the 12 ..., no it's the 3, yeah.
Interviewer: And, would you say that uh 5 is a divisor of 12?
Kate: It can be.
Interviewer: In what way?

Kate: That it can divide 12, the only thing is, it would have a remainder. So it's whatever number you're trying to divide the number by, yeah, the divisor is the number that you use to divide.

Interviewer: How about the term divisor, what does it mean to you to say that one number is a divisor of another number?

Mary: (pause) Divisor is, is when something is divided by in order to get the quotient, you've got the dividend and the divisor, dividend divided by the divisor equals the quotient ...

Interviewer: Hmm ...

Mary: That's I think in my mind what I see ... OK, so here we've got the quotient and we've got the divisor, so (pause) the, the divisor is, the divisor times, multiplied by the quotient equals the, the dividend.

Interviewer: How about a divisor? What does the word divisor mean to you?

Penny: (pause) I think of that as the process too, again as a label of, I can't remember which one it is, (laugh) I think it's the second one. Like you go, you're, how does it go ...

Interviewer: Could you illustrate to me on the paper, what you mean by the second one?

Penny: OK (laugh). You have A divided by B equals C, I think that this one's a divisor [points to B], I'm not sure, but that's the way I think of it.

In fact, 15 out of 19 interviewees defined divisor referring to the number's role in a division number sentence. In some cases, where the question posed by the interviewer was "What does the word divisor mean to you," this interpretation of a divisor was quite acceptable. However, in other cases, where the question was posed as "What does it mean to you to say that one number is a divisor of another number," there was an explicit contextual reference to the intended meaning of a divisor. In most cases this intention of the interviewer was not clear to the interviewee. Penny's reference to "the second one" and Mary's labeling of all the role-players in the division sentence indicated that the "role-player" meaning of divisor was more robust in students' minds than the "relation" meaningof the concepts used recently in their mathematics course. Several interviewees indicated that this meaning of divisor was part of their common knowledge from school.

All the examples of division provided by the interviewees involved natural number results. In these cases the conflict between the two meanings of the word divisor was not apparent. However, when confronted by the interviewer with an example of a division that does not result in a natural number, the students did not attempt to adjust their definitions. This was most apparent with Kate and Vickie above, who claimed that 5 was a divisor of 12 and 6 respectively. It has been suggested in the previous section that thinking of factors as a role numbers play in a number sentence, rather that a relation between numbers, did not result in mathematically wrong claims. However, as revealed in this section, the analogous view of a divisor resulted in erroneous claims.

Only four interviewees (out of 19) explained the meaning of a divisor as a relation between natural numbers, three claiming explicitly that a divisor was the same as a factor. The following excerpt with Ronnie shows that he has adopted a number-theoretical meaning for divisor.

Ronnie: Divisor it's the same as a factor. Like 3 is a divisor of 21. But 4 is not. For 3 if you do the division you end up with no remainders. Nice whole numbers. It's like putting equal pieces into equal groups, everything must fit perfectly.

Students describing divisors as a relationship between numbers were asked whether there was a difference between a factor and a divisor. The following are their responses.

Ronnie: Difference between a factor and a divisor. Um, just the relationship. If you're thinking in terms of division, then I would use divisor, I mean once again, bringing it back to concrete, if you're thinking about 21, 3 and 7, let's take those three, 3 and 7 are factors of 21, but if you sort of want to take a piece out of the big pie, and let's say this piece is 3, that, to me that would be a divisor.

Chris: Um hm, because the number is built up of factors, and factors, when you're dividing, break it down again. So it's just, one builds them up and then the divisor will break it down. So I can, because multiplication and division are kind of the corresponding operations, so factors is like multiplying and then divisors is dividing it down. So I can see the relationship.

Interviewer: You said multiplication and division are corresponding operations.

Chris: I don't know the special term is right, but like division and multiplication and addition and subtraction, you do the opposite.

Interviewer: Can you please show me? What do you mean when you say they do the opposite?

Chris: Like 6 divided by 3 equals 2, that's breaking it down smaller, into smaller, and then $2 \times 3 = 6$, you have the same numbers you're using, it's just, one of them puts them together as a whole, and the other one splits them apart.

Sarah: Um hm, oh um, I think the difference between a divisor and a factor would be that a factor is a prime number, but, like the factor tree, you look at the factor tree, and eventually you derive prime numbers, so the factors are all the numbers that can divide into a specific number, whereas a divisor is just a number that divides into the number given. Um, (pause) like with a factor tree, (pause) the factors eventually you derive at prime numbers, whereas a divisor is just choosing a number that derives the answer.

Liz: (Sigh) Um, I see a factor as being a whole number that divides perfectly. Well, I don't know if perfectly is the right word, into a number, like there's another whole number that's going to be the result. But, a divisor to me can be any number, even if doesn't go in perfectly, like it may not give you a whole number, it could give you a remainder, or a decimal pointed number—it doesn't go in perfectly. Or, it could too, but (laugh)...

Ronnie and Chris discussed the relationship between a factor and a divisor referring to the relationship between multiplication and division. Sarah faced the equivalence of factor and divisor in her definition, and became troubled by that. Her way to resolve the problem of having two different words for the same idea was to define factors as "prime factors," whereas divisors for her were not necessarily prime. Liz also realized the equivalence of factors and divisors in her initial explanation, so in order to resolve this she changed her view of divisor from the relation between numbers to the number's role in division. Although the interviewer's question intended to establish equivalence between a factor and a divisor, it is possible that these students were seeking a difference between the two concepts just because they were asked about it. However, an attempt to create a distinction can be also seen as an indication of a student's weak link between these concepts.

In summary, for the majority of students a "divisor" was linked to a division number sentence and a number we divide by. Similarly to the case of a factor, the essential links to the natural or whole numbers context as well as to relations among numbers appeared weak or missing in most interviews. The link of equivalence between factors and divisors appeared weak, students referring to this link sought a distinction between the two concepts.

4.3. Assigning Meaning to a Multiple. Multiple appeared to be the most problematic concept. Only four students provided correct examples for a multiple. Another four students used the word "multiple" as a synonym for the word "product." Eleven students identified explicitly or implicitly a multiple and a factor. This is consistent with the findings of Nicholson (1977), who investigated the understanding of mathematical vocabulary among students entering secondary school. Only 21 out of 185 participants in Nicholson's study provided an acceptable answer to the item "Give one example of a multiple of 30." Ninety-two students gave an example of a factor of 30 in their responses.

Chris was among the few who seemed to have a grasp of a concept of a multiple. However, later in the interview, as the following excerpt demonstrates, Chris testified to her confusion.

Chris: And factor is um, it's like, for example, 2 would be a factor of 8, because 2 is, ... 8 is divisible by 2, so 2 is multiplied by another whole number will give you 8 and, um multiple, um, (pause) I think that would be 6 is a multiple of 2, would be a number that could be formed by multiplying another number.

Interviewer: 6 is a multiple of 2, as you mentioned, can you please give another example?

Chris: 12 would be a multiple of 6.

In the same interview, 15 minutes later:

Interviewer: Do you think 117 is a multiple of 26?
Chris: I always get confused with multiple, that word, like I always think
               multiple is a smaller one, but it's really the bigger number, so . . .
Interviewer: It is very interesting to me, so I am interrupting you here . . .
Chris: Um hm . . .
Interviewer: Why do you want to think that it is smaller and you force yourself
               to think that it is bigger? What is it? I see some conflict here.
Chris: Because when you're multiplying, you're dealing with smaller
               numbers and I think of the word multiple, multiply, and I think of
               getting a bigger, bigger number, instead of having that bigger
               number and calling it a multiple.
Interviewer: Um hm . . .
Chris: So I get confused that way. So the terminology, is it a multiple of,
               like multiple of 26, I'm confused as to whether 26 is that bigger
               number, or, or is it a number that is bigger than 26
               that 26 divides into.

Chris appeared to be at the stage where her intuition about the concepts contradicted the mathematical meanings acquired in the course. Her thinking of "smaller" and "bigger" numbers helped her gain the appropriate terminology. Thinking of, in Chris's words, "getting a bigger number, instead of having that bigger number" appears natural in this group of participants. In the next excerpt Ronnie explains this line of reasoning:

Interviewer: OK, alright. How about multiple? What does that word mean
               to you?
Ronnie: I, I mean even without understanding the concept or anything else,
               had I not taken any mathematics, multiple means that you times
               it, you would mean that you times something.
Interviewer: Hmm, could you give me an example?
Ronnie: So when you say multiple, I want to say $2 \times 3 = 6$, I want to say 3
               is a multiple . . .
Interviewer: Um hm.
Ronnie: That's to me what it is, but I could be wrong.

Ronnie exemplified a multiple as a factor. Furthermore, in the following excerpts with Amy, Anne, Penny and Wendy the students' explanation for a multiple is matching the role of a factor in a multiplication number sentence. Anne and Wendy explicitly mentioned the equivalence between a factor and a multiple. Penny mentioned a connection between factor and multiple, but also pointed out the difference.

Amy: And a multiple is a number um, (pause) a number that you times
               another number by to get the other number (laugh) . . .
Interviewer: Can you give me an example?
Amy: OK, the multiple, well I guess if you have 6 times something
               equals 12, the multiple would be 2, is it? I don't know (laugh). I'm
               not sure.

Anne: And a multiple is hmm, I think the same as factor. It's, I think it's the same as the factor.

Interviewer: So you think that the word factor and the word multiple mean the same?

Anne: I think so.

Interviewer: Can you please give me an example?

Anne: So like 3 and 2, well, yeah, 3 and 2, $3 \times 2 = 6$, and 3 and 2 are both factors and they're both, both multiples of 6.

Interviewer: How about a multiple? What does a multiple mean to you?

Wendy: (laugh) The same as a factor.

Interviewer: Uh huh.

Wendy: Um, two numbers, when multiplied together to get a product.

Interviewer: How about a multiple? What does the word multiple mean to you?

Penny: Hmm, (pause) is that like, I think of a multiple, sort of the same way as factor, like, but it doesn't, factor I think, you've got to get it as low as you can, whereas a multiple doesn't have to be. It can be, like say with the 12 again, (laugh) multiples would be 4 and 3.

In the above excerpts Anne and Wendy believed that the words multiple and factor have the same meaning. Penny pointed out the resemblance of a factor and a multiple by thinking of these concepts as "sort of the same," but also found a difference between the two by claiming that factors should be prime, whereas multiples could be not prime. In the following excerpts students think of a multiple as a product.

Interviewer: Will it be correct to say that 117 is a multiple of 13? Is it a correct thing to say?

Sarah: No, it would have to be 13 and 9.

Interviewer: Will it be correct to say that 117 is a multiple of 3?

Sarah: No. It has to be 3 and 39.

Interviewer: So you would say that 117 is a multiple of 3 and 39.

Sarah: Um hm.

Interviewer: How about a multiple? What does it mean to you to say that one number is a multiple of another number?

Mary: Hmm, (pause) it's the product of, of multiplying it, something over and over again. (pause)

Sarah did not mention the word "product," however her perception of a multiple as a product was suggested in her rejection of the claim "117 is a multiple of 3" and acceptance of the claim "117 is a multiple of 3 and 39." Mary explicitly referred to a product in explaining the meaning of a multiple. Mary's reference to "over and over again" may be an indication of an additive perception of a multiple, a perception that is expressed explicitly in the following excerpts with Kate and Marty.

Interviewer: OK, alright. And, what does it mean to you to say that one
               number is a multiple of another number?
       Kate: If, that means that if you take the original number and you add
               itself continuously.
Interviewer: Um hm . . .
       Kate: Like if you have 6 and you add another 6, that would be 12, so 12
               would be a multiple of 6.

Interviewer: And what would it mean to you to say that one number is a
               multiple of another number?
      Marty: A multiple of another number? I just think these are multiples, if
               I go 2, 4, 6, 8, 10, these are all multiples of 2, so I'm going plus 2,
               plus 2, plus 2, that's what a multiple is to me.

This additive view of a multiple can be seen as a student's need to think of a
process. For example, Marty did not tell us what a multiple was, she explained
how multiples were generated.

   While it was stated earlier that for a majority of students the link between
a factor and a divisor was weak, the evidence in this section shows that the link
between a factor and a multiple was erroneous in the majority of cases. Unlike
a divisor and a factor, which present ambiguity by having a double meaning as
mathematical concepts, the concept of a multiple has only one meaning with respect
to natural numbers. Therefore it is particularly disturbing that the majority of
students in this group of interviewees did not have a grasp of the meaning of the
concept of a multiple. It may be the case that students were searching for a meaning
of the concept as a role player in a number sentence, and since such a meaning was
missing, they assigned it as it appeared reasonable for them. As Ronnie explained
in one of the above excerpts, the word multiple brings up the idea of multiplication.
Then, thinking about a multiplication number sentence, there are two possible roles
for its elements: a factor and a product. This may explain why the majority of
students saw a multiple as either one of these roles.

   Another explanation for the majority view of a multiple as a factor can be found,
as was suggested by one of the participants, in the structure and sound of the word
itself. In analogy to "addends," that is, numbers that are added in the addition
operation, we could have had "multiples," that is, numbers that are multiplied in
the multiplication operation. Unfortunately, the conventional understanding of a
multiple shared by mathematicians does not follow this analogy.

   4.4. Identifying Factors. In this section I will discuss the first part of Question Set 2, in which students were presented with a number $117 = 3^2 \times 13$ and asked to determine the prime factors and all the factors of the number. All students correctly identified the prime factors, however for some this wasn't an immediate response based on the $3^2 \times 13$ as the prime decomposition of the number. Two students (out of 19) built factor trees and two others performed a variety of division operations with the calculator in order to find the prime factors of 117. The following excerpts with Marty and Darlene exemplify these two approaches.

Interviewer: Can you list all of the prime factors of 117?
      Marty: I need a calculator.
Interviewer: And if you had a calculator, what would you do with it?

Marty: I would just (pause), is, is 9×13, 117?

Interviewer: Um hm.

Marty: It is. So I would just start plugging in numbers, like I'd, I'd say, 3 would probably be the first thing that I'd do, and I'd go, in my calculator, I'd go 117 divided by 3, and then that would give me an answer, if it goes into it evenly, then I know it is a multiple of it.

Interviewer: Would you please now look at a different number. The number is 117.

Darlene: Uh huh.

Interviewer: And it equals $3^2 \times 13$, I've already calculated it.

Darlene: OK.

Interviewer: So I'll ask you questions about this number 117.

Darlene: Um hm.

Interviewer: What are the prime factors of 117?

Darlene: The prime factors.

Interviewer: Um hm.

[Darlene found out by making a factor tree that $117 = 3 \times 3 \times 13$.]

Interviewer: And you have found 3, 3 and 13 and you have written it down as $3^2 \times 13$. Now I'm asking you, what I wrote here before, when I gave you the number 117, I wrote it here, it is $3^2 \times 13$.

Darlene: Um hm.

Interviewer: Could you usethis?

Darlene: As an explanation for this? [Darlene pointed to her factor tree]

Interviewer: Not explanation, maybe a hint.

Darlene: Oh no, I, like just looking at that, like I didn't, I wouldn't understand that.

Both Marty and Darlene ignored the information provided by the prime decomposition of 117. Since the strategies suggested by these students were time- and effort-consuming, it is reasonable to assume that they didn't know how to utilize the information. Marty suggested "plugging in numbers," where Darlene used a systematic approach of building a factor tree. When the interviewer pointed out to Darlene that what she found after building the factor tree was actually given to her earlier, Darlene admitted that without going through the process she wouldn't be able to use the information. However, in another case a student's reaction to the result of building her factor trees was quite different. The student felt she should have noticed and used the decomposition presented to her.

A request to list all the factors presented an unexpected challenge. Only four students (out of 19) generated all the factors of 117 by considering combinations of its prime factors. Other students relied on factor trees or a calculator as a means of determining the factors.

In the following excerpt Amy responded to the interviewer's request to list all the factors of 117, after the prime factors of 117 had been identified.

Interviewer: Now I'm asking you to give me a list of all the possible factors.

Amy: Right. So I would just go through on my calculator and divide them all. So I'd go, divided by $2 =$, it wouldn't be anything, and then I'd go divided by (pause), I would just go through like that and find all its factors ...

Interviewer: You're dividing 117 by 2 and by 5.

Amy: Yeah ...

Interviewer: To find out other factors?

Amy: Yeah. Well I just could go through the whole system ...

Interviewer: OK, and the fact that the number 117 is given to you as $3^2 \times 13$, does it help you in any way?

Amy: Well it helped me with the primes, because I know that 13 is prime and 3 is prime ...
              [...]

Interviewer: OK, so you've divided 117 here by 2 and by 5, how would you go on?

Amy: I would just keep going up the, like $1, 2, \ldots$, well I know it wouldn't be even numbers because this is, this one's odd, so I'd just do, just go on up there (laugh), I don't know.

Although Amy described her approach as "go[ing] through the whole system," she demonstrated some explicit beliefs about the factors. Amy knew that an even number cannot be a factor of an odd 117, therefore she didn't attempt to divide 117 by even numbers to find the factors. This even-odd consideration, exemplified again in the next excerpt from the interview with Robin, appeared to be rather typical of this group of participants. However, Amy, as well as her classmates, did not generalize this observation to multiples of 5 or any other non-factor of 117. Students' treatment of the property of parity versus the property of divisibility is discussed in detail in Zazkis (1998a).

Robin has built a factor tree for 117 in order to list its factors.

Interviewer: Can you find all of the factors of 117?

Robin: Well, you can see it's divisible by both 3 and 9, because the sum has the 3 digits, 117, $1 + 1 + 7 = 9$, so that tells a person that it's divisible by both 3 and 9, because 9 goes into 9 and 3 goes into 9 also, um, so that would show you your factors are there. One always goes into 117, because it's one of those things (laugh) ...

Interviewer: Um hm ...

Robin: Um, it's odd, it's an odd number at the end, so 2 wouldn't go into it, nor 4, nor 6, that sort of thing, so that leaves you with 3 and 9 and um you can find your multiples of either 3 or 9 that would give you 117, you know what I mean?

Interviewer: Um hm, yeah. So how many factors in total would there be for 117?

Robin: (pause) In total? Well, I'd have to figure that out (laugh), well factor ...

Interviewer: That's alright, take your time. (pause) OK Robin, you've worked out your factor trees and come up with how many?

Robin: Numbers that will go into it? Six. 1 and 117, 3 and 39, 9 and 13.

Earlier in the interview Robin had pointed out that 3 and 13 are prime factors of 117. However, Robin's deliberation on the divisibility of 117 by 3 and 9 by considering the sum of the digits, although correct, was absolutely unnecessary in this case. So was Robin's attention to the fact that 117 was odd and therefore could not have an even factor. These considerations show Robin's understanding and her ability to apply the recently learned or reviewed concepts and procedures. Nevertheless her attention to divisibility rules and not taking advantage of the prime decomposition suggests that she hadn't yet established the link between the prime decomposition and the factors of a number. Robin was very careful in looking for pairs of factors and listed all six of them after considering the factor tree. This can be seen as another application of her knowledge. However, in five other cases students' consideration of a factor tree resulted in only partial lists of factors. Numbers that were not written on the "branches" of factor trees constructed by students were not included in the list of factors. For example, when a factor tree of 117 was built by decomposing the number as 13×9 and then decomposing 9 as 3×3, the number 39 was not listed on the branches of the factor tree and therefore was frequently overlooked in listing all the factors.

Anne relied on the prime decomposition of 117 to generate the list of factors. She considered multiples of prime factors as potential factors and made decisions by performing division with her calculator.

Interviewer:	Can you please list all the factors. Not only prime factors, all the factors.
Anne:	Wait a second, (pause), it's hard, I can't really tell, I'd have to use a calculator I think ...
Interviewer:	OK, please use a calculator.
Anne:	(pause) I think that's all of them.
Interviewer:	So what is on your list? What are the factors of 117? I see here 1, 117 ...
Anne:	3 and 39, 9 and 13, and, well I think, maybe 27, no, (pause) ...
Interviewer:	You worked for a couple of seconds with your calculator, can you please tell me what were you checking?
Anne:	I was checking if there was another product of 13, like $13 \times$um, just the 13×3 and seeing if that was a factor of 117. You're just using different, different factors. Not factors, products of 9 and 3, seeing if they worked.
Interviewer:	So what product of 3 have you checked?
Anne:	27.
Interviewer:	And it didn't work.
Anne:	No. Well it, I think it was 4.5 or something.
Interviewer:	And you checked also 18. Does it work?
Anne:	No, it doesn't work.
Interviewer:	Do you have any explanation why 27 didn't work?
Anne:	Just because it, it would've worked if it had been a little bit smaller. If it had been half as much (pause) and, but then it would've been an uneven number, it would've been .5, because it only went in 4.5 times ... (pause)
Interviewer:	It was too hard for me ...
Anne:	Well 27 just goes into 117, 4.5 times, so it doesn't work.
Interviewer:	OK, but you said you were checking multiples of 9 and you were checking multiples of 3 (pause), were you checking something with 13?
Anne:	Um hm, that's how I got 39.

Anne seemed to grasp the idea of using numbers prime factors as building blocks in generating other factors. However, the understanding that these were the only building blocks allowed seemed to be missing, which resulted in unnecessary reliance on calculations and calculator.

The task of identifying factors demonstrated that the students knew what they were looking for, however, in most cases they didn't know an efficient way to find it. They demonstrated understanding of the meaning of the concept of a factor and the ability to apply this understanding. However, students' links between factors, prime decomposition, and prime factors appeared to be incomplete.

4.5. Identifying Divisors. Having identified all the factors of 117 the students were asked to make a list of all the divisors of 117. In this case the majority of students (12 out of 19) claimed that factors and divisors were the same numbers. Given the fact that in the beginning of their interviews only four students thought of divisors as relations between numbers, this finding can be interpreted optimistically as a growth in students' knowledge during the interview and the importance of the clinical interview not only as a data collection tool but also as a learning experience. This frequency can also be interpreted pessimistically as students' confusion and uncertainty about the ambiguity in meaning of the concepts and choosing a view which is convenient for a given problem solving situation.

Kate was confident in her view of divisors. For her the divisors of 117 were "whatever you choose for them to be." She disregarded the explicit clue to the meaning of divisor provided by the question requesting divisors *of* 117.

Interviewer:	What would the divisors of 117 be?
Kate:	Whatever you choose for them to be.
Interviewer:	Can you think of any factor of 117 that wouldn't be a divisor of 117?
Kate:	No.
Interviewer:	Can you think of any divisor of 117 that would not be a factor?
Kate:	No. No wait, yes I can, that I can. I can think of something that will divide into 117 that wouldn't create a whole number in the finish.
Interviewer:	Hmm, OK. An example would be?
Kate:	117 divided by 2.
Interviewer:	And in that case, 2 wouldn't be a factor ...
Kate:	No ...
Interviewer:	But it would be a divisor of 117.
Kate:	Yeah.

Amy and Robin in the excerpts below appear to be at the stage of resolving the conflict between the meaning they wish to apply for a divisor and the meaning suggested by the interviewer. Amy requested an additional reminder of the meaning of the word divisor. Robin asked whether the result of division had to be a whole number, a question which could indicate for her what meaning of divisor had to be utilized.

Interviewer:	OK, good. So this is your list of factors.
Amy:	Yeah.
Interviewer:	Now I'm asking you, what are the divisors of 117?
Amy:	The divisors are these numbers, all of these numbers are.
Interviewer:	All of these numbers?
Amy:	Yeah.
Interviewer:	So, can you think of a factor which is not a divisor?
Amy:	Um, (pause) what's a divisor? I get these all mixed up, what's a divisor again?
Interviewer:	Divisor is the number by which another is divisible by. Like we said 6 is a divisor of 12, but 5 is not a divisor of 12.
Amy:	Right, (pause) so I think these, these can all go into 117.
Interviewer:	Um hm ...
Amy:	So they're all divisors.
Interviewer:	OK, good. So I have here, are you telling me that divisors and factors of 117 are the same?
Amy:	In this case they are, but I don't know if in all cases they are.
Interviewer:	OK, so you mean that for 117 divisors and factors are the same?
Amy:	Yeah ...
Interviewer:	But if I give you another number, like 525, you are not sure whether it's the same?
Amy:	I think it probably would be, I think it would be.
Interviewer:	And what makes you think it's that way?
Amy:	Because if you break it all down into its factors, the way you find, like 3 goes into 117, 39 times, or ... , 13 times, right, or, well whatever ...
Interviewer:	You calculated the factors for 117.
Robin:	Um hm ...
Interviewer:	My question is, can you think of a divisor that is not a factor of 117.
Robin:	That would go into it. Well 2 wouldn't be, because it's an even number. Well it could be, but it would give, it wouldn't give you a whole number as your total ...
Interviewer:	Hmm ...
Robin:	Or is that what you're asking me and I'm just miss, I'm missing your point? (laugh)
Interviewer:	No, no I don't, I don't think so, basically the question is, can you think of a divisor of 117 that is not a factor of 117?
Robin:	Oh, OK, does it have to be a whole number?
Interviewer:	Does what have to be a whole number?
Robin:	Your end result. Like if you did 117 divided by 2, you would have either a fraction or a decimal. Yes, it can be divided, but it's not a, 2 is not a factor.
Interviewer:	OK, but you'd call 2 a divisor of 117.
Robin:	Yeah.

In these students' responses we can identify their effort to cope with the meaning of the concept of a divisor and with the dissonance presented when the meaning suggested by the interviewer did not fit their prior knowledge. Amy figured out that the divisors of 117 were the same numbers as the factors of 117, but she was not confident that this pattern would extend to other numbers as well. Attending to a variety of other examples could help her generalize this link. Although Robin would erroneously call 2 a divisor of 117, her earlier deliberation on whether the result of division had to be a whole number is an indication of her awareness that this result is crucial in considering divisors. However, no student in this group went through a process of calculating divisors, a finding that can be interpreted as a beginning of students' establishing of a solid link between factors and divisors.

4.6. Identifying Multiples. In students' responses to the request to list all the numbers of which 117 is a multiple, we recognize students' struggle with the meaning of the concept of a multiple, as well as students' links between multiple and factor/divisor.

Penny realized that there was a relationship between the numbers 117 and 39 which did not exist between 117 and 26. However, she described this relationship as "39 is a multiple of 117," suggesting further that "39 and 3 are multiples."

Interviewer: Would you say that 117 is a multiple of 26?
Penny: No.
Interviewer: OK. Um, of 39?
Penny: OK, 117 is a multiple of 39? I guess, yeah, see I, I'm not sure if I understand if multiple means (pause), actually no, I don't think, (pause) no, I wouldn't say that 117 is a multiple of 39. I'd say 39 is a multiple of 117.
Interviewer: Um hm ...
Penny: Because I think, yeah, that's what I think. Because multiples are the parts of, like it can't be bigger than the numbers, like it's a part of 100, like 39 is a part of 117, if you, like 39 and 3 are multiples and you multiply them together.
Interviewer: So why, why do you think that mathematicians would use the same word for multiple as for, say, factor?
Penny: Well it's just the context that you use it in ...

This excerpt with Penny is a further evidence of students' difficulties with the concept of a multiple and their disposition to identify a multiple and a factor and also to consider pairs of numbers as factors/multiples. This tendency has been described in a previous section.

The mathematical link between the concepts of a factor/divisor and a multiple is apparent: *B is a multiple of A if and only if A is a factor of B*. However, some students did not take advantage of this connection, which indicated that the meaning of this link has yet to be constructed. Both Kate and Kevin performed division, however having performed the division, their reactions were quite different.

Interviewer: Would you say that 117 is a multiple of 9?
Kate: Um, (pause) yeah, I think so. Wait just a sec. (pause) Yeah, it's a multiple, why did I have to do that (laugh) ...
Interviewer: And you, you did the long division.
Kate: Yeah.

Interviewer: And found that ...

Kate: 9 went into 117 (laugh) 13 times.

Interviewer: 13 times. And now you say, why, why did I have to do that.

Kate: Well because it was already in front of my face.

Interviewer: In what way would you say it was in front of your face?

Kate: Because I know that the prime factors were 13, 3 and 3 and 3 times 3 gives 9, so ...

Interviewer: OK. Would you say that 117 is a multiple of 26?

Kevin: (pause) Yes.

Interviewer: And why do you think so?

Kevin: Um, 117 is, 117 is um, (pause) 9, $= 9\times$ (pause), I thought it was, because 26 was just a double of 13.

Interviewer: Um hm ...

Kevin: (pause) And I thought that if I could reduce my 9 into half, then it would be a divisor, but because 9, half of 9 is a, is not a whole number, it's $4\frac{1}{2}$, or 4.5, then 26 is not a divisor of 117.

Interviewer: OK, thank you. Would you say that 117 is a multiple of 39?

Kevin: (pause) (laugh) [Kevin opens the calculator]

Interviewer: I'll let you use your calculator in a moment. Before you are using it, please tell me what you're, what you're thinking about.

Kevin: That was 39?

Interviewer: Yeah. And, once again, try to answer without doing division ...

Kevin: OK ...

Interviewer: But if you feel you need to do it, do it.

Kevin: OK. 39, (pause) I would say (pause), I would say yes, (pause) ...

Interviewer: Because ...

Kevin: I would say yes because, because I did the division in my head ...

Interviewer: (laugh)

Kevin: But I don't know if it's right. (pause) It is right. Um, 117 is a multiple of 39, 39×3.

Interviewer: OK ...

Kevin: I don't, I don't know how else I would have um, without doing the division, I don't know how else I would have decided whether that was or wasn't.

Interviewer: Do you think it is possible in the case of 26 to make a decision, without doing division with calculator?

Kevin: (pause) No. I did the 117, the 39 into, divided into 117 division, uh just to see if anything multiplied by the, the 9 in 39 would yield an ending of 7.

Interviewer: Um hm ...

Kevin: And then that's how I decided to continue with the division in my head. With 26, if I didn't divide, I couldn't just look at the number and decide whether or not it was a multiple without actually doing division. That's what multiple is, it's whether it can be divided.

After facing the evidence, Kate made the connection between a factor and a multiple. However, Kevin, having performed the division, claimed that the decision was impossible without it. He made an interesting observation about the last digit in multiplication, which suggested to him that it was possible for 117 to be a multiple of 39, so he was able to carry out a mental division. However, Kevin stated that in the case of 26 he "couldn't just look at the number and decide whether or not it was a multiple without actually doing [the] division."

Amy, when asked of which numbers 117 was a multiple, pointed to the list of factors of 117. Amy's claim "these are multiples," as well as her agreement with the interviewer that "117 is a multiple of all these" shows that her meaning of a multiple was not yet well established. However, she had already constructed some connections between factors and multiples.

Interviewer:	What numbers is 117 a multiple of?
Amy:	(pause) 117 is a multiple of 39 ...
Interviewer:	Good ...
Amy:	Because the answer would be 3, and 117 is a multiple of 13, I think all of these are multiples. [Amy points to the list of factors made earlier.]
Interviewer:	117 is a multiple of all of these?
Amy:	Yeah.
Interviewer:	OK. Now would you please look at 117 again and would you please tell me, is 117 a multiple of 26?
Amy:	Sure ...
Interviewer:	Is 117 a multiple of 26?
Amy:	OK. Can I use the calculator?
Interviewer:	Sure.
Amy:	OK. 26 (pause), oh, 26, (pause) no.
Interviewer:	Would you please tell me what you were doing with your calculator?
Amy:	I was trying a whole bunch of numbers, I was timing 26 by whole numbers to try and find 117.
Interviewer:	And?
Amy:	I couldn't find one.
Interviewer:	OK. Does it mean that there's no such number or does it mean that you haven't found one?
Amy:	(pause) It means I haven't found one (laugh), but maybe there's no such number ...
Interviewer:	What makes you think so?
Amy:	Trial and error.
Interviewer:	Trial and error. (pause) Can you check by looking at these numbers, by looking at 26 and by looking at 117, could you, without trial and error, could you immediately say, no it is impossible, 117 cannot be a multiple of 26?
Amy:	Well I think because this isn't even number, that, (pause) I think because this is odd and that's even ...
Interviewer:	So?
Amy:	(pause) When you times this by a number, that number is going to be even or something, I don't know ...
Interviewer:	I think you know.
Amy:	(laugh)

Interviewer: You said it absolutely correctly, you said 26, when you multiply it by another number it's going to be even.

Amy: Well, because I just was trying a couple in my head and they were all even, but I hadn't tried them all, so I wasn't completely sure.

Interviewer: So in order to be completely sure ...

Amy: I'd check it (laugh), check them all.

Amy was able to identify that 117 was a multiple of all its factors. However, she didn't seem to understand that this list was exhaustive, that is, there was no other number that 117 was a multiple of. In deciding whether 117 was a multiple of 26, Amy used an empirical approach, searching for whole numbers that when multiplied by 26 resulted in 117. Although this approach is consistent with the definition of a multiple, experimenting with performing various multiplications and not one single division can be seen as a lack of connection between the view of a multiple as relation in terms of division and in terms of multiplication. After being prompted several times and encouraged to apply deductive reasoning, Amy built her reasoning on empirical evidence and "wasn't completely sure" in her arguments without checking all the possibilities. This desire to check is an indication of a weak link between the concepts.

The following conversation with Chris took place after it has been established that 117 was a multiple of all the numbers listed previously as factors/divisors of 117.

Interviewer: Would you say that 117 is a multiple of 26?

Chris: Umm, (pause) well $2 \times 13 = 26$, but when I tried $2 \times 9 = 18$, it didn't work, so I'd guess, I don't think it would be.

Interviewer: You don't think it would be. And why don't you think it would be? Can you please share your thoughts?

Chris: Oh, because if we look at our factor tree, we don't have any to use in here, and the only way to get 26 is to have a 2, for example, 2×13, so I'd assume that wouldn't be, and also when I tried 2×9, for example, it didn't work as 18 into that.

Chris noticed that 26 had a factor of 2, and that 2 didn't appear on the factor tree for 117. Chris referred to her previous experience, when trying to extend the list of factors of 117 she "tried" 18. And since she had checked and concluded that 18 was not a divisor of 117, she easily concluded, without calculation, that 117 was not a multiple of 26, and was confident in her conclusion. Although her reasoning was sound, it could have been simplified considering the fact that a compete list of factors of 117 had been discussed just a few minutes earlier.

Chris and Amy used the connection between a factor and a multiple, deriving that 117 was a multiple of all the factors of 117. However, the inference that 117 cannot be a multiple of a number which is not its factor, was not utilized and the decision was reached considering factor trees or using trial and error with a calculator. This indicates that students' mental web of connection between factors and multiples appears to be much more complex than that suggested by the mathematical definition.

4.7. Relying on School Practices. Students' dependence on knowledge and practices acquired in elementary school was mentioned when the meaning assigned to a divisor and a factor was based for the majority on memory from school and not on what was recently learnt. Additive view of a multiple is another reference to a frequent school exercise of listing the multiples of a given number.

In this section I wish to bring additional examples that demonstrate students' reliance on school practices, such as reliance on calculations and reliance on multiplication tables.

4.7.1. *Relying on Calculations.* Chris claimed that 18 was not a factor of 117 because it equaled 2×9 and 2 was not a factor of 117. However, after applying this efficient reasoning, Chris reached for a calculator and performed division of 117 by 18. The interviewer asked her to explain her reliance on the calculator.

Interviewer:	You first gave an interesting explanation, 2×9, and you circled here 2 and 3^2, and then you plugged it fast, fast, fast in your calculator. Do you feel better after doing it?
Chris:	Yeah (laugh) ... I rely a lot on the calculator unfortunately, but ...
Interviewer:	Can you please tell me more about why do you rely a lot on your calculator?
Chris:	I think it's, it's, like the school system, like, that's why I'm really enjoying this class I'm taking right now, because it encourages you to think in a more broad conceptual sense, whereas in the school system you're always with formulas and using your calculator all the time, and it's all these complex calculations, and you don't really have time to rely on your brains, so I guess you get lazy in that sense.

Chris admitted to "getting lazy," however, most students, when stopped and asked about the purpose of using a calculator when it was unnecessary from the interviewer's perspective, replied that they wanted "to check" or "to make sure." Some referred to the habit from their school days of "check all your answers." However, it appeared that checking with a calculator gave students more assurance than an application of any mathematical reasoning.

4.7.2. *Relying on multiplication tables.* Many people mistakenly believe that 91 is a prime number, because its prime decomposition as 7×13 is not among the "basic facts" of multiplication memorized in early grades (Zazkis & Campbell, 1996a). In this set of interviews a belief that 39 was a prime number was expressed by three students, relying on their knowledge of multiplication tables. Divisibility of 39 by 3 is "obvious" considering the divisibility rule for 3 and the division of 39 by 3 (in comparison to the division of 91 by 7) is so simple that can be performed mentally without much effort. Therefore students' belief that 39 is prime is a very strong indication of students' being bound by the limits of their school knowledge. The following excerpts are students' testimony.

Interviewer:	And, how can you determine that 39 is a prime?
Liz:	I guess I could determine it the long way, but just because I know my multiplication tables, I know that nothing goes into 39.

Interviewer: So you've multiplied 3×39 in order to get 117. Do you feel as though they're both prime factors?

Penny: Yeah, I think 39 is a prime factor.

Interviewer: And how would you determine or convince yourself that 39 is indeed a prime?

Penny: Well I'd have to make sure that nothing else goes into it, other than 1 and itself. (pause) And I can't think of anything. (pause)

Interviewer: Um hm ...

Penny: OK, think, I know, well I can't think of anything at the time that goes into it, so then I go like 11, 12, 13, and then that's getting, that, I don't think there is anything, going in my head ...

Interviewer: Um, now I'm going to ask you to try 3.
[...]

Interviewer: I'd like to ask you why, why you feel as though 39 is prime? You, you seemed for a moment to be quite convinced ...

Penny: Because, I think, yeah I was, because (pause) like, I know I'm confident with, well maybe not, (laugh) I should have, but I thought I was confident with my multiples up to 10, you know, like, like to have something times each other that's under 10 ...

Indeed, 39 is not the result of any multiplication fact memorized as multiplication tables up to 10 or 12. Searching for a number decomposition using known multiplication facts seems to be a reasonable approach. However, Penny's being "confident with [my] multiples" and Liz's knowing her "multiplication tables" led them both to a wrong conclusion. Both Penny and Liz worked under an implicit assumption that numbers smaller than 100, when decomposed as a product of two factors, should have both factors smaller than 10 (or 12).

5. Summary and Conclusion

This study focused on factors, divisors, and multiples investigating the meaning of these concepts as constructed by preservice elementary school teachers, and the links among the three concepts as well as the links to other related concepts such as prime numbers and prime decomposition. The findings suggest that the links that appear clear and simple mathematically present a compound web for learners. Students' choices of problem solving approaches demonstrate that they have not fully utilized mathematical connections among these concepts. The findings also highlight the influence of students' prior experience on their constructions.

In this study, the meaning assigned by students to a multiple was often confused with the meaning of a factor or a product. Meanings assigned to a factor and a divisor were often inconsistent with the meaning intended in the context of introductory number theory. Existence of different mathematical meanings for factors and divisors contributed to students' confusion. From the perspective of number theory, factors and divisors are perceived as synonyms and describe a relationship among natural numbers. From the perspective of arithmetic operations, factors and divisors describe the role numbers play in multiplication or division number sentences. These two meanings, although related, can interfere in a problem situation. When the meaning assigned by a student is different from the meaning intended (by a teacher, an interviewer or a textbook) the ground is open for errors. For the three weeks of the mathematics course prior to the interviews the participants studied a unit on elementary number theory. The relation meaning of factor and divisor was utilized in the classroom during these three weeks. However, as the results suggest, during the interview most students explained these concepts by referring to their role-player meaning.

The intended meaning can be easily revealed using contextual and lexical clues (Zazkis, 1998b). In the context of introductory number theory, in the discussion of prime numbers and multiples, the relation meaning of a factor and a divisor is intended. The lexical clue to this meaning is in attending to a factor or divisor *of* a number rather than to a factor or divisor *in* a numerical equation. It is extremely important that preservice elementary school teachers learn to distinguish between the two meanings and to apply the meaning appropriate in a particular context. Since the role-player meaning is the one students were most likely accustomed to during their school years, the effort of constructing a new meaning appears to be greater than one could expect. Textbooks usually provide rigorous definitions followed by examples, but avoid any reference to what could be students' prior knowledge. At times, textbooks provide different definitions for the same term on different pages, but avoid mentioning the connection. Therefore the role of instructor in helping students construct the meaning of concepts provided by the underlying structure of the web of mathematics knowledge is crucial.

I suggest that explicitly addressing the issue of the ambiguity of meaning could be a promising strategy. A possible approach to have the relation view of divisors and factors to coexist and where necessary take over the role-player view is to actively address the ambiguity of meaning and the meaning intended in a specific situation. This can be achieved by engaging students in a situation in which different interpretations will lead to different results. (An example of such activity can be found in Zazkis, 1998b). A following classroom discussion could focus on resolving the disequilibration by attending to the possibility of different interpretations and seeking a desired interpretation through contextual clues and grammatical structure. Of course, empirical verification is needed to decide to what degree the suggested approach is beneficial.

"Building relationships between pieces of information does not always occur spontaneously" (Hiebert & Lefevre, 1986, p.16). Similarly, construction of links between mathematical concepts does not always occur spontaneously. This study highlighted an old wisdom of education—the importance of relating to students' prior knowledge. It showed that students' existing perceptions are much stronger than newly learned concepts and practices. An attempt to have students construct new connections among mathematical concepts might be more successful if those concepts were linked to their prior knowledge.

However, the notion of "prior knowledge" in the case of preservice elementary school teachers may be different from the prior knowledge of children or mathematics majors. What often occurs in a content course for preservice elementary school teachers is not construction of new meanings or concepts, but reconstruction of previously constructed meanings. This brings the discussion back to the first problem revealed by prior research—the problem of re-learning. In what way is re-learning different from learning? Can existing learning theories accommodate and explain re-learning or do we need to reconsider and extend them to generate "a theory of re-learning"? The idea of schema (Asiala et al., 1996) will most likely serve as a theoretical perspective attempting to accommodate and clarify re-learning as well as to sharpen the notion of "links" and their role in learning mathematics. Such theoretical advancement must be given consideration in future research.

"Mathematics educators must find an appropriate balance between attention to what prospective teachers should know and what they do know" (Simon, 1993, p. 253). Attention to the prior knowledge that prospective teachers bring to their mathematics courses may be the key to reducing the gap between the "should" and the "do."

6. References

Asiala, M., Brown, A., DeVries, D., Dubinsky, E., Mathews, D., & Thomas, K. (1992). A framework for research and curriculum development in undergraduate mathematics education. In J. Kaput, A. H. Schoenfeld, & E. Dubinsky (Eds.), *Research in collegiate mathematics education* (pp. 1–32). Providence, RI: American Mathematical Society.

Brown, A. (1999). *Patterns of thought and prime factorization.* Manuscript submitted for publication.

Brown, A., Thomas, K., & Tolias, G. (1999). *Conceptions of divisibility: Success and understanding.* Manuscript submitted for publication.

Campbell, S., & Zazkis, R. (1994). Distributive flaws: Latent conceptual gaps in preservice teachers' understanding of the property relating multiplication to addition. In D. Kirshner (Ed.), *Proceedings of the Sixteenth Annual Meeting of the North American Chapter of the International Group for the Psychology of Mathematics Education* (pp. 268–274). Baton Rouge, Louisiana: Louisiana State University.

R. Davis, R., Maher, C. A., & Noddings, N. (Eds.). (1990). *Constructivist views on the teaching and learning of mathematics,* Journal for Research in Mathematics Education. Monograph Series Number 4. Reston, VA: National Council of teachers of Mathematics.

Ferrari, P. L. (1997). Action-based strategies in advanced algebraic problem solving. *Proceedings of the 21st Conference of the International Group for the Psychology of Mathematics Education (Vol. 2, pp. 257–264).* Lahti, Finland: University of Helsinki.

Ginsburg, H. (1981). The clinical interview in psychological research on mathematical thinking: Aims, rationales, techniques. *For the Learning of Mathematics, 1*(3), 4–12.

Ginsburg, H. (1997). *Entering the child's mind: The clinical interview in psychological research and practice.* Cambridge, Cambridge University Press.

Hiebert, J. & Lefevre, P. (1986). Conceptual and procedural knowledge in mathematics: An introductory analysis. In J. Hiebert (Ed.), *Conceptual and procedural knowledge: The case of mathematics* (pp. 1–27). Hillsdale, NJ: Lawrence Erlbaum Associates.

Nicholson, A. R. (1977). Mathematics and language. *Mathematics in School 6*(5), 32–34.

Noss, R. & Hoyles, C. (1996). *Windows on mathematical meaning: Learning cultures and computers.* Dordrecht: Kluwer Academic Publishers.

Simon, M. (1993). Prospective elementary teachers' knowledge of division. *Journal for Research in Mathematics Education, 24*(3), 233–254.

Zazkis, R. (1998a). Odds and ends of odds and evens: An inquiry into stu-
    dents' understanding of even and odd numbers. *Educational Studies in
    Mathematics 36*(1), 73–89.

Zazkis, R. (1998b). Quotients and divisors: Acknowledging polysemy. *For the
    Learning of Mathematics, 18*(3), 27–30.

R. Zazkis, & Campbell, S. (1996a). Divisibility and multiplicative structure of
    natural numbers: Preservice teachers' understanding. *Journal for Re-
    search in Mathematics Education, 27*(5), 540–563.

Zazkis, R. & Campbell, S. (1996b). Prime decomposition: Understanding unique-
    ness. *Journal of Mathematical Behavior, 15*(2), 207–218.

Simon Fraser University

CBMS Issues in Mathematics Education
Volume **8**, 2000

On Student Understanding of AE and EA Quantification

Ed Dubinsky and Olga Yiparaki

ABSTRACT. We discuss college students' understanding of AE ("For all ...
there exists ...") and EA statements ("There exists ... for all ...") in natural
language and in mathematics. College students were given a questionnaire
with eleven declarative statements and asked to decide whether the statements
were true or false and explain their answers. We discuss conclusions based
on their written responses and on subsequent interviews with some of the
students. We found that students are not inclined to use the syntax of a
statement in order to interpret it, particularly if they do not understand it
very well; rather, they use the context of a statement to discuss their own
opinions about the topic in general, not the actual statement. We also found
that students are more inclined to interpret English statements as AE rather
than EA. Most students in this study could not distinguish between AE and
EA statements in mathematics and did not seem to be aware of the standard
mathematical conventions for parsing statements. Finally, we discuss the use of
quantifier games as a pedagogical tool to help students understand statements
with multiple quantifiers.

1. Introduction

> We do not see things as they are,
> we see things as we are.
> *—Anais Nin*

Does language govern our thoughts? Or do we form language so that we can
express our thoughts? Can a concept exist if we do not have a name for it? Or is
the creation of a name required to create a concept?

We are interested in the difficulties students have understanding statements in
mathematics that involve existential and universal quantifiers. Our starting point
was the idea that students can make sense out of statements in natural language
about everyday situations even when there is quantification. Our hope was that if
we could learn something about how people make sense of quantified statements in
everyday discourse, that we could use what may be natural modes of thinking to
help them understand similarly structured statements in mathematics.

©2000 American Mathematical Society

Our results suggest that the depth of student understanding of quantification in everyday discourse is not very strong and may not be a powerful resource for helping students understand quantified statements in mathematical contexts. Moreover, our results suggest that the understandings students do appear to have of quantified statements in everyday situations may not transfer very easily to mathematical situations. We will argue that the language used in mathematics obeys certain rigid rules that our students do not necessarily pick up on their own, and hence we need to help our students learn how the language of mathematics works in order to communicate with them. Thus we will argue that as teachers, instead of trying to make everyday life analogies between ordinary English statements and mathematical statements, perhaps we should remain in the mathematical contexts and concentrate our efforts directly on helping students understand mathematical statements in their natural mathematical habitats. We will further argue that one ends up making analogies even inadvertently, simply by using natural language to communicate a mathematics statement, and that it is for this reason that we need to teach our students that the language of mathematics works differently from natural language.

Throughout this paper, when we refer to *syntax* we will mean *the arrangement of and relationships among words, phrases, and clauses forming sentences*; when we refer to *semantics*, we will mean *the interpretation or meaning of the words in a sentence and the sentence as a whole*. We are interested in the extent to which students are inclined to use the syntax of natural language in order to understand mathematical language. Perhaps from the point of view of a mathematical logician there is little distinction between mathematics and formal language, but from a mathematics education perspective it is important to draw a distinction between semantics and syntax.

Most mathematics students (at least in a typical U.S. university or college) do not take a course in mathematical logic, or one that directly addresses the syntax of mathematical discourse, or indeed any course that tries to focus on making sense out of complex mathematical statements. Mathematicians would probably agree that we cannot hope to teach our undergraduate students much mathematics (which includes understanding and doing proofs), if they do not know how to read and interpret the language of mathematics. Yet there seems to be a substantial gap in communicating mathematics to our students. While struggling to follow a proof or to come up with one, our students often fail and get frustrated, or worse: they do not even know whether or not they have succeeded.

Our study was motivated by our desire to understand some of the reasons responsible for this gap in communication. Also, using students with diverse mathematical backgrounds, we intend to address some of the universal difficulties they experience.

We argue from our own mathematical experiences that one powerful tool a mathematician uses in understanding a complex statement is an analysis of the semantics based on the syntax of the language in which the statement is given. Is this a "natural" thing to do? In this report, we present data indicating that in the case of a student trying to learn mathematics, the most natural direction is to go from semantics to syntax and that students often are not able to use the syntax as a tool for understanding. Our data suggest that what students do is respond to a statement more or less globally and imagine a context or "world" which the statement is taken to describe. Our results are consistent with cognitive science research into logical thinking (see for example, Johnson-Laird 1983, 1995, and the Web page http://www.cogsci.princeton.edu/~phil/) where it is argued that students think about a syllogistic reasoning task in terms of a mental model. This model must be very close to the concrete experiences of the student who uses it in place of the syntax to make sense of the statement.

Not only is it possible to go from syntax to semantics, it is also formalizable. The theory of Denotational Semantics, inspired by Strachey's 1966 work and developed by Scott 1971, provides a framework for constructing mathematical models for programming languages. Using semantic equations, one translates a formal sentence; this translation is driven solely by syntax and it produces semantic objects (numbers, functions, tuples, etc.) for all sentences. Although, of course, most mathematicians probably do not even come close to using such a formal approach, they routinely construct "semantic objects" or mathematical worlds when attempting to support or refute a given statement. We claim that one of the important skills a mathematician employs regularly is the ability to go from syntax to semantics with an open and creative mind, without necessarily prejudging how reasonable or unreasonable a statement appears to be, based on mathematical worlds that are already known, established, and familiar.

The problem with novices then is that their restriction to familiar models is limiting and is not sufficient to understand very much mathematics once it goes beyond the most elementary. Imagine, for example, a student in a first course in analysis (or a bridge course) trying to understand the theorem that a continuous function on a closed finite interval is uniformly continuous. It is first necessary for the student to understand the distinction between continuity and uniform continuity. Epsilons, deltas and choosing them have traditionally been difficult for students. In part, this is perhaps because many students are uncomfortable with algebraic manipulations, but our data show that students have difficulties understanding the statements themselves (such as the mathematical statements defining continuity and uniform continuity). How do we help our students? It is not clear that examples and analogies will always help. It may be that the student must learn to use the structures of formal statements to make sense out of mathematical situations in contexts that are new and unfamiliar.

This study concerns a very special case of syntax difficulties: quantifiers. We were motivated by the following: It seems that practically all interesting mathematical statements have at least one universal and at least one existential quantifier—and the order in which they appear is crucial. Students have great difficulty understanding such statements. Regarding the latter, although it seems to be a part of "conventional wisdom," there is not much in the way of research data to support it. One of the results of this paper is to provide such data.

1.1. AE and EA statements in mathematics and similar natural language statements.

A formal statement of the type $(\forall x)\,(\exists y)\,R(x, y)$, where R is a quantifier-free binary predicate, is called an AE statement. Similarly, a formal statement of the type $(\exists c)\,(\forall d)\,S(c, d)$, where S is a quantifier-free binary predicate, is called an EA statement.

Throughout this paper we refer to AE and EA statements. However, most of these statements are not formal. When we say a natural language statement is AE (or EA), we mean that if the statement were to be formalized, then it would be an

AE statement (or EA statement). For example, we refer to the statement *every pot has a cover* as AE because we would formalize it as

$$(\forall p)\ (\exists c)\ R(p, c)$$

where p ranges over all *pots*, c ranges over all *covers* and $R(p, c)$ means c *covers* p.

Of course, some natural language statements can be interpreted as AE or EA. For example, the statement *Someone is kind and considerate to everyone* can be interpreted as "There exists one person who is kind and considerate to everyone" or as "For every person there is someone who is kind and considerate to that person." The first interpretation yields an EA formalization,

$$(\exists p)\ (\forall q)\ K(p, q)$$

where p and q range over the set of all people and $K(p, q)$ means p *is kind and considerate to* q. The second interpretation yields an AE formalization. One of the reasons for using formal mathematical language is to eliminate such ambiguity. One of our interests in this study is to see how students deal with such ambiguity in natural language.

1.2. Description of the paper. We begin this report with a description of the study that we conducted, the students involved and the instruments used. Then we present our results concerning the knowledge and understandings students appeared to have. We continue with a discussion of one device we did use to try to improve student understanding—asking them to play a mental game related to a statement. We present the results of this activity as well. Finally, we summarize our conclusions and make some pedagogical suggestions.

2. Description of Study

Our study has two components: a questionnaire and follow-up interviews. We describe these two instruments and the students to whom they were administered.

2.1. The students. Sixty-three college freshmen, sophomores, juniors, and seniors responded to our questionnaire. All were in some mathematics class: some were taking a first-semester linear algebra course, some multivariable calculus, some a junior-level class as preparation to teach middle school mathematics. Almost all were mathematics or mathematics education majors; the rest were science or engineering students. About half were students in one of two large public universities and the other half were attending a small liberal arts college. Finally, nine were in an advanced senior level/beginning graduate level abstract algebra class. We will refer to this small group as "the graduate students." We will refer to the remainder as "the undergraduate students."

2.2. Questionnaire. We devised eleven declarative statements and asked students to decide whether each statement was true or false. Students were told that there are no correct or incorrect answers; all they had to do is decide for themselves the truth or falsity of each statement. They were given some space below each statement to write a few sentences explaining their reasoning. (See Appendix A for questionnaire text.)

    (1) (AE) Everyone hates somebody.
    (2) (AE) Every pot has a cover.
    (3) (EA) Someone is kind and considerate to everyone.

(4) (EA) There is a mother for all children.

(5) (AE) All good things must come to an end.

(6) (EA) There is a magic key that unlocks everyone's heart.

(7) (AE) All medieval Greek poems described a war legend.

(8) (EA) There is a perfect gift for every child.

(9) (EA) There is a fertilizer for all plants.

(10) (AE) For every positive number a there exists a positive number b such that $b < a$.

(11) (EA) There exists a positive number b such that for every positive number a $b < a$.[1]

FIGURE 1. A condensed version of the statements with their labels. Justifications for the labels are given below.

Given that this study intended to investigate students' understanding of quantifiers, we did not want to complicate the statements any further. Thus, we avoided implications, logical conjunctions, and logical disjunctions within each statement, and so on. Furthermore, we tried to choose statements that would be short and sound reasonably natural in ordinary English (Statements 1 through 9).

The natural language statements (1 through 9) were intended to provide some variety of familiar statements (such as Statement 2 and 5 that are almost colloquial proverbs) as well as some variety of familiar notions involving debatable statements (such as Statement 1, for example, where the notion of *hate* is familiar, but opinions on the truth or falsity of the statement would be expected to vary). Moreover, Statement 7 was chosen to involve the familiar notions of *poem*, *war legend*, etc., but our hope was that most students would not be able to rely on any factual knowledge about medieval Greek poetry, given the obscurity of the topic. This statement was intended to be easy to understand and harder to assess.

Finally, the only two mathematical statements, numbers 10 and 11, were chosen to meet the above criteria of simplicity in logical structure and brevity; they were intended to rely on practically no mathematical knowledge from any particular course other than some understanding of some concept of *positive number* and the notion of *less than*. (Statements 10 and 11 have also appeared in Dubinsky, 1991.)

2.2.1. *Language conventions and the authors' AE and EA labels.* Why did we assign the AE and EA labels to the eleven questionnaire statements as mentioned earlier? Perhaps not many people would argue with the label "AE" to classify Statement 1, or the label "EA" to classify Statement 3 or Statement 11. We would expect that more people would argue that the label 'EA' is not necessarily appropriate for Statement 4. After all, perhaps in everyday speech, we frequently use the statements *There is a mother for all children* and *All children have a mother* interchangeably, yet in this study we would label the first statement as EA and the second as AE. Why? It is not that we wanted to ignore the speech of everyday language, quite the contrary.

---

[1] Our original statement for 11 was *There exists a positive number b such that, for every positive number a, $b < a$.* In the end, however, we became concerned that Statement 11 would have additional commas that Statement 10 did not have. We wondered whether the punctuation difference would lead students down the wrong path when trying to compare the two statements. So we decided to delete the commas and have Statement 11 exactly as it appears in the Appendix.

As mathematicians, when we talk to each other about the weather, sports, politics, gardening, etc., we almost certainly use the conventions of natural language. When we start discussing mathematics, however, we almost certainly use the conventions of mathematics. We switch between the two conventions with ease and we do not bother to signal explicitly that we are switching conventions, despite the fact that we are probably using the very same natural language in both cases. As mathematics educators, we almost always maintain the same behavior: we switch back and forth, but we don't tell our students that we are doing this. We expect them to catch on to this fact on their own. Our students in the meantime, are in the process of learning the material at hand. We believe that by failing to tell our students that the language conventions are different, we frequently fail to communicate the very topic of our focus: the material at hand. This is not limited to verbal communication, of course. The vast majority of mathematical statements our students encounter are not written in formal language, they are expressed in the "natural" language of the textbooks they read. These textbooks use the special conventions of mathematics.

One of these conventions is that the order of appearance (reading from left to right) of the quantifiers in a sentence makes a big difference. This important distinction (as well as many others) can slip by unnoticed merely because it is not usually addressed explicitly. Our interest in this study was to see to what extent our students are aware of this convention and if so, to what extent they are likely to apply it (that is, to use the syntax) in any context, mathematical or not. Thus, for the purposes of this study, we classify the natural language statements (Statements 1–9) following the mathematics convention of left-to-right parsing. How appropriate is it for this study to label and treat the natural language statements according to mathematics conventions? How appropriate is it for all of us, as mathematics teachers, to use natural language following mathematics conventions?

We chose natural language statements that would be at least to some extent ambiguous in the sense that many of them have two "reasonable" interpretations: AE and EA. We are aware, of course, that under ordinary natural language conventions, one of the two interpretations is more likely. Statement 4 is perhaps the clearest example of this, as it was intended. We anticipated that Statement 4 would be much more likely to interpreted as an AE statement. Thus, we agree with colleagues who have expressed criticism about treating Statement 4 as unequivocally EA. It is not surprising to us that, as we shall see later, the students in our study did not follow the mathematics conventions to interpret Statements 1–9. But what we believe is interesting is that students did not seem to be aware that they were using conventions—any conventions at all. In our view, this is the root of the problem with interpretations.

One last comment on statement interpretations. Throughout this paper we refer to students' tendencies or preferencies to interpret statements one way or another. These are not intended as value judgments on a student's reasoning or interpretive abilities, especially if the interpretation is the opposite of our label (AE or EA). We are not suggesting that our label for a statement is "the correct one." We are merely interpreting and labeling statements according to mathematics conventions and subsequently we are discussing when, where, and how often students appear to follow this set of conventions.

2.2.2. *Coding the questionnaires.* Every student was asked to provide an explanation for the decision (true or false) he or she made for each statement. We used the student's explanation to decide whether the student interpreted the statement as AE or EA, and given the student's interpretation, to assess the validity of her or his argument. We then "graded" each response with an ordered triple. The first component of the triple represents our understanding of a student's interpretation of the statement at hand: whether the student appeared to understand the statement as AE or as EA. The second component of the triple assesses whether the student's reasoning is valid, given her or his interpretation. These assessments were made by one of the authors and checked by the other. Finally, the third mark keeps track of the decision the student made: True or False.

For example, consider Statement 3: *There is a perfect gift for every child.* We believe that for most people there are only two reasonable interpretations: *For every child in the world, one can find a gift that is perfect for that child* (AE), and *There is some gift that would be perfect for each and every child in this world* (EA). Thus, when evaluating a student's answer (AE vs. EA), we compared each response to these two possibilities only; there were very few cases for which it seemed that the student did not have one of these two possibilities in mind.

Here are some examples of the triples we assigned.

- A student circled True, and responded with: *You can always find a gift to give a child that is more desired than any other. The gift may vary from child to child, but there is always something (not necessarily material).* We believe this student interpreted the statement as AE and we assigned the triple (AE, valid, T).
- Another student circled False, and responded with *I'm not sure what is meant by perfect gift. If it is a treasure of some kind, tangible or not, or if it is a skill, either way I could not limit life to perfectness for every child. I do, however, think that every child possesses a unique talent/skill, individual to the child, but not necessarily extraordinary.* We believe that this student understood this statement as AE and we assigned the ordered triple (AE, valid, F).
- Another student circled True, and responded with: *The perfect gift for every child is to love and nurture them. Any other material gift they will almost always grow out of.* We believe this student interpreted the statement as EA and we assigned the triple (EA, valid, T).
- Another student circled False, and responded with: *Some kids hate everything you give them, even if they asked for it, because they are spoiled.* It is not clear whether this student interpreted the statement as AE or EA because the response could be addressing either interpretation equally well. In fact, the response provides a valid reason for deciding that the statement is false in both cases (both possible interpretations). In this case we assigned the triple (unclear, valid, F).
- Finally, another student circled False, and responded with: *There is more than one perfect gift. Children like lots of different things.* It is unclear whether this student was thinking of the statement as AE or EA, and in either case, the response does not provide a valid reason for deciding the statement is false, in fact it seems to provide a reason that the statement

(in either interpretation) is a weaker truth, based on the student's viewpoint. So, we assigned the triple (unclear, not valid, F). (There were very
few of this case.)

Fifteen students misread (interpreted) Statement 11 as *There exists a positive
number b such that, for every positive number a, $a \cdot b < a$*. But of course we were
not interested in this interpretation. As it turned out, all students who claimed
Statements 10 and 11 were the same (or reworded) were among those who did not
interpret Statement 11 with a and b multiplied. Thus, despite this mishap, we have
been able to make sense out the data we collected for just this reason, as we were
interested only in those students who interpreted the quantifier-free part ($a < b$) as
the same in both mathematics statements in order to see how they distinguished (or
failed to distinguish) between the two statements. When we looked at how students
compared Statements 10 and 11, we considered only the responses of the forty-eight
students who interpreted Statement 11 without multiplication, as indicated in table
5 of Appendix B.

2.3. The Interviews. After coding the questionnaires we divided the students into very rough equivalence classes based on patterns in their responses. We
came up with five different classes, and selected members of these classes for our
interviews. However, because the selections were made to assure simple representation of each class, the set of students we interviewed does not reflect the difficulties
the students had in the same proportions as the set of students who responded
to our questionnaire. In particular, 57% of the students we interviewed were able
to assess the truth value of Statement 10 correctly, whereas only 49% of all the
students who responded to the questionnaire got statement 10 right.

We interviewed fourteen undergraduates. Each interview was conducted by
one of the authors with one student at a time, audiotaped, and later transcribed.
Interview length averaged 45 minutes.

During interviews students were given the opportunity to discuss the thoughts
and reasons that led to each decision about the truth or falsity of each statement
(while looking at her or his written responses to the questionnaire). Students were
also given the opportunity to revise any written answer—and some students did
revise answers upon rethinking some of the statements.

At the beginning of each interview, we explained that we wanted to focus on
the given statements and to discuss each student's reasoning that made her or him
decide whether a statement was true or false. We made it clear that explaining their
reasoning would necessarily involve personal opinions about the topic at hand, and
so we would ask them about their opinions, but that we would not debate any
opinions. We would question them on their reasoning to support or oppose the
statements printed on the questionnaire, no other statements.

The interviews focused only on Statements 1, 3, 6, 9, 10, and 11. Besides asking
students to reiterate or expand upon their reasoning for each of these statements,
the main questions we asked during these interviews were:

- For Statement 1, replace the word *somebody* with *at least one person* and
  with *one but no more than one person*. Replace the word *everyone* with
  many, with *most*, and with *a whole lot of people*. All these replacements
  were done one at a time, not several at once. In each case we asked whether

the truth value of the statement changed and whether the statement itself changed in meaning.

- For Statement 3, make similar replacements as in Statement 1 for the words *someone* and *everyone*. Then we asked students to describe how circumstances or the world might change in order for the very same statement to change its truth value. However, we did not use expressions such as "change the truth value." Here is an example of a typical scenario. A student has decided Statement 3 is false and has provided some reasons for this decision. We would then ask *What would it take for you to decide that this same statement is true? Can you describe the circumstances or the world situation that would be necessary for you to decide that this statement is now true?* In the case where a student had decided this statement was true, we would replace the word *true* by the word *false* in the above questions.

- For Statement 6, we asked students to explain their reasoning. We then asked whether the statement means there is a key that is intended to be common to all people, or whether the statement allows for different people to have different keys.

- For Statement 9, after asking students to explain their reasoning, we introduced a game and played variations of this game with the student. The game is described in detail in Section 4.

- For Statements 10 and 11, we first discussed the student's basic reasoning that was written on the questionnaire. We then played games analogous to the one introduced with Statement 9. See Section 4 for further details.

2.4. Methodological issues. The full set of questionnaires is available from the authors for any reader who would like to make an independent determination.

The interpretations of interview excerpts are entirely the judgements of the authors. We have tried to present enough of each excerpt and an explanation of our interpretation in each case so that the reader can make an indpendent judgement. Full transcripts are available on request.

The main thing that is lacking in our methodology for analyzing the transcripts, as well as, to some extent, the assessment of the validity of the students' explanations, is a theory. It would have been better to begin with a theoretical description of how people think about statements with two different quantifications and then analyze the student responses in terms of this theory. In fact, the present research is preliminary to developing such a theory. It is a first attempt at exploring data on student thinking about these statements. The next step will be to develop a theory and then conduct a similar study that involves this theory both in the design of the study and the analysis of the data. Our hope for the present work is that it will inform such future research.

3. Results

In this section we discuss the performance of the students on the questionnaire and in the interviews. We consider this performance in terms of several issues that seem to be salient in our results: tendencies of students to interpret the EA statements as AE; the contrast between student performance on statements in natural language about everyday situations and statements about mathematical situations; how students deal with a situation when they do not have enough information to decide whether it is true or false; tendencies of students to ignore quantifiers; abilities of students to negate statements or imagine situations in which the truth value of a statement changes; and some examples in which students were comfortable enough with a statement to be creative about it in the form of making a joke.

Although there were only nine graduate students in this study, there are some comparisons we can make with the undergraduate students. The graduate students had more brief and more succinct answers in their questionnaires (regardless of the validity of the reasons they presented). Furthermore, graduate students were somewhat more likely to give valid reasons for their answers.

Nevertheless, the success rate of even the graduate students was not ideal: two (out of nine) graduate students did not get Statement 10 right, only five of the nine got Statement 11 right. Indeed the overall patterns of responses for these students is sufficiently similar to those of the undergraduates that we do not distinguish between the two groups in the results that we present here. The breakdown of our numbers according to the two groups is available.

3.1. Interpreting EA statements as AE. On both the questionnaire and the interviews we found a strong tendency for students to favor an AE interpretation over an EA interpretation. As we indicated in section 2.2.2 we are aware of the fact that the statements we chose are to some degree ambiguous (though some more than others).

Though, as we noted above, we told students that the focus of the interviews was to discuss the questionnaire statements. Thus, given this focus was cast on the statements, we found it interesting that the students did not examine the statements more closely, even after our questions about whether some statements could be interpreted differently. In the following two sections we will provide some details regarding this observed tendency.

3.1.1. *AE over EA on the questionnaires.* In Table 2, we see that 94% of the students interpreted at least one EA statement as an AE. For the six EA statements, we see from Table 1 that the percentage of students who interpreted them as AE ranged from 11% to 81%. On the other hand, from Table 2 we see that only 5% interpreted at least one AE statement as an EA; the percentage of students who interpreted AE statements as EA ranged from 0% to 3%.

Finally, in Table 1 we see that when the student interpretation was clear, the chances of a valid argument were high. In other words, there is a close relationship between making a clear interpretation of a statement as AE or EA and giving a valid argument for the truth value assigned to it. Of course, this is, at least in part, due to the fact that we based our assessment of interpretation on the arguments students wrote to support their decisions (true or false) for each statement. On the other hand, if a student's interpretation of a statement's form was not clear from the arguments they presented, we gave them the benefit of the doubt: if the student's argument could support both the AE and the EA form of the statement (or their negations), we deemed the argument "valid" (see description of questionnaire coding, section 2.2.2). Hence, we believe it is fair to say that there is a close relationship between making a clear interpretation giving a valid argument.

3.1.2. *AE over EA in the interviews.* During the interviews we had the opportunity to confirm and question students about their interpretations of the statements. Students had the chance to revise any previous answers and to expand upon their thinking. Throughout this study, we welcomed any suggestion or even hint of a suggestion from a student that any of the given statements in our questionnaire was ambiguous. There were no such written responses on any of the questions indicating there was ambiguity. During the interviews, even when we explicitly probed to see whether a student who had interpreted Statements 6 and 9 as AE could see either of them as an EA, many did not. Almost all of the interviewees continued to see these statements as AE and saw no ambiguity. (There was only a single exception to this, where one student, during the interview, said that Statement 9 could be taken to be EA or AE, but then she decided that it was an AE statement.) To pursue this question further, we managed to get some students to agree that Statement 6 or 9 could be interpreted as an EA and then we asked them to suggest wording that could ensure this. In some cases, they gave us the same statement as the one in the questionnaire. We continued to see a strong tendency to interpret statements as AE, even when the interviewer suggested that the statement at hand may have an EA form. Some students avoided dealing with this issue directly, others simply rejected the possibility, and others were not really able to resolve the (possible) ambiguity. In fact, many times during the interviews, students appeared to be expressing their point of view about the topic at hand, not really focusing on the particular statement. For example, a student would look at Statement 1 and discuss the nature of hate among people without necessarily discussing what Statement 1 *says* about such hate. During these discussions, the tendency toward AE interpretations was strong. Here are some examples.

We begin with an example[2] of a student, AND,[3] who is asked for a view of the world that would make Statement 3 true. Her response is actually to change the statement by reversing the original predicate relation. The result is indeed an AE statement which agrees more with the world view that she then describes. The interviewer repeats Statement 3 in its original form, but this does not seem to affect the student. She seems to be talking more about the topic of kindness and less about what the statement says.

> I: OK, good. OK, let's move on to number 3. Ahm, and here again I want you to tell me how you were thinking. Someone is kind and considerate to everyone. And you said that that's true. So, what view of the world do you have to say that that's true?
>
> AND: For everyone is kind to somebody?
>
> I: Right.
>
> AND: Just ... I mean I know that people ... I am talking from experience, there are people that think they have done something ... or they might not even realize that sometimes, I don't think that people sometimes realize that they have done something, but what they have done could have been so nice to me, that I would have been like "oh, my gosh!!" ...

---

[2]Throughout the interview excerpts, we use the following conventions. The letter "I" indicates the interviewer is speaking; (P), (LP), and (VLP) indicate a pause, a long pause, and a very long pause, respectively.

[3]The actual names of students have been replaced by three-letter codes throughout this paper.

I: Hm-hm. Now I am interested in the someone and the everyone. What do you say? You are saying that "someone is kind and considerate to everyone"—you say that's true. What is ... what is it that makes that true?

AND: (P) Oh ... I guess someone is not kind ... well, every person has had some kind of a kind act done towards them. Now everyone hasn't done that act to everyone [student speaks very deliberately here] ... Because I don't ... I can't do a kind act to people that I have never met. I mean like ... or then again, I suppose I could ... I could give money to the poor countries, or you know things like that. But I mean there is no way that everybody can help every other person. I mean I am sure that things I have done have possibly helped people that I don't know, but there is no way it has helped everybody.

I: So, what is it that's happening here? Because you said that's true.

AND: Right. Because I mean someone ... everybody has had something nice. Somebody ... everybody has had something considerate done ... Whether it'd be a word of encouragement or, ... a pat in the back, or they'd been given food, or shelter. I mean everybody has had something nice done for them or to them.

Later, when AND is asked about Statement 9, she seems to respond by focusing on her opinion about the general topic rather than why, in her view, the statement is true. At first, she is not paying any attention to quantifiers at all. Then when the interviewer focuses her on the statement, she seems to be thinking about various situations with various plants and what the fertilizer would be—which is much closer to an AE interpretation than EA.

I: OK. Let's go over to 9. OK ... What was your thinking on this one?

AND: Ahm, that God really makes the true ... I mean, true there is the ammonia, and nitrates, that fertilize, but without God's nutrients and soil and the rain there wouldn't be anything to grow ... and therefore that is the real fertilizer of it.

I: OK. Suppose we consider that all of these things, the nutrients and the soil, came from God, ah ... but let's not think about that. Let's think about the nutrients and the soil, and so forth. That's the version of fertilizers. Not where they came from. The statement that there is a fertilizer for all plants, is that true?

AND: True. Because the nutrients and the soil fertilize. I mean they help it grow. Obviously if you just leave the same plant or crop in the same ground, you are gonna deplete the nutrients, so it won't have a fertilizer ... but unless you just do that, it will have its natural fertilizer from the nutrients in the soil.

I: OK. And would that fertilizer in that single plant work for all plants?

AND: No. I mean there are different areas of the world that are more suited to growing different types of ... crops... I mean part of it is the climate ... but part of it too is the soil is different. So it wouldn't have the same amount of nutrients so the crops wouldn't get as much of the nutrients as they needed.

It could be that AND had difficulty even understanding an EA version of Statement 9. This is not the case with JUL who clearly understands both possible interpretations and chooses the AE version because the statement says "a fertilizer" rather than "one fertilizer."

I: OK. Great. Take a look at number 9. What about that one? What were you thinking when you did that one?

JUL: (laughs) Ahm ... I don't know, I really didn't know quite how to ... what to think about this problem—or this question. A statement, I guess. Ahm, (P) ... This one is harder for me. I took it on face value and didn't read into it. I don't know how to really read into it I guess.

I: I am not sure I understand what you mean by face value and how to read into it.

JUL: Well, I mean ... I didn't know if there was a certain like meaning behind it—kind of like in the magic key problem, there is not obviously a certain actual key, so I don't know if you meant an actual fertilizer for all plants or if it was ... if that was just kind of (P) I don't know ...

I: One possibility then would be that there would be one fertilizer for all plants ...

JUL: Hm-hm.

I: What's the other ... what's another possibility?

JUL: That's what I didn't know. Yeah.

I: OK. So, is that the interpretation you gave it?

JUL: That there was ... that there was one fertilizer for all plants?

I: OK ...

JUL: Yeah.

I: And you thought that was true.

JUL: Yeah. (P)

I: Can you tell me why that's true?

JUL: Ahm ... That there's ... not necessarily the same fertilizer for all plants, but there is a certain like kind of fertilizer for all plants.

I: But not necessarily the same you say?

JUL: Right.

I: Oh. So that's not ...

JUL: Oh, is that ...

I: ... well ...

JUL: OK, no! I guess ... I see a difference between: there is *a* fertilizer and there is one fertilizer ... OK, so now ...

I: Right...

JUL: OK, so that's what ... I ... if it said *one* I would say false. Because it might ... that I would take that as the exact same fertilizer for all plants.

I: And you don't think that's true.

JUL: No.

I: Why not?

JUL: Ahm ... plants need fertilizers different fertilizers ...

I: Different plants need different fertilizers?

JUL: Yeah.

I: OK. So what about the other way now, the other interpretation?

JUL: There is *a* fertilizer ... That there is ... there is *a* fertilizer not necessarily the same one for all the plants, but there is a fertilizer made for all plants ...

Finally, we have an example from CAR who, in discussing Statement 6 appears to have an EA interpretation—water being the key for everyone. But she shies away from this, pointing out that it affects everyone differently and although the water may be the same for all, the glass is different. When the interviewer tries to get her to clarify more (with a tendency towards EA by mentioning only the water), the student comes down in favor of the AE interpretation.

I: OK, OK. That makes sense. Now let me ask you something about your interpretation of the statement. You were saying earlier that you saw this as implying that there is something out there, right? That there is this external key.

CAR: Hm-hm.

I: Did you interpret the statement as saying there is this external key that is supposed to be working for each and every person, sort of like a common thing that is supposed to work for everybody, or . . . did you interpret it to be ahm ... that for every person there would have to be *some* external key?

CAR: Yeah, I saw it as like an individual thing. Like, um, like in my mind when I read that question I saw something like ... like like, maybe like ... like water, or something, you know, like everyone can have a glass of the water. And that water will like affect them in a different way. You know? Because there is such a mass quantity of it. And every glass is different. That's how I saw it. And when I thought of it that way that like what's inside that cup, I mean like the magic key ... you know ... I saw key as like being as like that force. You know? [Laughs nervously.]

I: I think you *are* making sense. What I am trying to understand is whether you thought that the water, or whatever, would be ... ahm ... or *might be* different for every person ...

CAR: Yeah, I think it would be different. It would be the *same* like, in like, a large like scheme of things. Like the same concept, but then it would have different effects on every person. You know?

I: OK, OK, all right. Fair enough. Great. Let's move on to number 9.

3.2. Reasons for AE tendency. We consider several possible reasons for the observed tendency towards AE interpretations: likelihood of truth of a statement; cognitive difficulties; specific wording of a statement in general; and, the use of the term *all* as opposed to *every*.

3.2.1. *Truth of a statement.* One reason for an AE preference might be that more often than not, the statements we make in everyday life are AE and not EA (whenever they involve two different quantifiers, of course).

Another reason might be that people tend to interpret ambiguous statements so as to make them true. Given an AE statement and its EA variation, the EA version, of course, implies the AE version. So the AE version is much more likely to be true, and can be true while the EA version might be false; on the other hand, if the AE version is false then so is the EA version. So a tendency to favor truth would result in a tendency to favor an AE intrepretation.

Contrast the following (see Appendix B, Truth and Falsity): 30% of the AE statements were declared True, but only 10% of the EA statements were declared True. So, it could be that there is a preference for true statements, hence the preference for the AE interpretation which is more likely to make a statement true. Further support for a tendency toward true statements can be given from observing the following discrepancy: while only 11% of the students interpreted Statement 3 as an AE statement, Statement 4 was interpreted as AE by 81%. Statement 4 would be truly absurd (not "merely false") if interpreted as EA, whereas Statement 3 is not so absurd as an EA statement (Appendix B, Table 1).

3.2.2. *Cognitive difficulties.* Suppose we are given an AE statement, $\forall x\ \exists y\ R(x,y)$, and a corresponding variation $\exists y\ \forall x\ R(x,y)$. Proving an AE statement is false is equivalent to proving an EA statement is true (since $\neg(\exists y\ \forall x\ R(x,y))$ is equivalent to $\forall y\ \exists x\ \neg R(x,y)$), and vice versa. So, from a mathematical point of view, AE and EA statements have the same degree of complexity.

Despite this similarity between AE and EA statements, a student trying to assess the truth value of an AE statement, versus that of an EA, may encounter more cognitive difficulties in the EA case. Why? Let us contrast the sequence of steps one has to take in order to decide whether an AE statement is true with the corresponding sequence of steps taken for an EA statement. (In each of these two cases, we are not interested in examining the polished and final product of a proper proof here. We are interested in an approximate outline of the mental steps one might go through in order to come up with the ideas that will, perhaps, lead to a proof. Thus we ignore the possibility that we may have infinitely many xs and ys to consider.)

Suppose the AE statement at hand is

$$\forall x\ \exists y\ R(x,y).$$

We might start by considering mentally a single x_0 and asking ourselves whether there is some y_0 (of the appropriate type, depending on the context) such that the relation $R(x_0, y_0)$ holds. Then we iterate this process. As long as the answer to each individual question is "yes," we continue, until we exhaust (mentally) the whole set over which x ranges. If the answer turns out to be "no" at some point, we stop and declare the statement False.

On the other hand, suppose the EA statement at hand is

$$\exists y\ \forall x\ R(x,y).$$

To begin evaluating this statement, we might start with a y_0 and ask ourselves whether all the xs, simultaneously, satisfy $R(x, y_0)$. If the answer is "yes," we stop and declare the statement True. If not, we need to go back and consider another choice for y. Iterate this process until we reach a "yes"; if after exhausting all the possible ys we have answered "no," we say the statement is False.

Though both the above sequences of mental steps one might take involve some back-and-forth checking, they differ in one important way. The AE verification process remains at the x-element and y-element level. That is, each comparison is done with one pair (x_0, y_0) at a time. On the other hand, the EA verification process pushes us one level higher: each time we consider a y_0, we need to consider all the xs at once, the whole set of xs that are available. Thus, in set-theoretic terms, the *rank* of the sets has increased and in cognitive terms, it is necessary to see the set of all xs as an object of consideration. Perhaps this explains why one might first give an AE interpretation to a statement involving one existential and one universal quantifier: such an interpretation is easier to understand and easier to assess.

Obviously, as mentioned above, if the AE statement at hand is false, the student ends up verifying that an EA statement is true. But the point is that without knowing the answer in advance, a student can find an easy place to start the verification process. In other words, AE statements, thanks to the order of quantifiers, provide a starting point, a strategy, for a student who wants to begin assessing the truth value of the statement.

EA statements on the other hand do not provide such an obvious strategy. A student who needs to assess the truth value of $\exists y \; \forall x \; R(x, y)$ needs to look for y_0 that works for all the xs simultaneously. If we think of the search for such a y_0 as done totally blindly, this quickly becomes frustrating. One needs to look for a y_0 that is likely to work. But how do we choose such a y_0? It may seem that we are looking for something without knowing in advance what we are looking for. Perhaps this is a difficulty our students experience in dealing with EA statements in mathematics.

There are at least three obvious objections to the above discussion. Consider the EA statement $\exists y \; \forall x \; R(x, y)$. If, in fact, there is a unique y_0 such that $\forall x \; R(x, y_0)$ holds, and this y_0 happens to be the only reasonable choice among the other ys, then a student is more likely to find it without great difficulty. Another way a student may bypass the initial difficulty of not having a beginning strategy for assessing an EA statement is by having enough mathematical maturity to know that existential statements do not require constructive proofs. Finally, yet another way is for the student to think of assessing the negation of the given EA statement instead. These objections, however, do not really address the target population of this study. We believe that for the student who is new at dealing with AE and EA statements (and needs to construct proofs involving such statements), the cognitive difficulties described above are real.

3.2.3. *Effect of the specific wording.* It is possible that the difficulties the students had in interpreting the statements in the study are dependent on the specific wording that was used. For example, JUL interpreted Statement 6 as an AE on the questionnaire and maintains this view in her explanation during the interview. However, the interviewer is able to reword the statement in such a way that the student sees it as an EA.

I: OK, great. Have a look at number 6. What do you think about that? Cheesy [fate?] (both laugh)

JUL: I don't know, I guess I meant by that ... that there is kind of that in the whole fate of ... there is this person for everybody in the world and some people might not ... I think at one point in everyone's life they ... even if they don't get married, or whatever, there is someone that they ... not necessarily that they fall in love with or anything like that, just that changes their life or makes a big difference in their life ... Someone they always remember or treasure ... so ... I think that's what I meant.

I: OK. Suppose we change that statement ... So, for you the magic key would be a person—right?

JUL: Yeah. Or not necessarily, maybe a certain event or something, yeah ... it doesn't necessarily have to be a person.

I: OK. Now suppose we change the statement to read "there is a magic key common to all [persons]"?

JUL: ... common to all ... I would take that to mean that it is the same key I guess I would take that to mean that's the same key ...

I: The same key ...

JUL: To ... for everyone's heart, if it's common to all hearts, that it would be the same ... key whatever it was. Let's say something like love was that key.

I: I see.

JUL: Then love would be the only thing that could unlock everyone's heart I guess. But I wouldn't think that ... that would have to be necessarily true.

I: So you think that's false?

JUL: Yeah.

I: OK . And why?

JUL: Because I don't think that there is one particular thing that would ahm ... that would unlock everyone's heart because everybody is different. And so, I mean people with different views ... different I don't know, different things are important to people in their lives. And so it's not going to be the same.

Thus we see that this student is capable of interpreting a statement of this kind as an EA but either because of the wording or some other reason, she chose to interpret Statement 6 as an AE.

It is certainly true that the use of extremely specific wording, such as "common to all" or as another student suggested for this same statement, "There is one magic key," could influence students to give EA interpretations of statements. But it may be that this masks the tendency rather than explains it. For example, how does the rewording argument explain the fact that so many more students interpreted EA statements as AE than made the reverse? It is also the case that some students showed in the interviews a strong resistance to interpret a statement as an EA, no matter what rewording the interviewer used. See, for example, the last excerpt in Section 3.1.2.

3.2.4. *All and Every.* One specific kind of wording that one might consider as important for distinguishing between AE and EA statements is the choice between

using the word *every* (or one of its derivatives) and the word *all*. Perhaps this can
make the difference between the AE or EA interpretation of a statement. The idea
is that the word *all* seems to be collective, and it makes a statement understood as
EA, whereas the word *every* appears to address each case individually, and hence
would suggest that the statement is AE. Our evidence, however, does not support
this suggestion. This can be seen by comparing Statement 4 (which uses the word
all) with Statement 8 (which uses the word *every*): 81% of the students interpreted
Statement 4 as AE, whereas only about 40% interpreted Statement 8 as AE.

Another possibility is that the use of the word *every* (or one of its derivatives)
will signal the reader to interpret a statement as EA. Again, our data do not support
this. This can be seen by comparing Statement 3 (which uses the word *everyone*)
and Statement 6 (which also uses the word *everyone*): only 11% of the students
interpreted Statement 3 as AE, while more than three times as many students
(37%) interpreted Statement 6 as AE.

There are many other pairwise comparisons one can make (for example Statement 4 and Statement 9), but it seems that in all such comparisons, the mere
choice of words does not make the difference. Perhaps the context is most important. This could explain why 81% interpreted Statement 4 as AE, for example: the
EA version of the statement seems absurd. Yet only 38% interpreted Statement 9
as AE, despite the fact that Statement 4 and Statement 9 have identical syntactical
structure. In the context of Statement 9 neither the AE interpretation nor the EA
interpretation seem as absurd.

It makes sense that our students would favor the context of a statement over
its syntax. This is the convention we use in everyday, natural language. We tend
to favor interpretations of statements that "make sense" in the world in which we
live. A consequence of this principle is that we often end up not being fully aware
of the statement at hand, but rather of our interpretation of the statement. This
is what we observed in this study. Our students were not so much conscious of the
written statement on the questionnaire per se. It was as though the statement was
a window from which they were looking out. The students described what they
saw looking out the window, but they did not see the window itself. When coming
up with an interpretation of a statement, they were not aware of the process they
followed in order to reach their interpretation.

If context is indeed so important in determining one's interpretation of a statement, more so than syntax, then we see why we run into problems with the mathematics statements (10 and 11). We will see that students are on much shakier
ground because they do not have as good an understanding of the world of numbers as, say, pots and covers, and they do not have any everyday-type of conventions
to go on in order to interpret these statements. In addition, they typically have not
been taught the conventions for interpreting the order of quantifiers in mathematical statements.

3.3. Quantified statements in natural language and in mathematics.
Our data indicate that students were much more capable of handling the natural
language statements than the mathematics statements. Only 49% of the students
got Statement 10 right, and only 14% got Statement 11 right. (See Tables 1 and 5.
If we count correct interpretations and justifications of these statements when students considered a and b as ranging over the integers rather than the real numbers,
then 53% and 18% of the students got Statements 10 and 11 right, respectively.

Even so, these are not high rates of success with Statements 10 and 11.) Yet, 78% of the students gave a valid argument for seven out of the nine natural language statements (Table 3). We will try to look more deeply at this phenomenon by considering some excerpts from the interviews, first in everyday contexts and then in a mathematical context.

3.3.1. *Statements in everyday contexts.* Many students approached a strong understanding of simple statements from everyday life. Nevertheless, it is interesting to note that only 25% of the students gave valid arguments for all nine English language statements, and during the interviews, they did not show as strong an understanding on very many statements. We believe that overall, their understanding of the English statements was not as solid as the questionnaire data would seem to suggest.

During the interviews students were reasonably good at expressing their own viewpoints regarding the topic at hand, but did not appear to focus primarily on the given statement. The context of the statement was their focus. Furthermore, if a student had decided a statement was true, he or she was often unable, unwilling, or just resistant to discuss what it would take for the statement to be false. We view this as an indication of partial understanding, much the same way that in order for one to understand the notions of *chocolate* and *linear*, one needs to understand the notion of *non-chocolate* and *non-linear* as well. A second reason why we think students' understanding of the English statements is not so strong is that the students seemed to be unaware of the (possible) ambiguities of the natural language. The statements that we consider to be of EA form were, in many cases, regarded as unambiguously AE. When faced with the possibility that it might be otherwise, students did not entertain the thought and did not go back to re-read the statement in question. They either continued to talk about their viewpoints or they left the ambiguity unresolved. We believe all this is evidence of only partial understanding of the given statements.

3.3.2. *Statements in a mathematical context.* The performance dropped sharply when it came to the mathematical situations, Statements 10 and 11. Using the conventions of mathematics to assess the students answers, we required that the student interpret each statement correctly (Statement 10 as AE and Statement 11 as EA). Thus if the student was thinking of real numbers, he or she would have to provide a valid reason why Statement 10 is true and Statement 11 is false. On the other hand, if a student was thinking of integers only (as many students were), he or she would have to provide a valid reason why Statement 10 is false and Statement 11 is true. As it turned out, very few students thought only of integers as "numbers" and got Statement 10 or Statement 11 right. In other words, based on our data, it is not the density of the real line that caused the apparent difficulty students had with these statements.

In addition, three students indicated some confusion about 0 and considered that it could be considered to be a positive number. In two of these cases, because of other aspects of their response, our assessment would be same whether or not this error was reflected. In the other case, this was the only error the student made and we consider the argument to be invalid.

Table 1 shows the percent of students who gave a valid argument for each of the eleven statements (see % valid). In the case of Statements 10 and 11, this refers to those students who interpreted "number" as "real number." If we consider those who interpreted "number" as "integer," the % valid goes up from 49% to 53% in the case of Statement 10, and it goes from 19% to 23% in the case of Statement 11 (see Table 4 of Appendix B). Since the success rate with these statements was not substantially improved when looking at integers, we chose to list the success rate of those who made the 'real number' interpretation. All the remaining calculations were done with this assumption. A total of 50% were unsuccessful for both statements. The percentage of students successful on Statement 10 but not Statement 11 was 37%, successful on Statement 11 but not Statement 10 was 2% and successful on both was only 12%. (See Appendix B, Table 4.)

The difficulties students had with the statements in a mathematical context appeared both in their responses on the questionnaire and were maintained strongly in the interviews. We can provide a number of examples of this.

Here are some questionnaire responses given for Statement 10:

- *Eventually a will be small enough that b will have to be a nonpositive number, making the statement false.*
- *True, if zero is considered a positive number.*
- *False: a = b.*
- *True: b can be a number less than a.*
- *False: b might be a larger positive value than a.*
- *False. If a = 1 then b = 0. Zero is neither positive nor negative.*
- *False. a = 2, b = 1, 1 < 2.*
- *False. If a = 1 and b = 0. Zero isn't a positive or negative number, it is itself.*

Here are some questionnaire responses given for Statement 11:

- *Statement 11 is a reworded equivalent of statement 10.*
- *True: b can be less than a in all situations.*
- *This is the same thing as 10, just rearranged.*
- *True. Same reasoning as in 10: numbers grow and shrink to infinity.*
- *Same statement as 10, just syntax is different.*
- *Same logic as 10.*
- *True. If b is 1, then a will always be greater.*

Now, for the interviews, we begin with some positive responses. During the interviews, several students showed they had a good understanding of at least some of the mathematical statements. Here, for example, HAR came to the interview with a good understanding of Statement 10:

HAR: (rereads problem)
   I: OK. Tell me what you were thinking. Why do you think it's true?
HAR: Because for any number a, there will always be a number that is smaller than that ... By ... by ... there are an infinite number, an infinite amount of numbers, and no matter what number you take, however small it is, there will always be a number smaller.

Here is another example, from the interview with SOU:

   I: OK, good. Let's move on to number 10. You had said that that's true. Can you tell me why?

SOU: ... (LP) ... (VLP) ... because if you pick one ... a is a big positive number ... and there is always going to be a positive number that is smaller ... all the way down to if you pick a small fraction, there is always going to be a fraction that is smaller than this, but yet bigger than zero.

I: OK.

SOU: I think ...

I: OK, it makes sense.

Here is an example from the interview with LER, who started out thinking that Statements 10 and 11 were the same but, during the course of the conversation with the interviewer, changed her mind about Statement 11.

LER: ... (P) ... I think this is clearer ... number 10 is clearer than number 11, because ... it's ... I don't know ... [reads it mumbling] ... this kind of makes it so ... there is a number which is less than the original ... —this is hard to explain—this makes more sense ...

I: Number 10?

LER: Yes. Because ... You say there is a number that has this characteristic. And then here you are saying there is a number who ... for the other ... wait ... (P) Oh no ... it would make more sense to write this as $a > b$.

I: $a > b$?

LER: Because the last one ... like on here, you say b last, and then you define a characteristic of b. And here you say a last, and then you still define a characteristic of b.

I: For number 11?

LER: yes.

I: I see. So what if we change number 11 to say "there is a positive number b such that for every positive number a, $a > b$"?

LER: ... (LP) ... That would ... see they both make sense ...

I: Hm-hm, sure.

LER: And ... ahm ...

I: But would that change your opinion about the statements being the same or different?

LER: I think they would still be the same ... [reads it again] ... (LP) ...

I: Ahm, so would you say that it's true—number 11 is true, if we rephrase it that way?

LER: Hm-hm.

I: OK. Can we find that number b, if that's true?

LER: There is a number b such that for every a, $a > b$?

I: Hm-hm.

LER: ... (P) ... OK, well then it would still be the same thing because you would be saying ... so you would start off with ... so you can still lose. So it's still kind of . . . it's the same as that one. As this was before here. I don't know why this is ... this makes ... this works out, and I can't really figure out why.

I: Number 10 works out?

LER: Hm-hm. So if you change it, there is a positive number a ... such that for every positive number b ... $a > b$, would that work? ... I guess it's just what order you are using ...

I: So, if it is true, that there is a number b such that for every a, a is bigger than b, can we find that b?

LER: There is a number b ... for every ... OK can you say that again?

I: Sure. There is a number b such that that for every positive a, b ... excuse me, a is greater than b.

LER: ... (LP) ... No ... The word every is throwing me off there. There is a number b such that for every a, $a > b$ ... It's the word every and I can't explain why this is ... (P) there is a number ... (P) ...

I: Take your time.

LER: OK! ... (LP) ... NO, that's not true, because . . . OK. Then the b would have to be the smallest number in the world. Because for every a, a would be bigger than b. OK, it's not true! ... So that would be false.

I: That would be false.

LER: Hm-hm.

I: Now this rewording that we did, versus the original number 11, are they different or are they the same?

LER: It's different because ... let's see if I can explain ... You are defining a as like every positive number ... You are not really ... It's different somehow ... (P) ... This one you name a number a and you always find a b smaller than it—on the original. If you change it around, ... your b has to be determined by the a.

I: I see. OK.

LER: And the other way, your a is determined by the b.

I: OK. OK. All right! That's it.

Finally, here is MAT, the student who was capable of understanding both mathematical statements on his own. (He had originally interpreted Statement 11 with multiplication, but after the interviewer clarified the point, the student clearly did not need any prompting to assess the statement correctly.)

I: OK, great. Ahm, let's move on to 10. Here you said that statement is true. Can you explain what you were thinking here?

MAT: (P) ahm, it says for every positive number a, there is a positive number b such that b is less than a. So I took as in ... so obviously working between zero and ... I [...] zero is your fiction point and you can pick any object ... or any number a and no matter how close a is than zero or how far it is from zero there is going to be like a b which is the midpoint between zero and a. So no matter what a is, you can always find a b that is always less than that ... [mumbles]

I: OK, I would just like to say this for the tape—that as you were saying that, you were using your two hands to express a number line, and points on the number line. Now take a look at number 11. You say that one is true as well.

MAT: (P) [mumbles, whispers] (P) Yeah. OK, I think. So I first I took b to be a number between 0 and 1, so it's a fraction, so then, a fraction of any positive number is going to be less than the number itself.

I: OK—now I don't know that we wrote this down exactly the way you wanted to write it, so let me read this to you using intonation that you may find changes the meaning.

MAT: OK.

I: There is a positive number b such that for every positive number a, b is less than a.

MAT: OH! You put a comma or something between ...

I: Right, for instance we put a comma. Would that be true or false?

MAT: [mumbles] (P) I think that would be false. Because if you fixed b, there is a positive number b, which means you fixed b, ... then you can't say every positive number a is going to be greater than b.

I: Greater than b ...

MAT: Greater than b. Because there will be some positive a that will be less than b also ...

I: Could you say one for sure?

MAT: Ah, you can almost go back to 10. And use the midpoint between 0 and b.

I: Why almost?

MAT: Well, I guess you could! (laughs) For sure. Yeah.

We do not have any more examples of students expressing good understanding of Statements 10 or 11 on their own. Several students did make progress, however, during the course of the interview, particularly after the games (see the next section for details).

About half of the students we interviewed made explicit remarks to the effect that Statements 10 and 11 are the same or equivalent (a few students suggested that Statement 11 is harder to understand, but it is not any different than 10). Here is an example of student remarks comparing Statements 10 and 11.

I: OK. OK, great. Very good. Ahm, last question. Number 11. Let's see ... the statement says there is a positive number b such that for every a, b is less than a. And you have decided that that's true. So, can you explain to me why?

CPE: ... (P) ... Because I ... I thought it to be the exact same thing as the above. You just switch the ... letters around ...

I: The a and the b?

CPE: Hm-hm. ... Is that right though??? ... (P) (LP) OK. It switches it around ... but it's true ... (P) because again, you can always find a number that's less than another.

I: OK. Ahm, ... so what is the difference that you see between 10 and 11?

CPE: Well, 10 is less than 11. Eleven would be the a, and 10 would be the b.

I: No, I meant, the statements ...

CPE: OH!!! excuse me! OK ... I really don't see a difference at all.

I: Do you see them as reworded, or ...?

CPE: Hm-hm. Yeah.

Here is an example of a student, SOH, who first sees no difference between Statements 10 and 11, but then makes some progress toward understanding Statement 11 better.

SOH: How is this different than that one?

I: That was my question! Do you see them as different? Number 10 and number 11.

SOH: ... (LP) ... There ... it's possible that they differ ... I am trying to think exactly how, I mean I see that there is a confusion, when you say in number 10 that for every a there is a positive number b, and here you say ... there is a b ... well ... they are the same ... (P) They are ... yeah ... they are just placed differently. I mean you could say for every positive a, there is a positive number b ... (LP) ... This just ... when you put it that way, you might read it and think ... there is a positive number b such that for every positive number a ... It seems to be more like the controlling factor is b since it was said first. And so it makes it look like you have this b, and say you say it's 0.5, and then you, and then you read such that for every positive number a, b is less than a. And you are thinking, well, ... for every positive number a, well, a could be 0.2 instead of 0.5. And that way I mean it gets confusing, because you have like a control factor in ... that usually when you read it you think there is a positive number b, and there is the control. And then you say, such that for every positive number a, b is less than a. Where you might fill in examples, and say that that doesn't really work.

I: OK, you gave an important example. You said what if b is 0.5 ... then what did you say?

SOH: Such that for every positive number a, b is less than a. What if a was 0.2?

I: Yeah.

SOH: So ... (P) I guess ... they are different! I guess I would have to change that to false!

I: So, would you say that that's false?

SOH: Hm-hm. Yup.

I: And why would you say that's false?

SOH: ... (P) ... It's just different, the way you think about it in your head, like b being the constant, and then ... a being ... it could be whatever you wanted to be. It could be equal or less than ... so ... Now I am looking back at 10. Ahm ... 10 still seems right to me.

3.4. The unknown in mathematical and in non-mathematical situations.

One possible explanation for the drop in performance in going from everyday situations to mathematics is that, perhaps, for many students, the mathematical context simply represents an unknown domain and this increases their difficulty in making sense out of the structure induced by the quantifiers. Aside from the fact that comparing two real numbers for size (which is the mathematical context of Statements 10, 11) should not really be very much of an unfamiliar context for very many students, we tried to find out about this possibility through Statement 7.

Here are some examples of student responses to Statement 7, taken from the questionnaires:

- *There had to be some poems about love, romance, and/or religion.*
- *I have no idea! If I were to guess, I would say it's false; it sounds too stereotypical.*
- *True. That's all the Greeks seemed to do: fight in wars.*
- *I am not so sure, but it would not be surprising if it's true, considering the history of those times, fights, etc.*
- *I honestly don't have enough experience here to make a good call either way. But I can't imagine all Greek life was predominantly war.*

Considering the above responses, it is apparent these students did not know the answer for sure. In addition, their decisions disagree: some thought Statement 7 was true, others that it is false. Yet the common theme in most of the 63 responses we got was that they were able to make guesses that they could support with arguments that sounded reasonable. They were able to present reasonable arguments about an unknown situation. Many students certainly appeared to be able to deal with the structure of the statement.

Yet, when students were faced with presumably unknown facts about numbers in Statement 10 and Statement 11, their ability to provide valid arguments dropped sharply. One might reasonably expect that students who are already in some college-level mathematics class ought to have *some* understanding of the notions of *positive number* and *less than*. Yet, the students' ability to reason about the unknown in the context of Statement 7 did not transfer to the context of mathematics.

3.5. Attention to quantifiers.

In this section we give some examples regarding the tendency of students to ignore the quantifiers in a statement, including those who made this error only for the mathematical Statements 10 and 11.

During the interviews, when asked explicitly about quantifiers such as *everyone, for all, at least one, most, a whole lot,* most students expressed their meaning well. When viewing quantifiers in isolation, the students did not have difficulties understanding and expressing the differences and similarities between these quantifiers.

However, when students discussed their thinking process for deciding whether a statement was true or not, we had the opportunity to observe the degree to which they paid attention to quantifiers in a statement. We noticed two main trends: those students who did not pay attention to quantifiers in the natural language statements, continued to ignore quantifiers in the mathematics statements; and, some students who had attended to quantifiers in the natural language statements failed to do so in the mathematics statements. There were no students who attended to quantifiers in mathematics but failed to notice them in the natural language statements.

Here is an example from the interview with GIT who seems to ignore the quantifiers in Statement 6:

I: OK, all right. Good. Let's move on to number 6. The magic key. You have decided that that's true. Can you tell me a little bit about how you were thinking about it.

GIT: OK ... OK. Ahm ... Somebody might be sensitive, . . . or ... OK let's say that somebody might be sensitive, I mean might hold it inside him ... So, ... but ... there might be a time where he expresses his feelings. So he unlocks his feelings (laughs).

I: Hm-hm, that's good.

GIT: So, he unlocks his heart. Where he opens to talk about it. So that's the magic key. He opens his heart sometime. And talks about it.

I: OK, now when you were talking about someone, was that a particular person you had in mind or were you talking about people in general?

GIT: No, people in general.

I: OK. Everybody, or some people?

GIT: (laughs) ... Ahmmmmm, maybe some people, not everybody ... Well, I don't know if everybody is sensitive, or you know has other characteristics ...

I: Sure.

GIT: But I know some that are sensitive. I know from my knowledge ...

I: Yeah. Hm ... I guess what I am asking about is what about sensitive or non-sensitive, what about people in general justifies that this statement given in number 6 is true? You know, because we can talk about some people being sensitive, and some people not being sensitive,

GIT: Yes, yeah.

I: ... and so on. But what facts do we know that will justify that this statement is true?

GIT: ... (P) ... Mmmm ... well, I think everybody has his own . . . characteristics ... So everybody has something ... you know, inside him ... that is important to him. So, yeah. I think everyone does. Some feeling some characteristics that are staying inside.

I: Hm-hm, hm-hm ... And those would work as a key? Is that what you are saying? That the magic key, would those be the magic key??

GIT: ... (P) ... Maybe ... I think ... (P) ... Yeah ... I think everybody has a magic key that can unlock ...

I: Why, why?

GIT: Because everybody ... (P) ... OK, ahhhh ... Everybody might feel saaaad, or everybody might feel something, and ... when it comes to a situation, they might unlock that, and say it out loud. Yeah. So, everybody, I think everybody has ... characteristics ... [mumbles]

This student goes on to ignore quantifiers when asked to analyze Statement 10:

I: OK, all right. That makes sense. Let's move on to number 10 real quick.

GIT: OOOOhhhh... [mumbles] mathematics? (laughs)

I: You have decided that that's false. Can you tell me why?

GIT: Hm... (P) ... OK, ... Can I use an example?

I: Sure.

GIT: If say a is 2, ... and b is again a positive but 3, then 3 is not less than 2.

I: True.

GIT: SO that's why I said false. Because there might be a positive b which cannot be less than a.

I: Right. Now is that what our statement is saying?

GIT: ... for every positive number... [reads it mumbling] ... (P)... OK, it says "for every positive number a there is a positive number b." So that says that there might be a ... you know a b that would not work, and there might be a b that will work. It's because of the "a positive number b." And we know that ... a ... is a positive . . . like ... we know that ... OK that's a is every positive number. For every positive number a.

I: So, you are saying there are some b's that might work, and some b's that won't work. Like the example you gave, where a is 2 and b is 3.

GIT: Yes. Yes.

I: And that's not going to work.

GIT: Yes.

I: Is there a b that would work?

GIT: yeah. Let's say a is 2 again, b is 1.5. So the 1.5 is less than 2.

I: OK. So we have a b that will work and a b that won't.

GIT: Yes.

I: Does this help with the statement ? To decide if it's true or false?

GIT: ... (P) ... Well, I guess there might be a positive number a ... which will be less than b. Like it will not work for both cases. You know, if you have an a ... which is greater than b, then this will not work, because we say that b has to be less than a.

I: Hm–hm ...

GIT: So, there might be a number for a too that will not work.

I: OK... So, ... you are saying for an a there isn't a b that will work ... ? I guess I am asking you to rephrase, I am a little confused.

GIT: Yeah. There might be an a for which ... you know, let's say a is ... an example again, let's say a is like 2, and b is ... no. Let's say a is 3. And b is ... (P) ... b ... (P) ... OK, a is 3 and b is... (laughs) ...

I: Take your time ...

GIT: Let me see ... (P) ... One. or I guess ... (LP) OK, a is 3 and b is 4.

I: a is 3 and b is 4, hm-hm.

GIT: So, ... 4 is not less than 3.

I: Right.

GIT: Yeah ... (P) ...

I: OK, and that will make the statement false?

GIT: Hm-hm.

Here are two examples of two other students who, though they had been paying attention to quantifiers in Statements 1 through 9, appeared to ignore them in the mathematics statements.

I: What I want to do now is take a look at number 10. I hope you can read the copy there, it's pretty light. Oh, I see, you circled both true and false. On my copy I can't read what you said there, can you read it there?

ERI: It says that it is possible, but not always true. And could be true or ...

I: OK. So tell me your thinking there.

ERI: (P) It's possible for you to put 5 for a and 4 for b, and that means that a is larger than b. But it is possible it could be the other way. You could also put in a smaller number. For a. And a larger number for b, and that could be correct.

I: That would be correct? If you put a larger number for a and a smaller for b? But you circled both true and false.

ERI: (LP) That's because it is not always true. (LP) [mumbles] ... I don't know ... (LP) If you put in 4 for a and 5 for b ... or a smaller number ...

I: What about all these numbers: for every positive, there is a positive ... do they mean anything in this sentence?

ERI: (LP) I don't know... It's possible ... It depends what numbers you put in. It's possible. I think it's true that it's possible, but not always, depending on what numbers you put in.

Finally, we give AND's responses regarding Statement 10:

I: OK. Good. OK let's move on to number 10. Now this statement you say is false.

AND: (LP) Well, cause I mean it depends on where they are located. On the number line. I mean yeah, I mean I know in math they say a is less than b or b is less than a ... but I mean here you are saying for every positive number a there is a positive number b such that b is less than a?

I: Hm-hm.

AND: Not necessarily true. It depends on where those are located at
       the point. I mean who gets to pick where a and b are? I mean, I
       always put a less than b, so to me, I mean if I chose the points,
       I would pick a to be 1, and b to be 2. So it depends on who is
       picking the points or if you are saying . . . you know . . . this is
       where b is, this is where a is. Therefore, you know...

I: What about the words "for every" and "there is"—do they affect
       anything?

AND: (LP) Well I suppose . . . because I mean if you were to say for a
       positive number a there is a positive number b such that b is less
       than a, but for "every" . . . I mean you are saying in all cases,
       there is no exceptions to this rule. In every single case. I mean
       and in some cases, I mean it just depends on who picks where a
       and b go.

3.6. Difficulties with changing the truth value. Another indication that
students had trouble, even with the English language quantified statements, was
the difficulties they exhibited in trying to change the truth values of the statements.
Many students were unable to give any reasonable responses in the interviews regarding the possibility of keeping a statement unchanged but imagining a situation
(or a different world view) in which the statement would have the opposite truth
value. Others exhibited a tendency to exaggerate the "negation." That is, in order
to make a statement false, the student constructed a world or a statement in which
the statement is false for every instance of each variable.

We begin with an example of the latter. In constructing a world in which
Statement 1 is true, CAR, who had interpreted it as an AE, goes all the way to a
world in which everybody hates everybody.

I: I see. OK. OK, very good. Now let's go back to the original
       statement. "Everybody hates somebody." You decided for your
       reasons that that's false. OK, now, . . . can you imagine a situation
       where you would decide, with the same beliefs that you have, a
       different situation where you would decide that this statement is
       true. Can you describe such a situation? What would the world
       have to look like in order for that to happen?

CAR: OK, everyone hates somebody. When this statement would be
       true would be like when like . . . I see like, . . . brown . . . like every-
       thing being really drab and like no one like smiles anymore, and
       there is maybe there is it is like nuclear winter or something like
       that and I see it like I mean you know like they talked about like
       Europe after World War II, I see that where like everyone hates
       someone . . . I mean like where after a period where there's been
       just mass hysteria and everyone is confused and no one wants to
       talk to anyone else about anything at all.

This also occurred with statements that the student had interpreted as an EA
as in the following discussion of Statement 3, which this student had interpreted
as an EA and evaluated to be false. In this case, CPE first gives the negation (a

situation in which Statement 3 is true) as essentially an AA, but then responds to the prompt by reducing to a single person as the statement requires.

> I: OK. All right. Now once again, just like we did for number 1, is it possible for you to imagine a different situation in which you would decide that this very same statement is true? What would it take for it to be true?
>
> CPE: ... (P) ... We have to be on another planet I think! (both laugh) I just can't ... I can't imagine everyone being sincere, considerate, kind ... all the time, for every moment in their life.
>
> I: OK. Ahm, so in that weird planet, would everyone have to be kind and considerate? Or would some people have to be kind and considerate? For the statement to be true.
>
> CPE: (P) Just someone. Just one person. Would have to be kind and considerate for this to be true.

The next excerpt shows HAR, who has decided that Statement 3 is false but seems to have difficulty imagining a world in which this statement would be true. She eventually gets it, and, when asked, she shows she understands the syntax of the statement.

> I: OK, all right. Let me ask you something else. Is it possible for you to describe a situation—imaginary, right, because you have decided that that's false—is it possible for you to imagine a situation where you would decide that this is true? What would it take for that to happen?
>
> HAR: ... (LP) ... I am not sure ... (P) ... I don't know, I don't know. Because again, I relate it to all the time. Being kind and considerate all the time. So ... I mean it would take ... I guess my situation would be a case where you around every other people for, you know, a day, or ... something like that, I mean I think that it's possible to be kind and considerate to everybody for short periods of time. But when you extend that ... I don't know if I am answering your question ...
>
> I: You are getting towards that question ... Here is what I am trying to focus on. You are saying "you can be blah blah blah." The "you" there is generic, we are talking about people in general.
>
> HAR: Right, right.
>
> I: OK. So ... If the statement turned out to be true, for some reason, even though we are saying here it is not the case, but if it turned out to be true, would it have to be true for ... kind of the generic person in society, or would it be true for just one person, ... ?
>
> HAR: It would have to be a special person.
>
> I: It would have to be a special person. Would that special person, let's say this person existed, would that be enough of a reason to make the statement true, or would we need more?
>
> HAR: That would make it true.

I: That would?

HAR: Yes.

I: Just one person?

HAR: Yes.

I: Why?

HAR: Because you say someone, and the one means it only takes at least one person.

I: OK.

HAR: There can be more, but if there is at least one person, then this is true.

Finally, here is another student, LER, who had decided Statement 1 is false. When asked to describe what it would take to make this statement true, she was not even willing to entertain the possibility.

I: OK. All right. Let me ask you something else before we move on. Is it possible for you to imagine a situation that would make this original statement true? What would it take for it to be true?

LER: ... (P) ... I mean we would have to like live in different universe... Maybe I don't know, the world is taken over by evil or something. I mean just in practical terms, I think there is always going to be someone who doesn't hate people.

3.7. Creativity. The difficulty seen just above (or at least reluctance, in some cases) would suggest that students have a difficulty building a mental world unless they view it as realistic to begin with. This is very important because one of the skills one needs to have in order to do mathematics is to create new mathematical worlds without judging them in advance as realistic or not.

Students' written responses on the questionnaires often contained jokes. Yet they were not dismissing the statements: they provided a valid justification for their responses (True or False) within their joke. Here are some examples:

- A student declared Statement 2 to be False and responded with *This pot does not have a cover.* There was an arrow pointing to a sketch of a leaf next to it.
- Another student declared Statement 6 to be False and responded with *Yeah? Well, where are Kevin's and Mike's keys? Haven't found them yet ...*
- A student declared Statement 5 to be False and responded with *Nope, death is a good thing and hasn't ended yet.*

There were many more playful answers to the natural language statements, but none found in the students' responses to the mathematics statements. We are not suggesting that one needs to be humorous to justify Statements 10 or 11 (or any other mathematics statements for that matter). However, we claim that humor is a sign of creativity, that is, we see these humorous approaches as evidence of some creativity in devising rational explanations. This creativity was noticeably absent, however, in the justifications of Statements 10 and 11.

4. The Game

In the latter part of the interviews, the interviewer intervened by asking the student to play a certain game involving first Statement 9 and then Statements 10 and 11. In our opinion, this game expresses the logical statement we are trying to help a student understand. Moreover, our use of it is actually a pedagogical tool introduced in our investigation of student understanding. In other words, instead of just asking what the student understands and is capable of doing, we are asking what the student is ready to learn given a particular pedagogical treatment, such as playing a certain game. This could be related to the idea of a "zone of proximal development" put forward by Vygotsky (1986).

The notion of a game is not at all uncommon in mathematics. An example of this is the Axiom of Determinacy, which is an alternative to the Axiom of Choice (Moschovakis, 1980). Examples such as Ehrenfeucht games (Ebbinghaus, Flum, & Thomas, 1984) abound in logic, and in computer science (for example in Nerode, Yakhnis, & Yakhnis, 1992), the authors discuss how to interpret programs as game strategies).

Furthermore, the notion of a game based on a statement is not new as a pedagogical tool. For example, the games used in TARSKI'S WORLD, a software program designed to introduce students to the language of first-order logic, are essentially identical to the games we described to the students we interviewed. A final example is Courant and Robbins (1996, pp. 311–312) where the precise definition of continuity is expressed in terms of a contract in which one person agrees to produce a δ with a certain property whenever the other contractee produces an ϵ. We have known colleagues who sometimes employ the use of games similar to those described below in order to help students understand a statement. So, although we do not claim to be pioneering this idea, we are not aware of any studies of the use of such games as a pedagogical tool. Therefore we decided to incorporate the use of these games in our present study, by way of a first step in a systematic investigation of a potentially useful pedagogical strategy.

Our specific goals for using a game in this study are to specify precisely what we mean by a game and the conditions under which it can be used; to produce some first data on student reactions to a game; to see what results might be possible from students playing a game in an interview; and to generate tentative conclusions and conjectures for future study of the effects of a game.

4.1. The idea of a game. By "game" we mean exactly what is meant in the usual sense of the word in the English language. The game we have in mind in this context is much like ordinary games our students know (such as chess or tic-tac-toe) except that our games are based on some given formal (or formalizable) sentence with one existential and one universal quantifier.

Suppose we have some sentence S. In order to give a general description of the game based on S, we will assume that S is expressed formally (this is only for the convenience of the reader so that our description is uniform; once the game is understood, it will be clear that it is not necessary to formalize sentences in order to play their corresponding games). To describe this game based on a formal sentence, let us take a generic example of such a sentence to be

$$(\forall x \in B)(\exists y \in C)\ R(x,y).$$

The game has two players, the A-player and the E-player. The A-player chooses x's from the set B, while the E-player chooses y's from the set C. Who goes first is decided by the order of quantifiers in the sentence. So, in this case, the A-player will start the game by choosing an x from the set B. Then, the E-player will respond by choosing a y in the set C. The players go on like this, alternating. The E-player is trying to establish that the formal sentence is true. So, given some x_0 that was named by the A-player, the E-player tries to find some y_0 so that $R(x_0, y_0)$ holds. If the E-player can do this, no matter what x the A-player chooses, then the E-player wins. On the other hand, the A-player is trying to antagonize the formal sentence, that is, this player is trying to make the sentence false. The A-player tries to come up with a cleverly chosen x^* such that no matter what y the E-player responds with, $R(x^*, y)$ is false. If the A-player succeeds in finding such an x^*, then he or she wins. So, player A has a winning strategy if and only if the sentence is false.

Similarly, we can describe a game based on the generic formal sentence

$$(\exists y \in C)(\forall x \in B) \ R(x, y).$$

Once again, the game will have the two players: the A-player who chooses x's from the set B, and the E-player who chooses y's from the set C. Once again, the E-player's objective is to establish the sentence, while the A-player's objective is to antagonize, and prove the sentence false. The difference this time is that the E-player starts the game (since the existential quantifier occurs first in the formal sentence). So, the E-player tries to find some y_0 such that no matter what x the A-player will respond with, $R(x, y_0)$ will be true. If the E-player succeeds in finding such a y_0, he wins. On the other hand, once the E-player has named a y_0, the A-player tries to find an x_0 such that $R(x_0, y_0)$ is false. If the A-player succeeds, she wins. So, player A has a winning strategy if and only if the sentence is false.

4.2. What we found. Based on the transcripts of interviews with 14 students, we consider the following questions. Does playing the game affect student understanding of the statements we are dealing with? Under what conditions does this effect seem to occur? What are some effects other than directly on understanding the statements that may be present? Finally, we make a few miscellaneous comments.

4.2.1. *Effects on understanding.* The most common interchanges during the interviews were those in which the student appeared to be helped in understanding the statement as a result of playing the game. In some cases it did not seem that playing the game helped the student to understand. In a few cases it is possible that the student's understanding was reduced as a result of her or his experience with the game.

In the next few paragraphs we give some examples of these observations.

4.2.2. *Helping the student's understanding.* We begin with the example of SOH, who explains her determination that Statement 9 is false with a general discussion about what we do or do not know and concludes with the explanation that the statement is false because there could be plants for which no fertilizer exists. In other words, for her, the negation of a statement of the form "exists f such that for all p, $G(p, f)$" is a statement of the form, "exists p such that for all f, not $G(p, f)$." This is, of course, a correct negation of the Statement 9 interpreted as an AE statement.

I: Let me see. You have decided that that's false. OK, so can you
    ...

SOH: I guess I stopped being metaphorical here, I guess. Ahm, ... yeah,
      so, what do you want me to say?

I: Just go through the thinking. Why did you decide it's false?

SOH: Well ... I thought about all plants, and I thought about do we
      even know some of them, how could we even make fertilizers for
      them... But I guess there are natural fertilizers also, ... so . . .
      it might be possible, but ... I guess it's just too unknown for me
      to say that's true. So, ... I marked false because I wasn't sure, I
      mean if we don't even know about the plants ... and even though,
      I mean I don't even know much about plants, but there could be
      plants out there that nothing helps them, you know. I think I
      am being too realistic about it, you probably didn't mean it that
      way!!

The interviewer continues to discuss the statement in terms of fertilizers and
plants, trying to get the student to be cognizant of the quantifiers. She appears to
remain confused, however and talks in terms of a world she is imagining.

I: Now let's assume let's say that we have perfect knowledge about
    plants and fertilizers ... let's say we have a list of all the world's
    plants and a list of all the world's fertilizers (whether they are
    natural or chemical that someone has created). So, we have access
    to complete information.

SOH: OK.

I: How would you go about deciding if the statement is true or false?
    So, you didn't have any suspicions at this point, you could really
    check for sure.

SOH: I could check? ... I mean wouldn't that be evidence?

I: So, what would it take to check, that's what I mean.

SOH: Testing the fertilizers on the plants?

I: OK. So, what would you do? Just describe what would go first.
    How would you check?

SOH: Ahm ... well, I mean you would check off the ones that you know
      work. Like that are sold in the stores or whatever. I mean things
      that people already knew.

I: OK. The ones that ... when you say the "ones" what do you mean,
    the fertilizers or the plants?

SOH: The fertilizers ... I mean I don't know ... I mean I just don't know
      what it means having perfect knowledge because I just don't have
      it ...

I: Let's imagine we do.

SOH: Even if I knew about some of the plants, even if I was all-knowing about plants, I might not be all-knowing about the fertilizers that would go with that plant. Maybe if I knew everything about plants and knew how chemicals would react with other chemicals. But I guess I would have to experiment with all of them. Collect data, and find out ...

I: OK. And then what would it take for you to decide—let's say you went through your collection, your data, and all that. What would it take for you to decide that the statement is true?

SOH: I would just have to see it.

This continues through several attempts by the interviewer and the student displays varying degrees of understanding. Then the game is introduced to the student, and after some discussion, she appears to have come to an understanding of Statement 9.

I: OK, great. All right. So, now I want us to look at this data collection and all that, as some information that we have access to—that's what I meant earlier that we have perfect information. Imagine that we have done the experiment, or someone else has, imagine that we have a list of everything. Let's say we are playing a game, the two of us. And we take turns, and I choose a plant, and you get to choose a fertilizer ... And I get to go first. So I choose a plant, then you respond ...

SOH: This is like playing the game Memory?! (laughs)

I: I don't know that one.

SOH: OK, never mind.

I: I choose a plant, and you choose a fertilizer , and then I choose another plant and you choose another fertilizer , and we go back and forth like this. I try to win, and by this I mean I try to find a plant for which there is no fertilizer , so for which you can't come up with a response that will work for my plant. If I can't win, then you win.

SOH: OK.

I: Now, let's say that you decided that this statement is false. Who gets to win this game?

SOH: You win ... no ... Let me think of the game again! (P) You are trying to give me a plant that I can't find a fertilizer for. And if you win, the statement is false ... Which is kind of like saying there is a plant with no fertilizer ... Kind of a different statement ... but ... hm ... I haven't had logic or anything, but it seems to work that way ... seems logical to me.

I: OK, so in this case, if you circled "false," I would get to win this game. If you had circled "true," who would get to win this game?

SOH: ... (P) ... [mumbles]

I: Let's say the statement here, that there is a fertilizer for all plants was true.

SOH: Is true?

I: Hm-hm. Let's say we decided that that's a true statement.

SOH: And so the object of the game was for you to prove that's true?

I: No, the object of the game is once again: we each are trying to beat each other, right? Win. So, I go first, I choose a plant, then you go and you say a fertilizer, and then I choose a plant, and you choose a fertilizer, and so on. I win if ...

SOH: If you say a plant and I can't name a fertilizer.

I: Right. Otherwise, you win.

SOH: Right.

I: So, let's say the statement here is true. Who wins the game?

SOH: I do.

I: You do. Why?

SOH: Because I would have an answer for you. Every time, if there was really a fertilizer.

I: OK. Good. So, now let's do a variation on this game. And the difference is ... you still get to choose a fertilizer and I still get to choose a plant, but we reverse the order. You go first: you choose a fertilizer and then I choose a plant. Now ...

SOH: It's not the same, though. Because I would have a fertilizer but there is no plant for it ... so ...

I: OK. So let's say I would win in this case if ... once you come up with a fertilizer I can come up with a plant which doesn't like that fertilizer ...

SOH: OH!

I: Otherwise, you get to win.

SOH: That would be easy for you to find a plant that doesn't like that fertilizer because I think a lot of fertilizers can kill plants ... (laughs). So, what was the question?

I: So, this game would be won by a different person, right?

SOH: (P)

I: Who would win this game?

SOH: You would. Because you could find a plant that would not like a fertilizer. I don't think there is a universal fertilizer that for every plant would say you help me. You know!

Although SOH came to a better understanding of the statement after playing the game, there was also considerable discussion not related to the game and we cannot be sure that this did not help so the game may not have been the only reason for her improvement.

The following transcript is clearer. HAR begins with the serious error of thinking that Statement 11 is the same as Statement 10. She holds to this idea fairly strongly in discussion with the interviewer, so we might infer that this discussion did not help. Then the interviewer explains the idea of the game, the specific rules, and how it is determined who wins and who loses. After playing the game, HAR not only comes to a good understanding of Statement 11, but spontaneously expresses the view that it was playing the game that led to her understanding. In fact, it does appear that the student's understanding of the game contributed to her understanding of the statement because she revised her answer to agree with her decision about who wins the game, and not the other way around.

I: OK ... Let's go to the last one, number 11. You have decided that that's true. What is the reason?

HAR: MMM ... This one confused me ... (P) ... I didn't see any difference in Statement 10 and 11, none whatsoever.

I: OK, that was going to be my next question! Do you see any difference?

HAR: No ... I mean I looked at it, and I looked at it, and I ... I don't see any difference.

I: Ahhh, do you see them as slightly reworded, or ...

HAR: Yes. Because all you did was you put "for every number a" up there. I don't see any difference ... (P) ...

I: OK ... (P) ... Let me ask you something else ...

HAR: Am I supposed to be seeing a difference?

I: No, I am just interested in just how you see it—exactly. Let me ask you something else. Let's say we play the same, sort of the same game that we were describing before, for number 10. I was doing the a-choosing, right, I would choose the a, and then you would try to find a b and so on, we would go back and forth. Let's say I still do the a-choosing, and you do the b-choosing, but we reverse the order in which we do this. So, you get to go first. OK? So, You find a b, positive, some number, it's up to you to choose it. And then I respond with an a. Ahm, if your b is less than my a, you win, or if we continue like this, you survive, you win just like before. OK? You were trying to make your b's smaller than my a.

HAR: Right, right.

I: So, if your b is smaller than my a, you win.

HAR: But your goal is to make your a bigger than my b? Yes?

I: No. You win if ...

HAR: You are changing the rules!! (laughs)

I: No, no (both laugh)

HAR: Yes, you are!! Because you are ... you are changing the way that ... you are making it your advantage to win ...

I: The only thing I am changing is the order in which we are playing.

HAR: I know, because you have... OK, I ... I choose a number. And but I have to make sure that my number is smaller than your number. You know my number before you say your number.

I: Correct.

HAR: So therefore, you can always pick a number smaller than mine.

I: And thereby, who wins?

HAR: You.

I: I win.

HAR: But that's different! That's a different game than what we were playing in number 10.

I: Oh, absolutely. It is different, yeah. I am not suggesting it is the same game.

HAR: OK, right. Because you know my number I know your number. That's the difference.

I: Yes. So are you sure that I would always win?

HAR: Yes, for the same principle as that ... Because basically, when you ... the game that you just did, ... OK, in here, over here, you chose the a and I chose the b.

I: Hm-hm, hm-hm.

HAR: And you said that ... in this one, I'm gonna choose the b, but I am gonna choose the b first. So basically what you are doing is I ... the b becomes the a ...

I: OK...

HAR: And your ... and what you chose as your a ... becomes the b up here.

I: I see, hm-hm. OK. Now let me ask you this. Is it is conceivable that number 11 might be describing this game? The second game?

HAR: ... (LP) ... Yeah, I guess so ... Because it is saying ... because the way the statement is, you choose the b before you choose the a.

I: Where in the statement do you see that?

HAR: Oh, just in the order ...

I: In the order ...

HAR: Not what the word says, but ... but once you said that—I wouldn't have seen it ...

I: Sure ...

HAR: But once you said that, I can see the statement is telling you to choose b before you choose a. And so, because of that . . . I don't ... ahhhh ... see ... OK, me saying that you would win, on this, the game we just played, is based on the fact that I ... the assumption that you want to win. Do you see what I am saying? That you will manipulate the game so that you can win.

I: Hm-hm.

HAR: Because ... OK, you said that I would choose a b. And I could win if you chose an a that was larger than my b.

I: Ahm, but you said, you said something interesting. You said for the b that you may choose, there is going to be a's that will be bigger. Right? Which will help you win. But there will be a's that are going to be smaller.

HAR: Yes.

I: Ahm, what is the ... reference here to [aim?]   the statement number 11. Does ... how does it qualify the a's?

HAR: [mumbles] ... for every positive ... (P) ... (LP) ... AHHHHHH-HHHHHHH [in pain] (laughs) ... You are confusing me!!! (both laugh) ... (P) ... OK. I see what you are saying. Because by s ... I am being contradictory, because if what I said is every ... because a ... that if I can also choose an a that is less than b, then the "every" doesn't work ...

I: Hm-hm.

HAR: OK. So, it is a different statement.

I: Yeah ... would you say it's true or false? Number 11.

HAR: ... (P) ... False.

I: False?

HAR: Hm-hm ...

I: Why?

HAR: Because you ... you forced me to see that you choose b first ...
        And then, and then you choose a. And, whatever ... once you
        choose b ... you can choose a number that's larger than b or a
        number that's smaller than b. But in 10 it's different, because
        you say "every number a" and then you say a positive number b,
        and so that makes it go the other way.

I: Hm-hm. Ahm, now the reality is, of course, that it's like you said.
        You see no game. There is a statement and it's about numbers,
        you don't see a game. So is it possible to decide and answer like
        11 is false, like you said, without reference to a game?

HAR: Without you ... I wouldn't have noticed the difference without
        you saying so.

I: Oh, I understand that, that makes sense. But now that we have
        established that there is some difference.

HAR: Hm-hm.

I: ... is it possible to justify that number 11 is false without saying
        hey, we can talk about a game, and so on. Can we just talk about
        numbers and no games, no players, and still justify that number
        11 is false?

HAR: You mean come up with specific examples?

I: Ahm, no, just kind of argue why number 11 is false. What's wrong
        with it?

HAR: And not reference it to a game?

I: Right.

HAR: ... (P) ... Yeah, I think it's possible, I mean I think it ... I think
        takes someone knowing ... someone knowing the idea of this game
        before and helping somebody else through it ... (P) ...

I: Hm-hm. So, how might the answer go, how would we explain why
        11 is false?

HAR: ... (LP) ... By saying that you choose one b, and I see it as like
        the span, you know, like a number line ...

I: Hm-hm.

HAR: You have a b and for every, every number, means every single
        number would fall over here ... that a, every single number a
        would fall over here. And that's not true.

I: To the right of b?

HAR: Yeah. Because there are things that will be on the left of b. There
        are infinitely many things over here.

Having no effect on the student's understanding. It is not always clear that it
is the game that helps a student increase her or his understanding. For example,
SAN, said on the questionnaire that Statement 11 is true. Later, she changed her

mind, gave a good explanation of Statement 11, and seemed to understand that it is different from Statement 10.

Yet, all of this discussion took place with no mention of the game relative to Statement 11. The interchange occurred at the end of the interview in which the game had been used for other statements, in particular for Statement 10 where it did appear that the game helped her understand that statement. It could be that the effect of the game carried over, or it could be that for this statement the game did not have much effect.

Possibly reducing the student's understanding. Finally, we note that it is possible that playing the game can be counterproductive and increase a student's confusion. The following excerpt is from a student, JUL, who is struggling to understand Statement 9. She appears to be making some progress.

> JUL: I guess ... I see a difference between: there is *a* fertilizer and there is one fertilizer ... OK, so now ...
>
> I: Right...
>
> JUL: OK, so that's what ... I ... if it said "one" I would say false. Because it might ... that I would take that as the exact same fertilizer for all plants.
>
> I: And you don't think that's true.
>
> JUL: No.
>
> I: Why not?
>
> JUL: Ahm ... plants need fertilizers different fertilizers ...
>
> I: Different plants need different fertilizers?
>
> JUL: Yeah.
>
> I: OK. So what about the other way now, the other interpretation?
>
> JUL: There is *a* fertilizer ... That there is ... there is *a* fertilizer not necessarily the same one for all the plants, but there is a fertilizer made for all plants ...

Then the interviewer introduces the game. It may be that the student becomes more confused and at the very end, she indicates that, for her, the situation and the game are different.

> I: OK. OK. Let's suppose we play a little game with this one. Now in this game we have to pretend that you and I are both experts on plants and fertilizers—that is, we know everything. [Student agrees throughout.] So, here is how the game goes. I go first. OK. Then after I go, you go, and then I go, and then you go, and so forth as long as it keeps going. And what I do is I pick a particular plant. OK. Then your job is to find a particular fertilizer for that plant. OK? If you can't, then I win. If you can, then I go on, and name another plant, and you have to name a fertilizer for that, OK? So, we continue in this way. If we ever come to a time where you can't name a fertilizer then I win the game. If that never happens, and we continue doing that, then you win the game. You understand the game? [Student agrees throughout.] OK. Who wins the game?

JUL: (laughs) (P) Ahm... I would probably say that you would win. Because I think that you could probably name ... since I don't really ... I think there is a lot less fertilizers maybe than there is plants ...

I: You are allowed to use a fertilizer twice if it works.

JUL: (P) ahm ... I would still say that probably ... you would win... well ... yeah, I think it'd probably be easier to name plants than fertilizers ...

I: So what would happen at some point is that I would name a plant, and you would ...

JUL: ... Unless it was the same ... (P) I mean there are probably some plants that yeah would take the same fertilizer ... so ...

I: Right. You are allowed to do that. But you still think it's false.

JUL: Yeah.

I: And that means that there would come a time when I would name a plant and you couldn't name a fertilizer ...

JUL: Hm-hm.

I: But before you said that the statement "there is *a* fertilizer for all plants," you said that was true. Now what's that got to do with the game?

JUL: Then ... because if you can name a plant then I can't name a fertilizer ... if you ...if you win the game by naming a plant, I can't name a fertilizer ... Then I am saying that there is not a fertilizer for all plants.

I: So that statement would be ...

JUL: False.

I: But you said it was true before.

JUL: Yeah ...

I: OK. Which way do you think now? True or false?

JUL: (sighs ...) ...

I: I know this can be disconcerting ...

JUL: I know ... ahm (P) I guess now that I ... if I think of it in that way, like the game then it would be false. But . . . (P) ahm, but still ... see I guess I see it ... (P) ... I guess I see how it's the same thing how you are saying it's the same thing but yet when I look at that statement, it means something different than the game, I guess.

In this case, the student did not have a very good understanding of Statement 9 and it seems clear that the game did not help. It is possible, although we cannot be sure, that playing the game increased the student's confusion.

4.2.3. *Conditions for the effects.* Playing the game, like every pedagogical strategy, can have different kinds of effects for different students in different situations. There is need for careful studies to determine, as seems to be the case here, whether playing the game is more often helpful than neutral, and to make sure that it is not harmful. Another point of view is that, given the fact that playing the game can help the student understand, it is important to investigate the conditions under which it is most likely to be helpful.

To begin such an investigation, we offer the observation, which these interviews generally seem to support, that in order for the game to help, the student must understand the rules of the game and know exactly how it is determined who wins and who loses. If the student is not given this information, or if (as occurred in some cases) the student is asked to devise the game, then understanding may not be increased. This is not surprising. Our point is that the game may be an effective device for the interviewer to help the student construct an understanding that fits with the interviewer's understanding. This understanding is expressed in the game, so it is reasonable to insist that the game come from the interviewer, at least at first.

In the following excerpt from AND, the interviewer fully explains the game for an AE interpretation of Statement 9 and the student seems to understand the statement. Then, the interviewer asks the student to change the game (to get to an EA interpretation) but does not specify the conditions under which each player wins. The student's response indicates an unwillingness to accept an EA interpretation of the statement. This data are consistent with our suggestion of a strong connection between fully understanding the game and interpreting the corresponding statement.

> I: So then, ahm, would in your interpretation, is the statement true? "there is a fertilizer for all plants"? [mumbles] You still think it's true. OK. Let's play a little game with this statement, OK? And the game is the following. I go first, and first of all when we play this game, we will assume that you and I know everything there is to know about fertilizers and plants. OK, we know all the answers. So I go first and I pick a particular plant, and then your job ... you have to pick a fertilizer that's a good fertilizer for it. And then if you can't, I win.
>
> AND: If I can't?
>
> I: If you can't I win the game. If you can, then I get to pick another plant, and you try a different fertilizer . And if you cannot pick a fertilizer for this one, then I win. But if you can, then we go on. If ever I pick a plant that you can't pick a fertilizer, then I win the game. But if that never happens, then you win the game. OK, you understand the game?
>
> AND: OK.
>
> I: Who wins the game?
>
> AND: (P) Me.
>
> I: You would win this game. Because?
>
> AND: because all these nutrients in the soil, there is always going to be some type of fertilizer there as well as God to help it, too.
>
> I: OK, now I don't like that game, since you win so I am going to change the game, OK, see if I get a better chance, OK? This time, you go first. You pick the fertilizer and I have to pick a plant. Does that change anything? Who wins now?

AND: You will, because no matter what plant I pick, you can do the same the thing I would have done because there is always nutrients in the ground and God's help to make it grow. So therefore you'd win because whatever I would say, you'd be able to come up with some type of plant that could use God's help or could use those nutrients.

4.2.4. *Other effects.*

There were other effects that indicate that discussing the game can be a meaningful experience for students and that it can contribute to the learning/teaching enterprise in several ways. These include the fact that the students tend to adopt the language of the game, and that this can lead to improvements in the clarity of their explanations.

Adopting the language of the game. LER provides an example of the many cases we saw in which students used the language of the game to discuss the statement. Here, in finally realizing that Statements 10, 11 are different, she uses language in which two people are active in choosing and naming numbers. She did not speak this way before playing the game.

I: Now this rewording that we did, versus the original number 11, are they different or are they the same?

LER: It's different because ... let's see if I can explain . . . You are defining a as like every positive number ... You are not really ... It's different somehow ... (P) ... This one you name a number a and you always find a b smaller than it—on the original. If you change it around ... your b has to be determined by the a.

I: I see. OK.

LER: And the other way, your a is determined by the b.

The use of game-like language indicates that the student understands the connection between the statement and the game. This is very likely an important step in developing an understanding of the statement.

Improving the clarity of explanations. Aside from indications we have seen in which the game appears to help the student directly to understand the statement, we saw examples in which the understanding may have been fine before playing the game so there was little room for improvement. Nevertheless, the student's explanations in some cases became richer and clearer. At times the game appeared to provide students with a clear mechanism by which they could explain their answers.

4.2.5. *Final comments.* We end this section with two general comments about the game and its effects. As we have seen elsewhere in this study, students have greater difficulty with EA statements than with AE statements. This might at first appear surprising to those mathematicians and logicians who understand such statements so well that each appears equally simple. However, the difference is profound for students. We find that this difference persists in the context of the game, whether or not the game is helping the student to understand, and we take this persistence as additional evidence for the depth of the difficulties students have with quantified statements.

5. Discussion

In this section we summarize the conclusions which can be drawn from the results presented and then discuss some pedagogical suggestions that arise from this study.

5.1. Conclusions. Following are the main conclusions that we feel can be drawn about the students in this study.

(1) The students did not appear to use mathematical conventions to interpret the questionnaire statements (the mathematical and the non-mathematical statements).

(2) The students did not appear to be aware of having engaged in *interpreting* the questionnaire statements. That is, aside from which interpretation conventions they might or might not have used, and aside from their own interpretation (AE or EA), they did not focus on the statements as entities on their own right. During the interviews, when they were told explicitly that the focus was on the statements, and that their opinions were needed solely for establishing the truth value of a given statement, the students still focused on their opinions about the general topic of a statement, rather than how these opinions concerned the statement's truth value. Even when we raised the possibility that a statement might be ambiguous or offered an alternate interpretation (such as an EA interpretation when the student's interpretation was AE), in almost all cases the students assessed this possibility by referring to their opinion about the statement's topic, rather than the statement itself. This behavior suggests to us that in general the students were not conscious that in they were interpreting statements or of any conventions they might have used in their interpretations.

(3) Continuing the trend described above, students did not appear to use the syntax of a statement to analyze it. Rather, they referred to a world they were already familiar with and they considered that the statement described that world. They interpreted the statement and decided on its truth or falsity based on the nature and properties of this world. If the statement described a world unfamiliar to the student, he or she was very likely not to exhibit much understanding of the statement. The most important point is that what the student can understand seems to depend on what worlds he or she is able to think of and this scope is not expanded by thinking about the statement.

(4) Given a natural language AE statement, these students seemed to be able to interpret it as such and give a valid argument for its truth or falsity.

(5) When faced with a possibly ambiguous statement, students tended to interpret it as an AE statement. Even when the possibility that a statement might be interpreted as EA was brought to their attention, almost all students either dismissed it quickly and without explicit reference to the statement, or they left the possibility unresolved.

(6) There was a close relationship between making some interpretation of a statement on the questionnaire as an EA or as an AE statement that was articulated clearly and giving a valid argument for the truth value they assigned to it.

(7) Students had great difficulties imagining a situation in which the truth value of a given statement changes. This is the case even for statements which students at first appeared to understand.

(8) The ability to interpret correctly and to give a valid argument for truth or falsity dropped significantly for both AE and EA statements when the situation is in a mathematical context.

It seems that the eighth item above points out a *semantic* difficulty on the students' part. Positive numbers are not as real to them as children, fertilizers, pots and covers, etc. Items 5 and 7 suggest that even with statements about situations familiar to students, their understanding of multiple quantification is not very strong. Thus, even if we were able to effect a transfer from everyday to mathematical contexts, the knowledge we are transferring may not be as strong as we might have hoped. But Item 3 above suggests it might be over-optimistic to expect that very much of this transfer will take place. Items 3 and 4 suggest that there is a *syntactic* difficulty in the sense that most students failed to make use of, and/or were unable to understand the structure of the statements. Moreover, they had difficulty viewing a statement as an entity on its own right, apart from any truth value, and apart from what it means in some particular world.

If indeed it is the case, as seems to be suggested by this study, that students have difficulty viewing a statement as an object in its own right, apart from what the statement "says" in the (only) world they have in mind, then the implications are many. This would point to reasons why, for example, our students often read (and memorize) a theorem that appears in a box of a textbook, and can only apply it as it is stated in the book. Applying the theorem by using its contrapositive form, for example, often seems to require a "different" theorem. The above observations could provide some of the reasons why many students have difficulty experimenting with different mathematical "worlds" in order to assess a mathematical statement, to come up with counterexamples, etc. This is consistent with Johnson-Laird's theories about human reasoning — that we use mental models to assess statements and draw conclusions (and, hence, that we take a semantic, not a syntactic, approach to reason). See (Johnson-Laird, 1983).

Finally, the substantial drop we observed in the students' ability to handle the given mathematics statements when compared to the natural language statements strongly suggests that the conventional wisdom to *teach by making analogies to the real world* can fail dramatically. Our study suggests that simple analogies don't work. Many of these students were able to produce valid arguments to support their answers to Statements 1 through 9 on their own. A major breakdown occurs when they move to the mathematical realm. Furthermore, even regarding the natural language statements, when we probed a little more, their understanding appeared to be more shallow than we first thought by reading the questionnaire responses. Thus, there are two arguments against making analogies to natural language when trying to help students understand mathematical statements. One is that we cannot take advantage of something that may not be there. Second, even if it is there, it is in a form that is not likely to transfer to the mathematical realm. This study would suggest that in order to help our students, we need to remain in the mathematical realm and approach the difficulties from an integrative viewpoint: combining the study of syntax with the study of semantics.

5.2. Pedagogical suggestions. One of the key issues of this research is the difference between two approaches to making sense of a complex statement. One is to imagine a world which the statement describes and to reason about that world. The other is to use rules of syntax as an aid in constructing meaning for the statement. Our position is that the latter approach is a powerful one that can be used to construct meaning when the sentence is too complicated to imagine a world directly. On the other hand, it seems that many students are restricted to the former approach and this might reduce the scope of statements they are capable of understanding. Therefore, our most important pedagogical suggestion is to find ways, perhaps including, but certainly not restricted to, the use of formal language, in which to help students learn to use the syntax of a statement as a tool for making sense out of it—in short, our advice is to *Teach the syntax, teach the conventions of mathematical discourse relating to syntax.*

In this section we consider two different approaches one can take in order to integrate the active and explicit study of syntax in the curriculum: computer activities and the game.

5.2.1. *Pedagogy related to computer activities: ISETL.* One approach has already been used and studied before: the incorporation of a programming language such as *ISETL*. *ISETL* is a language whose syntax closely imitates that of mathematics. In asking students to read and write code in *ISETL*, we ask them to observe, understand, and construct formal statements that express a precise idea. The *ISETL* approach is introduced and analyzed in the following references: Asiala, Dubinsky, Mathews, Morics, & Oktac, 1997, Dubinsky, 1995, and Dubinsky, 1997. *ISETL* is free and can be downloaded from the web. It is available for Unix, IBM-compatible computers (in DOS and Windows versions) as well as Macintosh computers. For more information on *ISETL* and download sites, see http://csis03.muc.edu/isetlw/about.htm.

TARSKI'S WORLD and HYPERPOOF. TARSKI'S WORLD and *HYPERPROOF* are software packages written by Jon Barwise and John Etchemendy, and are designed to teach students various aspects of mathematical syntax, logic, proof, and sound reasoning. They are innovative in part because they involve the rigorous use of visual information, not only symbolic statements. In other words, these packages represent an approach that goes well beyond the more narrow approach of traditional formal logic and the syntax of programming languages. Unlike *ISETL*, or another programming language, this method skillfully combines syntax and semantics in a natural and even playful way while maintaining a sharp distinction between syntax and semantics. *HYPERPROOF* in particular extends the traditional notions of syntax, semantics, logical consequence, and proof, in ways that are both rigorous and natural (see Allwein & Barwise, 1996 and Barwise & Etchemendy, 1989). These packages were awarded the 1997 Educom Medal for innovative software. For more information on these packages, and a paper by Barwise and Etchemendy describing their viewpoints on pedagogy and logic, see Barwise and Etchmendy (1991, 1994) and http://csli-www.stanford.edu/hp/index.html.

5.2.2. *Pedagogy related to the game.* Finally, another approach that appears to have some promise is the use of (mental) games based on a statement with at least two different quantifiers. The use of such games is illustrated in our interviews and what we obtained here can form the basis for designing, implementing and studying a pedagogical strategy in which such games play a significant role. This is a matter for future investigations.

In light of the fact that the pedagogical tools we mentioned above can provide substantial help for students with the syntax (and correct reading) of mathematical statements, it might appear that the pursuit of these "low-tech" mental games described in this paper is superfluous. Furthermore, as we said above, if *HYPER-PROOF* manages to extend the traditional notions of syntax, semantics, logical consequence, and proof, notions that go far beyond the mere action of reading mathematical statements correctly, why would we want to re-invent the wheel? We do not. We are interested in pursuing the approach of these games precisely because they are "low-tech" and unlimited in scope. They can be used anywhere, they can be applied to any setting.

A pedagogical suggestion to improve student understanding of multiple quantification arises out of our data relative to the game. It could be that, just as asking a student to express a statement as a computer program helps her or him make use of the syntax to analyze it, asking the student to play the game related to the statement leads to using the rules of the game as a syntax. Continuing our analogy, we can imagine that to the task of moving from the syntax of a programming language to the syntax of the statement and of mathematical discourse corresponds the task of moving from the rules of the game to the mathematical syntax.

A final pedagogical suggestion is related to the fact that we saw several examples in which a student showed a good understanding of the statement both within and without the context of the game. Our data do not permit us to decide in every case if one of these understandings contributed to the other. But even if it was the student's understanding of the statement that led to her or his understanding of the corresponding game, it would be possible to make pedagogical use out of the situation. It might be possible to use this "joint understanding" to transfer the student's understanding to other, similar but still troublesome, logical statements.

References

Allwein, G., & Barwise, J. (Eds.). (1996). *Logical reasoning with diagrams*. New York: Oxford University Press.

Asiala, M., Dubinsky, E., Mathews, D., Morics, S., & Oktac, A. (1997). Development of students' understanding of cosets, normality and quotient groups. *Journal of Mathematical Behavior, 16*(3).

Barwise, J., & Etchemendy, J. (1989). Visual information and valid reasoning. In J. Barwise & J. Etchemendy (Eds.), *Visualization in teaching and learning mathematics*. Washington, DC: Mathematical Association of America.

Barwise, J., & Etchemendy, J. (1991). *Tarski's world*. Stanford: CSLI, and Cambridge: Cambridge University Press.

Barwise, J., & Etchemendy, J. (1994). *Hyperproof*. Stanford: CSLI, and Cambridge: Cambridge University Press.

Courant, R., & Robbins, H. (1996). *What is mathematics*. New York: Oxford University Press.

Dubinsky, E. (1991). The constructive aspects of reflective abstraction in advanced mathematics. In L. P. Steffe (Ed.), *Epistemological foundations of mathematical experiences*. New York: Springer-Verlag.

Dubinsky, E. (1995). *ISETL*: A programming language for learning mathematics. *Communications in Pure and Applied Mathematics*, *48*, 1–25.

Dubinsky, E. (1997). On learning quantification. *Journal of Computers in Mathematics and Science Teaching*, *16*(2/3), 335–362.

Ebbinghaus, H. D., Flum, J., & Thomas, W. (1984). *Mathematical logic*. Springer-Verlag.

Johnson-Laird, P. N. (1983). *Mental models: Towards a cognitive science of language, inference, and consciousness*. Cambridge: Cambridge University Press.

Johnson-Laird, P. N. (1995). Inference and mental models. In S. E. Newstead & J. S. B. T. Evans (Eds.), *Perspectives on thinking and reasoning: Essays in honour of Peter Wason* (pp. 115–146). Hove, England: Lawrence Erlbaum Associates.

Moschovakis, Y. N. (1980). *Descriptive set theory*. New York: North-Holland.

Nerode, A., Yakhnis, A., & Yakhnis, V. (1992). Concurrent programs as strategies in games. In Y. Moschovakis (Ed.), *Logic from computer science*. Springer-Verlag.

Scott, D., & Strachey, C. (1971). Toward a mathematical semantics for computer languages. *Proceedings of the Symposium on Computers and Automata*, *21*, 19–46.

Strachey, C. (1966). Towards a formal semantics. In *Proceedings IFIP TC2 Working Conference on Formal Language Description Languages for Computer Programming* (pp. 198–220). North Holland.

Vygotsky, L. (1986). *Thought and language*. Cambridge, MA: MIT Press.

Appendix A. Questionnaire

Following is the questionnaire given to the students. It is an exact copy of the instrument that was used except that we omit here the space left for student responses.

Questionnaire

This questionnaire is to be filled out at one continuous sitting. There is no time limit. Please do not consult with anyone or any book (or other source) while you are still working on your answers.

Here are 11 statements. Decide for each one if it is True or if it is False, and indicate your answer by circling one of them. Then, underneath the statement, provide a brief explanation for your answer.

(1) Everyone hates somebody.   True   False
     Why?

(2) Every pot has a cover.   True   False
     Why?

(3) Someone is kind and considerate to everyone.   True   False
     Why?

(4) There is a mother for all children.   True   False
     Why?

(5) All good things must come to an end.   True   False
     Why?

(6) There is a magic key that unlocks everyone's heart.   True   False
     Why?

(7) All medieval Greek poems described a war legend.   True   False
     Why?

(8) There is a perfect gift for every child.   True   False
     Why?

(9) There is a fertilizer for all plants.   True   False
     Why?

(10) For every positive number a there is a positive number b
     such that $b < a$.   True   False
     Why?

(11) There is a positive number b such that for every
     positive number a $b < a$.   True   False
     Why?

How much time did you spend on this questionnaire? _____

Appendix B. Questionnaire Data

Number of students: 63.

Table 1

Statement

1	2	3	4	5	6	7	8	9	10	11

Label

AE	AE	EA	EA	AE	EA	AE	EA	EA	AE	EA	*All AEs*	*All EAs*

Percent consistent with label

84	92	65	16	83	22	76	27	29	59	19	79	30

Percent consistent with label and with valid argument

84	90	63	14	76	19	73	25	25	49	14	75	27

Percent reverse of label

0	0	11	81	1	37	3	40	38	1	22	1	38

Percent reverse of label and with valid argument

0	0	10	81	0	33	2	40	35	0	5	0.3	35

Percent unclear

16	8	24	3	16	41	21	33	33	40	57	20	32

Percent unclear but with valid argument

5	2	8	0	0	16	2	19	14	0	0	2	10

Percent with valid argument

89	92	81	95	76	68	77	84	74	49	19	76	70

Percent with "True" circled

24	40	60	83	40	54	14	65	79	68	46	37	65

Table 2. AE versus EA

	as EA	as AE
% stdts interpreting at least 1 EA	92	94
% stdts interpreting all 6 EAs	0	0
% stdts interpreting at least 1 AE	5	97
% stdts interpreting all 5 EAs	0	35

Table 3. Valid arguments in natural language.

n	1	2	3	4	5	6	7	8	9
% $C(n)$	100	98	98	98	97	92	78	50	25

$C(n)$ is the set of students who gave at least n valid arguments for natural language statements.

Truth and Falsity

- Percent of statements that were interpreted as AEs and were declared True: 30%.
- Percent of statements that were interpreted as EAs and were declared True: 10%.

Table 4. More on Statements 10 and 11.

% 10 right	49
% 11 right	14
% 10 & 11 right	12
% 10 right & 11 wrong	37
% 10 wrong & 11 right	2
% 10 & 11 wrong	50

Comments

(1) Percent of students who claim 10 and 11 are the same or reworded: 42.
(2) Percent of graduate students who claim 10 and 11 are same or reworded: 33.
(3) Percent of undergraduate students who claim 10 and 11 are same or reworded: 44.

(4) Number of students who claim 11 is a rewording of 10 and give 10, 11 different truth values: 0.

(5) Percent of students who explicitly state "they do not know" for Statement 7 but give a valid argument: 51.

(6) Percent of students who explicitly state "they do not know" for Statement 7 but give a valid argument for 7 and also claim Statement 10 is the same as Statement 11: 18.

Table 5. Mathematics vs. English

n	1	2	3	4	5	6	7	8	9
$R(n)$	92	92	92	92	92	93	92	94	94
$B(n)$	57	57	57	54	56	55	92	94	94

- $R(n)$ is the percent of students who got at least one of Statements 10 or 11 wrong or blank among those in $C(n)$.
- $B(n)$ is the percent of students who got both Statements 10 and 11 wrong or blank among those in $C(n)$.

265 NORTH WOODS ROAD, HERMON, NY 13652
E-mail address: edd@mcs.kent.edu

AGNES SCOTT COLLEGE, ATLANTA, GA 30030
Current address: IBM Corporation, 9000 S Rita Rd, Tucson, AZ 85744
E-mail address: yiparaki@us.ibm.com

Research in Collegiate Mathematics Education

Editorial Policy

The papers published in these volumes will serve both pure and applied purposes, contributing to the field of research in undergraduate mathematics education and informing the direct improvement of undergraduate mathematics instruction. The dual purposes imply dual but overlapping audiences and articles will vary in their relationship to these purposes. The best papers, however, will interest both audiences and serve both purposes.

Content. We invite papers reporting on research that addresses any and all aspects of undergraduate mathematics education. Research may focus on learning within particular mathematical domains. It may be concerned with more general cognitive processes such as problem solving, skill acquisition, conceptual development, mathematical creativity, cognitive styles, etc. Research reports may deal with issues associated with variations in teaching methods, classroom or laboratory contexts, or discourse patterns. More broadly, research may be concerned with institutional arrangements intended to support learning and teaching, e.g. curriculum design, assessment practices, or strategies for faculty development.

Method. We expect and encourage a broad spectrum of research methods ranging from traditional statistically-oriented studies of populations, or even surveys, to close studies of individuals, both short and long term. Empirical studies may well be supplemented by historical, ethnographic, or theoretical analyses focusing directly on the educational matter at hand. Theoretical analyses may illuminate or otherwise organize empirically based work by the author or that of others, or perhaps give specific direction to future work. In all cases, we expect that published work will acknowledge and build upon that of others—not necessarily to agree with or accept others' work, but to take that work into account as part of the process of building the integrated body of reliable knowledge, perspective and method that constitutes the field of research in undergraduate mathematics education.

Review procedures. All papers, including invited submissions, will be evaluated by a minimum of three referees, one of whom will be a Volume editor. Papers will be judged on the basis of their originality, intellectual quality, readability by a diverse audience, and the extent to which they serve the pure and applied purposes identified earlier.

any reasonable length will be considered, but the
e smaller for very large manuscripts.
script should be submitted. Manuscripts should be
ations and bibliographies according to the format of
Association as described in the fourth edition of the
ierican Psychological Association.
umes are produced for electronic submission to the
ed manuscripts will be prepared using AMS-LATEXand
vailable from the AMS Web site. Illustrations should
table for electronic submission (namely, encapsulated

Correspondence. All manuscripts and initial editorial correspondence should
be sent to the managing editor of *RCME*:

> Cathy Kessel
> School of Education, Tolman #1670
> University of California at Berkeley
> Berkeley, CA 94720-1670
> `kessel@soe.berkeley.edu`

Subequent correspondence may be with the managing editor, or with the Volume
editor who has been assigned primary responsibility for decisions regarding the
manuscript.